EXS 80

Frontiers in Biosensorics I

Fundamental Aspects

Edited by F. W. Scheller
F. Schubert
J. Fedrowitz

Birkhäuser Verlag
Basel · Boston · Berlin

Editors

Prof. Dr. F.W. Scheller
Institut für Biochemie
und Molekulare Physiologie
c/o Max-Delbrück-Center
für Molekulare Medizin
Robert-Rössle-Strasse 10
D-13 122 Berlin

Dr. J. Fedrowitz
c/o Centrum für Hochschulentwicklung
PO Box 105
D-33311 Gütersloh

Dr. F. Schubert
Physikalisch-Technische Bundesanstalt
Abbestrasse 2–12
D-10587 Berlin

Library of Congress Cataloging-in-Publication Data
A CIP catalogue record for this book is available from the library of Congress,
Washington D.C., USA

Deutsche Bibliothek Cataloging-in-Publication Data
EXS. – Basel; Boston; Berlin: Birkäuser.
Früher Schriftenreihe
Fortlaufende Beil. zu: Experientia
80. Frontiers in Biosensorics I. Fundamental aspects. – 1997
Frontiers in Biosensorics / ed. by. F. W. Scheller ... – Basel;
Boston; Berlin: Birkhäuser.
(EXS; 80)
ISBN 3-7643-5481-X (Basel ...)
ISBN 0-8176-5481-X (Boston)
NE: Scheller, Frieder [Hrsg.]
I Fundamental aspects. – 1996
Fundamental aspects. – Basel; Boston; Berlin: Birkäuser. 1997
(Frontiers in Biosensorics I) (EXS; 80)
ISBN 3-7643-5475-5 (Basel ...)
ISBN 0-8176-5475-5 (Boston)

© 1997 Birkhäuser Verlag, PO Box 133, CH-4010 Basel, Switzerland
Printed on acid-free paper produced from chlorine-free pulp. TCF ∞
Printed in Germany
ISBN 3-7643-5481-X (Volumes 1+2, Set)   ISBN 0-8176-5481-X (Volumes 1+2, Set)
ISBN 3-7643-5475-5 (Volume 1)   ISBN 0-8176-5475-5 (Volume 1)
ISBN 3-7643-5479-8 (Volume 2)   ISBN 0-8176-5475-8 (Volume 2)
9 8 7 6 5 4 3 2 1

Contents

Frontiers in Biosensorics I
Fundamental Aspects
ed. by F. W. Scheller, F. Schubert and J. Fedrowitz
© 1997 Birkhäuser Verlag Basel/Switzerland

Present state and frontiers in biosensorics

F. W. Scheller[1], F. Schubert[2] and J. Fedrowitz[3]

[1] *Analytcal Biochemistry, Institute of Biochemistry and Molecular Physiology, University of Potsdam, c/o Max-Delbrück-Center of Molecular Medicine, D-13122 Berlin Germany;*
[2] *c/o CHE Center for Higher Education Development, D-33311 Gütersloh, Germany;*
[3] *Physikalisch-Technische Bundesanstalt, D-10587 Berlin, Germany*

The field of bioanalytics

Nature offers an arsenal of interesting principles for the optimization of existing and the realization of novel technical processes. These principles include "classical" reactions of enzymes, antibodies, receptors or nonprotein macromolecules like nucleic acids, carbohydrates, DNA or RNA. This volume aims at indicating the frontiers in biosensorics by summarizing new recognition elements, new developments in thin layers and interfaces as well as by giving examples for recent applications of biosensors.

Molecular recognition

As far as analysis is concerned, *enzymes* and *antibodies* represent powerful tools that allow for sensitive and specific methods of detection and quantitation to be developed for a wide variety of substances.

The first step of the interaction between enzyme and substrate, or antibody and antigen, is the binding of the analyte to the complementary protein structure. The basic principle behind the high chemical selectivity of these biomacromolecules is the structural complementarity of the recognition elements and the target analyte. While the binding to the antibody normally does not lead to a chemical alteration of the antigen the formation of a complex between substrate and enzyme protein initiates the chemical conversion of the substrate. However, a strict differentiation between both reaction types is not very suitable since they merge into each other. In the absence of cosubstrate an enzyme acts as a binding protein for the substrate; the same is generally true for the binding of enzyme inhibitors and other effectors. Recently catalytic antibodies have been developed which are capable not only of binding the partner but also of catalyzing its chemical conversion.

Chemoreceptors located in biological membranes function like biomolecular devices driven by the cell metabolism and controlled by the presence of biologically active substances. They are extremely attractive tools for the selective recognition of toxins, hormones and drugs. Binding of the agonist either by the opening of ion channels (e.g. with the nicotinic receptor) or via enzyme cascades (e.g. in case of the β-adrenergic receptor) triggers a signal amplification of 4 to 9 orders of magnitude. The interrogation of the receptor function requires its integration into an artificial membrane; in other cases it can be performed by using (intact) chemosensing structures, e.g. antennae of crustaceans.

In addition to the proteic macromolecules *nucleic acids* and *carbohydrates* are increasingly used in specialized areas, for example for the sequencing of genomes and for cell surface characterization. The stereo-specific interactions of nucleotides – the building blocks of the nucleic acids – are primary for replication of DNA, synthesis of RNA in the transcription, and ribosomal protein synthesis. Structurally complementary single strands associate to double helical segments where the strength of interaction is determined by the degree of sequence homology. Nowadays, hybridization assays based on this interaction are a fundamental tool of molecular biology and gene identification.

Measurement and sequence analysis of DNA and RNA are not only the keys to analyzing the human genome but are unique instruments for genetic profiling susceptibility to diseases, prenatal diagnosis of genetic diseases, and research in molecular biology.

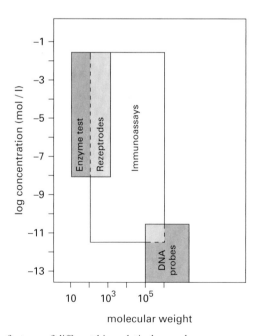

Figure 1. Specific features of different bioanalytical procedures.

The biological recognition elements presented so far cover almost all relevant analytes. As a first step in the design of a bioanalytical process one of them has to be chosen that performs the molecular recognition of the analyte of interest. This selection, however, is dictated by size and concentration of the analyte (Fig. 1).

Signal transduction and bioanalytical configurations

In order to obtain a quantifiable output, the interaction between analyte and biomolecule has to be transformed from the chemical into a physico-chemical signal. The transfer of information from the biochemical domain to human knowledge requires the command of the techniques for characterization of biochemical systems as well as the transducer principles and technologies. Furthermore, due to the increased use and availability of computers, it is generally desirable to obtain an electrical output. Thus, the electronic domain is common to all modern bioanalytical instruments.

The increasing availability of highly purified enzyme preparations and (monoclonal) antibodies induced the development of a wide range of methods based on these biochemical reagents. For their repeated use, as well as that of cells and other biologically active agents, such as receptors, in analytical devices, numerous techniques for fixing them to carrier materials have been developed. *Immobilization* of the protein, particularly of enzymes, brings about a number of further advantages for their application in analytical chemistry: (i) in many cases the protein is stabilized; (ii) the immobilizate may be easily separated from the sample and (iii) the stable and largely constant biomolecule activity makes the enzyme an integral part of the analytical instrument. Thus, application of immobilized enzymes in analytical chemistry has become common.

As early as 1956 the principle of the litmus paper used for pH measurement was adopted to simplify the enzymatic determination of glucose. By impregnating filter paper with the glucose-converting enzymes the "enzyme test strip" was invented. It can be regarded as the predecessor of optical biosensors and, at the same time, initiated the development of the socalled *dry chemistry*. Nowadays, highly sophisticated enzyme and immuno test strips are commercially available for the determination of about 15 low-molecular metabolites and drugs as well as the activities of 10 enzymes.

In parallel analytical enzyme- and immuno-*reactors* have been developed where the progress of the reaction is indicated in the reactor effluent colorimetrically or electrochemically. In *packed bed reactors* the enzyme-catalyzed reaction is carried out in a column of 100 µl–10 ml volume filled with tiny particles bearing the immobilized enzyme. In contrast, in *open tubular reactors* the enzyme is attached in a monolayer to the inner walls. Such reactors permit a higher measuring frequency.

An alternative route was opened in the early 60's by Leland C. Clark, the inventor of the Clark oxygen electrode. He arranged the enzyme solution directly in front of that electrode and avoided the mixing of the enzyme with the bulk solution by covering the reaction compartment with a semi-permeable membrane. Thus, a single enzyme preparation could be used for several samples. This measuring arrangement gave birth to a new sensor concept – the *biosensor*.

The first step in the biosensing process is the specific complex formation of the immobilized recognition element with the analyte. The biological part of a biosensor is often submitted to a conformational change in context with the binding of its partner. In nature this effect may immediately be used for transduction and amplification, e.g., in the ion channels of nerve tissue. The effects of interaction between the analyte molecule and the biological system are quantified by the transducer and electronic part of the biosensor. As transducers, chemical sensors, i.e., potentiometric, amperometric and impedimetric electrodes, optical detectors using indicator dyes, as well as physical sensors, such as piezoelectric crystals, thermistors, and plain optical sensors, have been combined with appropriate biocomponents (Tab. 1). In analogy to affinity chromatography, in so-called *binding* or *affinity sensors*, dyes, lectins, antibodies, or hormone receptors are being used in matrix-bound form for molecular recognition of enzymes, glycoproteins, antigens, and hormones. The complex formation changes the magnitude of physico-chemical parameters, such as layer thickness, refractive index, light absorption, or electrical charge, which may then be indicated by means of optical sensors, potentiometric electrodes, or field effect transistors. After the measurement the initial state is regenerated by splitting of the complex. On the other hand, the molecular recognition by enzymes, which can also be applied in the form of organelles, microorganisms and tissue slices, is accompanied by the chemical conversion of the analyte to the respective products. Therefore this type of sensor is termed *catalytic* or *metabolism sensor*. It usually returns to the initial state when the

Table 1. Principles of Biosensors

1. Bioaffinity sensors		2. Biocatalytic sensors	
receptor	analyte	receptor	analyte
dye	protein	enzyme	substrate
lectin	saccharide	organelle	cofactor
	glycoprotein	microbe	inhibitor
enzyme	substrate	tissue slice	activator
	inhibitor		enzyme activity
apoenzyme	prosthetic group		
antibody	antigen		
receptor	hormone		
transport system	substrate analogue		

Transducers

optoelectronic detectors, field effect transistors, semiconductor electrodes, potentiometric electrodes, amperometric electrodes, thermistors analyte conversion is completed. Under appropriate conditions catalytic biosensors are capable of determining cosubstrates, effectors, and enzyme activities via substrate determination. Amperometric and potentiometric electrodes, fiber optics, and thermistors are the preferred transducers here, the former being by far the most important.

Frontiers in biosensorics

Recently the International Union of Pure and Applied Chemistry issued recommendations on the definition, classification and nomenclature concerning (electrochemical) biosensors. The following proposals made by a group of experts have been published in Biosensors & Bioelectronics: *A biosensor is a self-contained integrated device which is capable of providing specific quantitative or semi-quantitative analytical information using a biological recognition element (biochemical receptor) which is in direct spatial contact with a transducer element. We recommend that a biosensor should be clearly distinguished from a bioanalytical system which requires additional processing steps, such as reagent addition. Furthermore, a biosensor should be distinguished from a bioprobe which is either disposable after one measurement, i.e. single use, or unable to continuously monitor the analyte concentration. (Biosensors and Bioelectronics 11, (1996), i).*

Unfortunately, this definition attempt reflects the state of development several years ago. Owing to rapid technological progress essential parts of it appear ambiguous today. Thus, in the light of the ability of miniaturized bioreactors to interact with transducers (without additional sample processing elements) on a single chip the term 'direct spatial contact' of the elements is amenable to different interpretations. Furthermore, the borderline between recognition tools that are biological in nature, and synthetic (organic) receptor molecules is no longer well definable; these two classes of recognition elements are merging. This is especially true for enzyme models, polymers imprinted by biomolecules and, although to a lesser extent, (synthetic) oligonucleotides and ionophores which mimic the function of channels.

Although at present routine application of biosensor technology is more or less restricted to enzyme electrodes, optical immunosensors and whole cell-based receptor assays, the forefront of biosensor resarch exploits results and principles of molecular biotechnology and nanotechnology for creating qualitatively new *molecular sensors*. These new approaches shall be highlighted in the introduction to this monograph on the frontiers in biosensorics.

New recognition elements

By mimicking the binding area of enzymes *host* molecules for binding of the respective *guests* have been developed. Alternatively, molecular model-

Biomolecular Recognition Elements

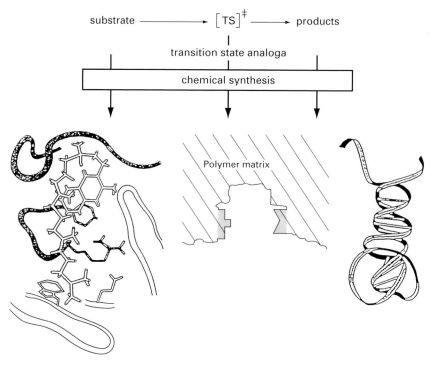

substrate ⟶ $[TS]^{\ddagger}$ ⟶ products

transition state analoga

chemical synthesis

Polymer matrix

Figure 2. Principles of generating binding or catalytically active antibodies, molecular imprints, and nucleic acids.

ing is used to design appropriate binding structures which are combined as building blocks for signal generation, for example in the sensing of glucose, ethanol and creatinine. Likewise, polymer chemistry, immunotechnology and RNA technology are contributing to the development of new molecular recognition elements (Fig. 2). *Molecular imprintings* can be synthesized by forming a polymer network around a template. In this way structures complementary to the analyte are created on the surface of the polymeric carrier. Another synthetic approach uses oligonucleotides (aptamers) generated by random synthesis. In the next step the library is screened for the binding to the immobilized target molecule. The appropriate sequence is then amplified by a combination of the well-established reverse transcription and polymerase chain reaction. Using this highly innovative approach a recognition partner has been synthesized that possesses a 10^4-fold higher affinity constant for theophylline than for the very similar caffeine molecule.

Molecularly imprinted polymers have been applied for an increasing number of analytes in (bio)sensors, e. g. of D- and L-amino acids, atrazine, cholesterol, ephedrine, diazepam, and morphine. On the other hand, the application of aptamers in sensor arrangements is at its very beginnings.

Whereas the binding imprints and aptamers mimic the action of anti-bodies, the catalytic function of enzymes may be simulated by an analogous concept. The basic dogma of catalysis is the increase of the reaction rate by stabilization of the transition state by the catalyst. By using (stable) analogues of the transition state as the template in the generation of antibodies, imprin-ted polymers and aptamers, in fact new catalytically active species have been generated. On the basis of this concept catalytic antibodies for the following reactions have been prepared: ester hydrolysis, amidohydrolysis, cyclization, amide bond formation, Claizen rearrangement, decarboxylation, peroxida-tion, cis-trans isomerization and β-elimination. Up to now, enzymes are the better catalysts, with typical turn-over numbers of several ten thousands per second; typical catalytic antibodies reach only one percent of this value. In contrast, antibodies exhibit higher affinity towards the antigen, as it is reflec-ted by affinity constants between 10^6 and 10^{12} M^{-1} compared to values for enzymes in the per-milli and micromolar range. The binding area of antibodies is not as appropriate as that of enzymes, the latter having been optimized during biological evolution. Furthermore, the analogue used for immunization may reflect an intermediate rather than the "real" short-lived transition state. Nevertheless, catalytic antibodies represent appropriate tools for substances not metabolized by enzymes.

The analytical performance of enzymes and antibodies may be improved by applying them in nonbiological environments, like organic solvents or the gas phase. In addition to a higher stability, specificity may be improved under these nonconventional conditions, and a simplified measuring pro-cess may be achieved. In this way bioanalysis might be carried out directly in extracts of soil or in the gas atmosphere.

Genetic engineering allows to optimize the properties of biomacro-molecules, such as stability and substrate specificity, by exchanging amino acids. Multifunctional proteins can be provided by fusing the struc-tural genes. The latter approach permits to combine both analyte recogni-tion and signal generation on the molecular level. Examples of this func-tional integration are fused proteins containing the binding fraction of antibodies and the marker enzyme on the same protein chain. An analo-gous approach couples a reporter gene, e.g. for luciferase or β-galactosi-dase, to a DNA region, the expression of which is controlled by the pre-sence of the analyte. When modified according to this concept, the cell acts like an bioindicator.

Signal processing

Sensor technology has an enormous influence on the progress in bio-senorics as well. Three directions shall be emphasized in this chapter:

(i)   direct (label-free) signal transfer from the biocomponent via trans-
        ducers to the electronics,

(ii) miniaturization enabling the indication and manipulation of single molecules,
(iii) high sample throughput, redundancy and multi-analyte determination by parallelizing.

Direct signal transfer
The direct transfer of the signal gained in the analyte recognition from the biocomponent to the transducer could provide distinct advantages over a process exploiting the mass transfer of one reaction partner. First, the loss of sensitivity by migration of the signal-mediating component into the bulk phase is minimized. Secondly, the direct transfer is restricted to the phase boundary at the transducer, thus restricting contributions of the bulk phase.

For electrochemical biosensors the realization of the (mediator-free) electron transfer from the analyte converting oxidoreductase to the electrode has been described for several PQQ-containing dehydrogenases. Recently, also for glucose oxidase covalently bound to the electrode surface, direct electron transfer has been demonstrated by the occurrence of catalytic currents in the presence of the substrate, glucose. Alternatively, oxidoreductases, e. g. peroxidase, have been 'wired' to the electrode via redox polymers or modified prosthetic groups. These approaches lead to a reagentless measuring procedure. A technological breakthrough has been achieved for direct immunosensors. Optical techniques, like surface plasmon resonance, reflectometric interference spectroscopy, and grating couplers, do not require the presence of a signal-generating labeled reaction partner. These techniques directly sense changes in the respective optical properties at the transducer surface. In this way label-free binding sensors have been sucessfully developed.

Piezoelectric detection, which normally indicates the surface coverage in the gas phase, has been adapted to aqueous media. In this manner the change of mass occurring during biospecific interaction can be translated into a sensor output.

The use of magnetic toroids and intact nerve tissue provides an approch for investigating neural processes. Unlike microelectrodes, which can damage the neural membranes when suction is applied, the toroids themselves are not in direct contact with the nerve tissue. Biomagnetic measurements could thus indeed be used as analytical sensors.

Detection of single molecules
Detection and identification of single particles is an important challenge faced by analytical chemists today. The 'particles' may be single cells or parts of their surfaces, cell organelles, viruses, single genes, proteins (enzymes, receptors), or even small molecular entities, such as peptide hormones or other oligomeric compounds. The ability to accomplish such measurements could lead to a beakthrough in DNA analysis, groundwater monitoring, immunoassay, and fundamental studies of physical and

chemical phenomena at the single-molecule level. Using fluorescence correlation techniques single-molecule detection by illuminating a very small volume ($\cong 1$ fl) in a drop of solution below the objective of a confocal microscope has been demonstrated. Fluorescence correlation spectroscopy, which records spatiotemporal correlations among fluctuating light signals, expands the horizon in molecular diagnostics by making it possible to monitor concentrations down to 10^{-15} M without amplification. Another novel, powerful analytical tool, the atomic force microscope, gives subnanometer resolution of biological objects. On this basis analyte quantitation can be performed in a discrete or 'digital' manner. This technique allows not only to count individual particles but also manipulate single molecules. Evidently, these capabilities might result in biotechnology on the molecular level.

Applications

Nowadays the techniques of enzymatic and immunochemical analysis are most widely used in clinical diagnostics. Other fields of application are food analysis, bioprocess control, environmental monitoring and, to a lesser extent, the cosmetics industry. Furthermore, a virtually unlimited area of application exists in molecular biology research.

The most successful principles of applying biomolecules for analytical purposes are enzyme test strips, especially for blood glucose measurements, and immunoassays. The world market for these applications may be estimated for test strips in the order of $ 3 billion and almost $ 10 billion for the 1 billion immunoassays carried out per year. The field of biosensor application has been expanded from clinical laboratories to the point of care. The commercialization of home glucose meters for patients' self control reflected this trend during the eighties. Recently, enzyme electrodes have been introduced into analyzers devoted to critical care medicine. The analyte combination of electrolytes, blood gases and key metabolites is highly valuable for diagnostics under critical metabolic circumstances. On line measurement of the glucose concentration in patients is a very rewarding task for biosensor development. Both, combination with dialysis or direct implantation of the sensor, are under intensive study. However, also noninvasive measurement by physical instrumentation may result in true solutions for the characterization of metabolic situations within the next ten years. In the light of such prognoses future application of biosensors could be envisaged as including *in vivo* detection of short-lived intermediates, e.g. superoxide or nitric oxide, and low concentration substances, e.g. neurotransmitters or hormones. The monitoring of processes on the level of nucleic acids, the action of drugs on receptors as well as transmitter-receptor interactions represent further applications at the frontiers of biosensorics.

New recognition elements

Frontiers in Biosensorics I
Fundamental Aspects
ed. by F. W. Scheller, F. Schubert and J. Fedrowitz
© 1997 Birkhäuser Verlag Basel/Switzerland

Imprinting techniques in synthetic polymers – new options for chemosensors

G. Wulff

Institute of Organic Chemistry and Macromolecular Chemistry, Heinrich-Heine-University of Düsseldorf, D-40225 Düsseldorf, Germany

Summary. Highly selective molecular recognition in synthetic polymers may be the basis for a new type of chemosensor. For the preparation of such polymers, a molecular imprinting procedure during crosslinking is used to generate, with the aid of template molecules, microcavities of specific shape and with a defined arrangement of functional groups. In this review the role of the polymer matrix and the type of binding site interaction is discussed in more detail. Recent attempts to use these polymers as the basis for chemosensors are highlighted.

Introduction

Biosensors are usually based upon defined biological-type recognition systems between immobilized biomolecules (enzymes, antibodies, nucleic acids etc.) and certain substrates or between immobilized substrates and biomolecules. The interaction between the substrates and the biomolecules is of a rather complex nature involving different types of noncovalent interactions. For a highly selective interaction, the orientation in space of the functional groups acting as binding sites is an important factor. This orientation facilitates a highly cooperative combination of interactions. In addition to the orientation of the functional groups, an exact steric fit of the two complementary compounds (or at least part of them) considerably improves selectivity (shape selectivity).

It was the purpose of our work to develop a chemoselective recognition system acting in a similar mode but based upon synthetic polymers as sensing materials.

This means that the orientation of the binding sites in the polymer has to be controlled during preparation in order to obtain certain cooperativities in binding and, in addition, shape selectivity has to be achieved. Since binding in biological systems is rather complex and often not known in detail, simplified binding mechanisms have to be developed for such an approach. Sensors based on synthetic polymers should be considerably more stable and should, therefore, have a longer lifetime.

Highly selective interactions in bioselective recognition are observed between antigens and antibodies. Antibodies can be generated in animals against substances acting as antigens. Quite some time ago, we introduced (Wulff and Sarhan, 1972; Wulff et al., 1973, 1977a; Wulff, 1986, 1995)

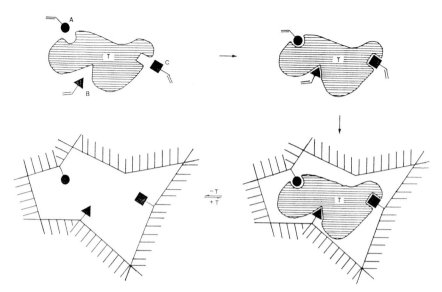

Figure 1. Schematic representation of the imprinting of specific cavities in a cross-linked polymer by a template (T) with three different binding groups.

a technique analogous to what was formerly thought to be the mechanism of the formation of antibodies (Pauling, 1940).

   In our case crosslinked polymers are formed around a molecule that acts as a template and this template is then removed. For this, polymerizable binding-site groups are bound by covalent or noncovalent interactions to a suitable template molecule (see Fig. 1). This template monomer is copoly-merized in the presence of a high amount of cross-linking agent. After extracting the template molecules from the polymer, micro-cavities with an imprinted shape and an arrangement of functional groups complementary to that of the template are generated. Such imprinted polymers show high selectivity in binding to their template molecules and find application in molecular separation, in catalysis, and in sensors. The method has been extensively used and refined over the past two decades by several research groups (reviews see: Mosbach, 1994; Shea, 1994; Steinke et al., 1995; Vidyasankar and Arnold, 1995; Wulff, 1986, 1995).

Examples for the imprinting procedure

As an example for the preparation of cavities with a defined arrangement of binding-site groups and with predetermined shape, the polymerization of **1** is considered. Phenyl-α-D-mannopyranoside (**2**) acts as a template. Two molecules of 4-vinylphenylboronic acid are bound to the template molecule by esterification with each pair of hydroxyl groups. The boronic acid was chosen as the binding-site group because it undergoes an easily reversible interaction with diol groupings. The template molecule **2** is chiral and optically active, and for this reason the cavities produced should be chiral as well. Therefore, after cleavage of the original template, the accuracy of the steric arrangement of the binding sites in the cavity could be tested by the ability of the polymer to resolve the racemate of the template. The monomer **1** has been extensively used for the optimization of the imprinting method (Wulff et al., 1977a, b, 1982, 1987; Wulff, 1986, 1995).

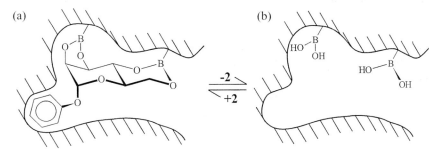

Monomer **1** was copolymerized by free-radical initiation in the presence of an inert solvent (porogenic agent) with a large amount of a bifunctional cross-linking agent. These conditions led to macroporous polymers possessing a permanent pore structure and a high inner surface area. These polymers exhibited good accessibility and low swelling characteristics and, hence, a limited mobility of the polymer chains.

The template can be split off by water or methanol to an extent of up to 95% (see Fig. 2). When this polymer is treated with the racemate of the template in a batch procedure under equilibrium conditions, the enantiomer that has been used as the template for the preparation of the polymer is taken up preferentially. The specificity is expressed by the separation factor α, which is the ratio of the distribution coefficients of D- and L-forms between solution and polymer. After optimization of the procedure, α-values between 3.5–6.0 were obtained (Wulff et al., 1982, 1987;

Figure 2. Schematic representation of a cavity (a) obtained by polymerization of **1**. The template **2** can be removed with water or methanol to give (b). Addition of **2** causes the cavity to be reoccupied, giving (a) again. In this case, the binding of the template is by covalent bonds (Wulff, 1986, 1995).

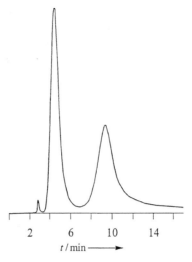

Figure 3. Chromatographic resolution of D, L-**2** on a polymer imprinted with **1**.

G. Wulff, K. Jacoby and J.H.G. Steinke, personal communication). With α-values of $\alpha = 5.0$ in the simple batch procedure we observed a maximal enrichment of the D-form at the polymer to the extent of 70–80%. This is an extremely high selectivity for racemic resolution that cannot be reached by most other methods.

Polymers prepared by this procedure can be used for the chromatographic separation of the racemates of the template molecules. The selectivity of the separation process is fairly high (separation factors up to $\alpha = 4.56$) and at higher temperatures with gradient elution resolution values of $R_s = 4.2$ with baseline separation have been achieved (see Fig. 3). These sorbents can be prepared conveniently and possess excellent thermomechanical stability.

Apart from using sugar derivatives as templates, this method permits the separation of amino acids, hydroxy carboxylic acids, diols and other racemates (Wulff, 1986, 1995). Furthermore, it has been possible to localize two amino groups in a defined distance at the surface of silica with this method (Wulff et al., 1986). These materials bind strongly by a cooperative binding dialdehydes or diacids in those cases where the functional groups have the correct distance from each other. The group of K. Mosbach (Sellergren et al., 1988) widened the scope of the imprinting procedure

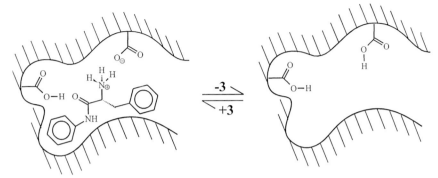

Figure 4. Schematic representation of cavity produced in the presence of **3**. Polymerization takes place in the presence of acrylic acid. Noncovalent electrostatic interactions and hydrogen bond formation occur (Sellergren et al., 1988).

considerably when they achieved an effective imprinting by using exclusively noncovalent interactions. Many investigations in this direction were carried out with L-phenylalanine anilide (**3**) as template. This was polymerized with acrylic acid as the comonomer under the polymerization conditions established for **1**. One acrylic acid unit forms an electrostatic bond with the template **3**, and another forms a hydrogen bond (Fig. 4). The research groups of K. Mosbach and K.J. Shea extensively optimized the process, and achieved high selectivities for resolution similar to those obtained with **2** (O'Shanessy et al., 1989; Sellergren and Shea, 1993).

The optimization of the polymer structure

The structure of the matrix is crucial in the imprinting process. As the specific structure of the cavity is not determined by low molecular weight molecules but by the fixed arrangement of the polymer chains, the optimization of the polymer structure is extremely important. The polymer should have the following properties (Wulff, 1986).

a) *Stiffness of the polymer structure* enables the cavities to retain their shape even after removal of the template, thus giving high selectivity.
b) *High flexibility* of the polymer structure works against the above but is essential for the kinetics to give rapid equilibrium with the substrate to be embedded.
c) *Good accessibility* of as many cavities as possible in the highly cross-linked polymer can be achieved by forming a particular polymer morphology.
d) *Mechanical stability* of the polymer particles is of great importance for many applications, for example, for use in an HPLC column at high pressure or as a catalyst in a stirred reactor.
e) *Thermal stability* of the polymers enables them to be used at higher temperatures, at which the kinetics are considerably more favorable.

From the beginning, macroporous structures have been used almost exclusively for imprinted polymers. Macroporous polymers are obtained if polymerization of the monomers is carried out with a relatively high content of cross-linking agent in the presence of inert solvents (also known as porogens). During the polymerization phase separations take place and, after removal of the porogen and drying, a permanent pore structure remains. The relatively large inner surface area ($50-600 \, \text{m}^2\text{g}^{-1}$) and large pores (about $10-60$ mm) ensure that the specific microcavities formed by the imprinting process (about $0.5-1.5$ nm in diameter) are readily accessible, and smaller molecules can diffuse freely inside the pores.

The enantiomer selectivity of these polymers is strongly dependent on the type and amount of cross-linking agent used during the polymerization (Wulff et al., 1982, 1987; Wulff, 1986, 1995). With ethylene dimethacrylate as the cross-linking agent, it was observed that for polymers containing < 10% cross-linking (Fig. 5) virtually no specificity was observed. Up to 50% cross-linking the α-value increases linearly to 1.50. From 50 to 66.7% cross-linking a dramatic increase in α-value from 1.50 to 3.04 was observed, thus implying a fourfold increase in selectivity over this range. With further increase in cross-linking up to 95% the specificity rises to $\alpha = 3.66$. On the other hand, with the use of tetramethylene dimethacrylate and especially p-divinylbenzene as cross-linking agents, a much lower specificity is observed as a function of cross-linking percentage.

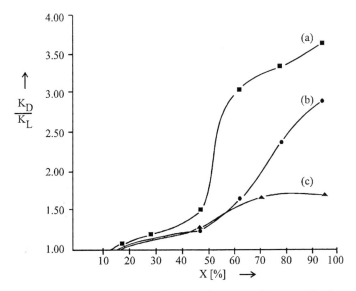

Figure 5. Selectivity of polymers as a function of the type and amount (X) of cross-linking agent (Wulff et al., 1982). The polymers were prepared in the presence of **1** with various proportions of the cross-linking agents ethylene dimethacrylate (a), tetramethylene dimethacrylate (b), and divinylbenzene (c). After removal of the template **2** the separation factor $\alpha = K_D/K_L$ was determined for the resolution of D,L-**2** in a batch process.

Polymers obtained with ethylene dimethacrylate as cross-linker retained their specificity for a long period. Even under high pressure in a high-performance liquid chromatography (HPLC) column, the activity remained for months. This was true even when the column was used at 70–80°C. On the other hand, polymers cross-linked with divinylbenzene gradually lost their specificity at higher temperatures. Interestingly, at 60°C the α-value for racemic resolution was further increased to 5.11. With relatively low ratios of substrate to polymer the selectivity further increases up to values $\alpha \cong 6.0$ (Wulff, 1995).

In this connection it must be mentioned that, as with polyclonal antibodies, the individual cavities have varying degrees of selectivity. The selectivity distribution was determined in a multisite model where association constants for each newly occupied binding site were calculated. This showed that a certain proportion of the cavities has relatively low selectivity (changes due to shrinkage or swelling, position on the surface), while the others have a distribution of selectivities around an average figure that can give very high selectivities (Wulff et al., 1977b).

The function of the binding sites

In the imprinting procedure the binding groups have several functions during the different steps of imprinting (Wulff, 1986, 1993, 1995). The requirements for the consecutive steps are quite different.

(1) During *polymerization* a stoichiometric, complete conversion between binding site and substrate functional group should be achieved to have a 1:1 proportion of both groups. The bonding should be stable under the polymerization condition. Furthermore, it is desirable that the bonding orientation is fixed in space.

(2) The *splitting* off of the templates after polymerization should be possible under mild conditions and should be as complete as possible.

(3) During *equilibration* with substrates binding and release should be very quick and reversible. It should be possible to adjust a favorable position of the binding site equilibrium in order to use it for different purposes and to control the selectivity by thermodynamic means. The binding site interaction should be as selective as possible, and it should be fixed in space.

Interactions can be covalent or noncovalent; both types show advantages and draw-backs. With covalent interactions the conditions under (1) can easily be met since this binding is stoichiometric, stable and fixed in space. Problematic are conditions (2) and (3). Only a few covalent interactions can be cleaved completely from the polymer and show an easily reversible interaction during rebinding.

A good example is binding *via* a boronic ester. In this case the formation of the ester according to Eq. (a) can easily be made quantitative by removal of the water formed. The bonding is stable during the polymerization and can be cleaved afterwards by addition of water or alcohol. For the equilibration it is important that the kinetics can be enhanced by more than 6 orders of magnitude by the addition of ammonia. Therefore, the attainment of equilibrium according to Eq. (b) is extremely quick and in the same order of magnitude as with many noncovalent interactions.

With noncovalent interactions (e.g. electrostatic or hydrogen bonding) during formation of the bond a considerable excess (usually at least fourfold) of binding groups is required in the polymerization mixture so that the binding sites in the template are completely saturated at equilibrium. In this way, a considerable proportion of the binding groups are incorporated randomly into the polymer, thus reducing selectivity. Furthermore, under these conditions only 10–15% of the cavities can be reoccupied since by swelling of the polymer matrix the cavities are partially shrunk in size (Sellergren and Shea, 1993b). With stoichiometric binding the great majority of the cavities can be reoccupied. On the other hand, the template is usually very easily removed after noncovalent binding, and fast and reversible interactions with substrates are possible in most cases. When producing chromatographic materials for analytical purposes, noncovalent interactions are usually preferred, as the materials are more readily obtainable and an excess of binding groups apparently does not have a detrimental effect on the separations. For the construction of catalysts, the orientation of the binding groups and catalytically active groups in the cavity are of greater significance, so that covalent interactions should be more advantageous here.

There is at present a lot of activity going on to develop new and better binding sites for imprinting. A very promising type of binding during polymerization and later in the final polymer can be achieved by coordinative bonds to metals. This type of bond is analogous to that used in ligand exchange chromatography. The advantage of this kind of interaction is that its strength can be controlled by experimental conditions. Definite interactions occur during the polymerization, and an excess of binding groups is not necessary. The subsequent binding of the substrate to the polymer is so rapid that in many cases even rapid chromatography is possible. This method has been used by Fujii et al. (1985) and worked out in great detail by Dhal and Arnold (1991) and Vidyasankar et al. (1995). Arnold's group used bonding of imidazole-containing compounds with Cu^{2+} complexes [Eq. (c)]. Model experiments were first carried out in which bisimidazoles with various distances between the imidazole groups were used as templates to position polymerizable iminodiacetate groups in the polymer. These experiments were aimed at developing an effective recognition pattern for proteins that depends on the correct spatial arrangement of a few binding sites on a polymer, recognizing the histidyl residues at the protein.

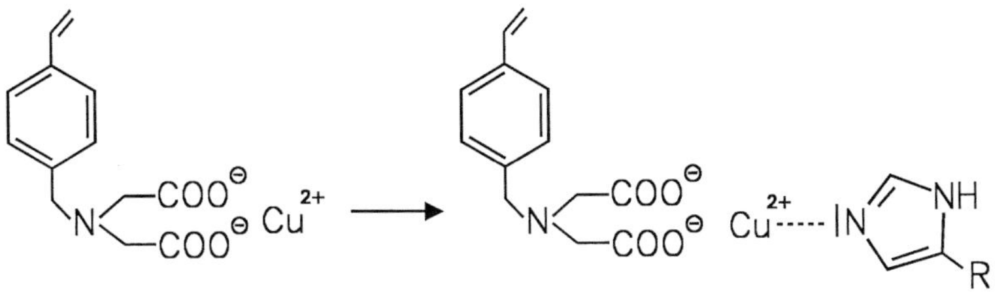

(c)

In order to meet the very different requirements of the binding sites during polymerization and the final equilibration, different types of interactions during polymerization and during equilibration were used. For example, during polymerization, a Schiff base interaction between a polymerizable amine and an aldehyde was used. Later, during equilibration, electrostatic interaction between the polymer-bound amine and a carboxyl group is used (Wulff et al., 1986). Another way to combine the advantages of covalent bonding during imprinting with noncovalent interaction during equilibration was reported by Whitcombe et al. (1995). They employed 4-vinylphenyl carbonate esters during imprinting. The ester is efficiently cleaved hydrolytically afterwards with a loss of CO_2. The resulting binding-site phenol is capable of interacting with an alcohol through hydrogen bonding.

It is to be expected that by far more and better binding interactions will be applied soon since a number of research groups are working in this direction.

Applications in sensors

Imprinted polymer membranes or polymer surfaces could be used as the basis of chemosensors. These are considerably more stable than biosensors and therefore have a longer lifetime. In principle, layers selective for many

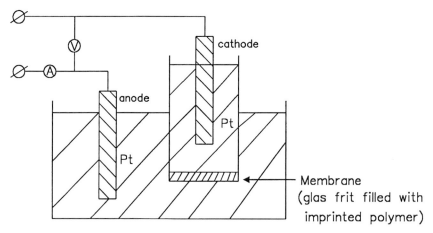

Figure 6. Scheme of setup for measurement of electrodialysis (Piletsky et al., 1992, 1994).

substances can be prepared by molecular imprinting. In practice, the preparation of effective and selective sensors is still beset with considerable difficulties. One problem is the development of a suitable detection system.

Piletsky et al. (1990, 1992, 1994, 1996) have investigated the development of sensors based on imprinted polymers. They prepared imprinted polymers in the form of free-standing membranes (Piletsky, 1990) or membranes supported on glass frits (Piletsky, 1992, 1994). The polymers were prepared from monomeric 2-diethylaminoethyl methacrylate and ethylene dimethacrylate, and were imprinted with adenosine monophosphate, amino acids, cholesterol, or artrazine by noncovalent interactions. Electrodialysis showed that these membranes were selectively permeable as represented in Fig. 6. Considerable selectivity for racemates was also obtained. The limit of detection in these systems is around 0.05 mMol/L. The mechanism of these remarkable separations does not yet seem to be fully explained.

A recent paper on sensors for sialic acid described the use of fluorescence spectroscopy for detection (Piletsky et al., 1996). Fluorescence seems a very promising detection technique due to the high sensitivity of the system in which 1 µMol/L sialic acid can be detected. Selectivity, however, was not very high in this system. Fluorescence has also been used by Kriz et al. (1995). In another experiment a membrane was imprinted with tryptophan, and this compound could be selectively transported through it, but the capacity for selective transport can be inhibited by photoisomerization of the polymer (Marx-Tibbon and Willner, 1994). This transport of amino acids through the polymer can thus be regulated by light. Sensors were also prepared by polymerization of an imprinted polymer layer on a silicon wafer with a thin coating of SiO_2 (Hedborg et al., 1993). This very refined design (see Fig. 7) gives a product with selectivity for various amino acids.

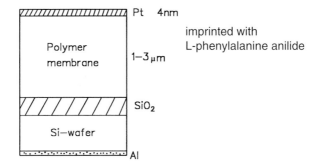

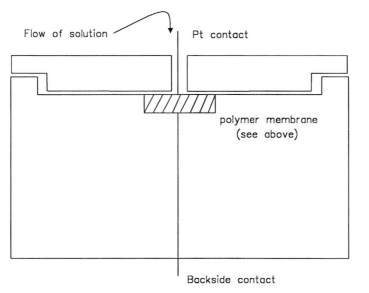

Figure 7. Experimental setup for capacitance-voltage measurements and a schematic cross section of the polymer membrane field-effect capacitor (Hedborg et al., 1993).

Differences in embedding of the substances on the polymer are measured by changes in the capacitance-voltage curve. The detection limit is around 2–5 mMol/L. This system seems promising but has to be improved considerably before routine use.

An interesting new route for the construction of sensors was found by D.C. Sherrington's group (Steinke et al., 1996). They succeeded in synthesizing imprinted macroporous polymers as completely transparent monoliths. The template molecules bound to the polymer (for example, Michler's ketone) are irradiated with polarized light, and the molecules lying in the direction of excitation react with the polymer matrix. After removal of the unchanged template molecules, the anisotropic polymer contains empty

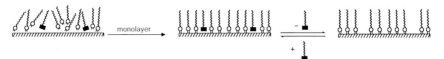

Figure 8. Schematic representation of the production of mixed monolayers with template molecules on glass. Components that are only physisorbed can be removed (Sagiv, 1979).

cavities that not only preferentially bind the templates but orient them in a particular direction. The polymers therefore show a pronounced dichroism in UV light, which enables nonselective binding to be distinguished.

A method of modifying surfaces with imprinting by self-assembling systems was used by J. Sagiv (1979). Mixed monolayers of trichloro-*n*-octadecylsilane and a dye modified with a detergent as the template were absorbed on glass and chemically bound as siloxanes. The modified dye molecule that was not covalently bound could then be removed by dissolution. This left holes within the stable network of chemisorbed and poly-condensed silane molecules. Such layers showed preferential adsorption of the template used (Fig. 8).

I. Tabushi et al. were the first to prepare sensors purely by surface imprinting (Yamamora et al., 1979). They used Sagiv's method and bound octadecylsilyl monolayers together with inert template hosts (*n*-hexadecane, 2-cholesteryl-3-6-dioxadecylcarbonate, or decyladamantane-1-carboxylate) covalently to tin dioxide or silicon dioxide layers. After extraction of the hosts, vitamin K_1, K_2, and E as well as cholesterol and adamantane were detected by strong electrochemical signals. Other possibilities for detection are surface-enhanced resonance Raman spectroscopy (Kim et al., 1988) and ellipsometry (Andersson et al., 1988). Silicon wafers with SiO_2 coatings function similarly (Starodub et al., 1991). In this case the templates (nucleotides, nucleosides, and amino acids) are adsorbed to the coatings, and the free silanol groups are then treated with trimethylchlorosilane or octadecyltrichlorosilane. After extraction of the template contours of the molecules are left behind on the surface of the silica. The same method has been used on the surface of glass electrodes.

Conclusion

By molecular imprinting highly selective adsorbants can be prepared. There is a lot of activity going on to develop molecular sensors on this basis that can change spectroscopical or electrochemical properties in response to the presence of biologically important chemical substances. Molecularly imprinted polymers promise a number of advantages over biological receptors. They possess higher thermal and mechanical stability. Therefore, their life expectancy should be high and they may even be sterilized for use, e.g., in fermentors. By tailor-making the polymer it should be possible to prepare sensors for all kinds of substances, and by using the correct templates one can prepare selective sensors for individual substances as well as for a certain group of substances with a characteristic structural feature. As synthetic polymers they can be fabricated with reproducible optical and electrical properties and can be processed as thin layers on a solid material, as membranes or as porous beads.

For routine application there are still a number of problems to overcome. Selective detection of substances on the surface of membranes while in contact with a solution is still a problem. Non-selective adsorption cannot easily be distinguished. There are some promising approaches under way so that interesting results can be expected in the near future.

References

Andersson, L.I., Mandenius, C.F. and Mosbach, K. (1988) Studies on guest selective molecular recognition on an octadecylsilylated silicon surface using ellipsometry. *Tetrahedron Lett.* 29:5437–5440.

Dhal, P.K. and Arnold, F.H. (1991) Template-mediated synthesis of metalcomplexing polymers for molecular recognition. *J. Am. Chem. Soc.* 113:7417–7418.

Fujii, Y., Matsotani, K. and Kikuchi, K. (1985) Formation of a specific coordination cavity for a chiral amino acid by template synthesis of a polymer Schiff base cobalt(III) complex. *J. Chem. Soc. Chem. Commun.* 415–417.

Hedborg, E., Winquist, F., Lundstrom, I., Andersson, L.I. and Mosbach, K. (1993) Some studies of molecularly-imprinted polymer membranes in combination with field-effect devices. *Sens. Actuators A* 37–38:796–799.

Kim, J.-H., Cotton, T.M. and Uphaus, R.A. (1988) Molecular recognition in monolayers and species detection by surface-enhanced resonance Raman spectroscopy. *Thin Solid Films* 160:389–397.

Kriz, D., Ramström, O., Svensson, A. and Mosbach, K. (1995) Introducing biomimetic sensors based on molecularly imprinted polymers as recognition elements. *Anal. Chem.* 67:2142–2144.

Marx-Tibbon, S. and Willner, I. (1994) Photostimulated imprinted polymers: A light-regulated medium for transport of amino acids. *J. Chem. Soc. Chem. Commun.* 1261–1262.

Mosbach, K. (1994) Molecular Imprinting. *Trends Biochem. Sci.* 19:9–14.

O'Shannessy, D.J., Ekberg, B., Andersson, L.I. and Mosbach, K. (1989) Recent advances in the preparation and use of molecularly imprinted polymers for enantiomeric resolution of amino acid derivatives. *J. Chromatogr.* 470:391–399.

Pauling, L. (1940) A theory of the structure and process of formation of antibodies. *J. Am. Chem. Soc.* 62:2643–2657.

Piletsky, S.A., Dubey, I.J., Fedoryak, D.M. and Kukhar, V.P. (1990) Substrate-selective polymeric membranes. Selective transfer of nucleic acids components. *Biopolm. Kletka* 6:55–58.

Piletsky, S.A., Butovic, I.A. and Kukhar, V.P. (1992) Design of molecular sensors based on substrate-selective polymer membranes. *Zh. Anal. Klim.* 47:1681–1684.

Piletsky, S.A., Parhometz, Y.P., Lauryk, N.V., Panasyuk, T.L. and El'skaya, A.V. (1994) Sensors for low molecular weight organic molecules based on molecular imprinting technique. *Sens. Actuators B* 18–19:629–631.

Piletsky, S.A., Piletskaya, E.V., Yano, K., Kugimiya, A., Elgersma, A.V., Levi, R., Kahlow, U., Takeuchi, T., Karube, I., Panasyuk, T.L. and El'skaya, A.V. (1996) A biomimetic receptor system for sialic acid based on molecular imprinting. *Anal. Lett.* 29:157–170

Sagiv, J. (1979) Organized monolayers by adsorption III. Irreversible adsorption and memory effects in skeletonized silane monolayers. *Isr. J. Chem.* 18:346–353.

Sellergren, B. and Shea, K.J. (1993a) Chiral ion-exchange chromatography. Correlation between solute retention and a theoretical ion-exchange model using imprinted polymers. *J. Chromatogr.* 654:17–28.

Sellergren, B. and Shea, K.J. (1993b) Influence of polymer morphology on the ability of imprinted network polymers to resolve enantiomers. *J. Chromatogr.* 635:31–49.

Sellergren, B., Lepistö, M. and Mosbach, K. (1988) Highly enantioselective and substrate-selective polymers obtained by molecular imprinting utilizing noncovalent interactions. NMR and chromatographic studies on the nature of recognition. *J. Am. Chem. Soc.* 110:5853–5860.

Shea, K.J. (1994) Molecular Imprinting of Synthetic Network Polymers: The *De Novo* Synthesis of Macromolecular Binding and Catalytic Sites. *Trends Polym. Sci.* 2:166–173.

Starodub, N.F., Piletsky, S.A., Lavryk, N.V. and El'skaya, A.V. (1993) Template sensor for low weight organic molecules based on SiO_2 surfaces. *Sens. Actuator B* 13–14:708–710.

Steinke, J.H.G., Dunkin, I.R. and Sherrington, D.C. (1995) Imprinting of Synthetic Polymers Using Molecular Templates. *Adv. Polym. Sci.* Vol. 123:81–126.

Steinke, J.H.G., Dunkin, I.R. and Sherrington, D.C. (1996) Molecularly imprinted anisotropic polymer monoliths. *Macromolecules* 29:407–415.

Vidyasankar, S. and Arnold, F.A. (1995) Molecular Imprinting: Selective materials for separation, sensors and catalysts. *Curr. Opin. Biotechnol.* 6:218–244.

Vidyasankar, S., Dhal, P.K., Plunkett, S.D. and Arnold, F.H. (1995) Selective ligand-exchange adsorbents prepared by template polymerization. *Biotechnol. Bioengin.* 48:413–436.

Whitcombe, M.J., Rodriguez, M.E., Villar, P., and Vulfson, E.N. (1995) A new method for the introduction of recognition site functionality into polymers prepared by molecular imprinting: Synthesis and characterisation of polymeric receptors for cholesterol. *J. Am. Chem. Soc.* 117:7105–7111.

Wulff, G. (1986) Molecular recognition in polymers prepared by imprinting with templates. *In:* W.T. Ford (ed.): Polymeric reagents and catalyst, ACS Symposium series, American Chemical Society, Washington, D.C., pp. 186–230.

Wulff, G. (1993) The role of the binding-site interactions in the molecular imprinting of polymers. *TIBTECH* 11:85–87.

Wulff, G. (1995) Molecular Imprinting in Cross-Linked Materials with the Aid of Molecular Templates – A Way towards Artificial Antibodies. *Angew. Chem. Int. Ed. Engl.* 34:1812–1832.

Wulff, G. and Minarik, M. (1990) Template imprinted polymers for h.p.l.c. separation of racemates. *J. Liq. Chromatogr.* 13:2987–3000.

Wulff, G. and Sarhan, A. (1972) Use of polymers with enzyme-analogous structures for the resolution of racemates. *Angew. Chem. Int. Ed. Engl.* 11:341.

Wulff, G., Sarhan, A. and Zabrocki, K. (1973) Enzyme-analogue built polymers and their use for the resolution of racemates. *Tetrahedron Lett.* 44:4329–4332.

Wulff, G., Vesper, W., Grobe-Einsler, R. and Sarhan, A. (1977a) Enzyme-analogue built polymers, IV. On the synthesis of polymers containing chiral cavities, and their use for the resolution of racemates. *Makromol. Chem.* 178:2799–2816.

Wulff, G., Grobe-Einsler, R., Vesper, W. and Sarhan, A. (1977b) Enzyme-analogue built polymers, V. On the specificity distribution of chiral cavities prepared in synthetic polymers. *Makromol. Chem.* 178:2817–2825.

Wulff, G., Kemmerer, R., Vietmeier, J. and Poll, H.G. (1982) Chirality of vinyl polymers. The preparation of chiral cavities in synthetic polymers. *Nouv. J. Chim.* 6:681–687.

Wulff, G., Heide, B. and Helfmeier, G. (1986) Molecular recognition through the exact placement of functional groups on rigid matrices via a template approach. *J. Am. Chem. Soc.* 108:1089–1091.

Wulff, G., Vietmeier, J. and Poll, H.G. (1987) Influence of the nature of the cross-linking agent on the performance of imprinted polymers in racemic resolution. *Makromol. Chem.* 188:731–740.

Yamamura, K., Hatakeyama, H., Naka, K., Tabushi, I. and Kurihara, K. (1988) Guest selective molecular recognition by an octadecylsilyl monolayer covalently bound on an SnO_2 electrode. *J. Chem. Soc. Chem. Commun.* 79–81.

Frontiers in Biosensorics I
Fundamental Aspects
ed. by F.W. Scheller, F. Schubert and J. Fedrowitz
© 1997 Birkhäuser Verlag Basel/Switzerland

Biomimetic recognition elements for sensor applications

U. E. Spichiger

Centre for Chemical Sensor/Biosensors and bioAnalytical Chemistry, Swiss Federal Institute of Technology (ETHZ-Technopark), CH-8005 Zurich, Switzerland

Summary. Ethanol and other alcohols with low molecular mass, glucose and creatinine are the analytes and/or substrates which have to be most frequently determined to obtain analytical informations in a variety of different process. The extensive literature on the development of biosensors is indicative of the interest in these sensors. In addition to enzyme-coupled recognition in biosensors, chemical host-guest interactions based on synthetic hosts are currently being investigated. Such a research project has applied an optical ethanol sensor to monitor ethanol generation in a bioreactor continuously. In this chapter, some examples of the state-of-the-art and recent results of these research projects are presented. Difficulties encountered in coupling a transducer to the synthetic host-guest recognition process, and in two-phase systems are described.

Introduction

The ideal chemical sensor can be regarded as a technical analogue of the human or animal sensory system. It responds continuously to the increasing and decreasing activities of a species. The term "selectivity" was coined in order to describe quantitatively the capability of a chemical sensor to prefer specific target analytes and to discriminate specific background species. The chemical sensor, characterized by the recognition process involved, must either respond reversibly or work under steady-state conditions in order to allow real-time continuous monitoring. The chemical recognition process usually known as *host-guest interaction* involves the fields of coordination chemistry, supramolecular chemistry, reversible chemical reactivity, receptor- and immuno-chemistry (for an overview, see Spichiger, 1994).

For specific substrates, such as monosaccharides (glucose), phenols, lactate, enzymes were extensively used as routine reagents and as host molecules in order to create biosensors. An enzyme presents a very well-defined binding site to the analyte. The selectivity of the catalytic activity makes it a suitable host molecule or reagent for chemical analysis. However, enzymes require special operating and storage conditions, which limit their activity and their widespead use. The lifetime of most enzymatic sensors is not comparable to that of chemical sensors used in ion-selective analysis (for overviews, see Gorton et al., 1991; Gilmartin and Hart, 1995). From this point of view, a synthetic approach in order to create analogues to the active site may well be promising.

The key to analyte-specific enzymatic activity is its specific reactivity and the typical conformation of the active site. Many enzymes require the presence of a cofactor which can be firmly attached to the protein core of the active site. The coenzyme defines the specific chemical reaction between the substrate or analyte and the enzymatic host. For a large number of substrates, a very limited number of chemical reactions are operative, which indicates that the environment within the active site and the specific conformation of the active site contribute considerably to the specificity. The catalyzed chemical reactions, as well as many of the enzyme conformations, have been documented (Brookhaven Protein Data Bank). This fact has stimulated two different approaches to creating synthetic analogues of the enzymatic host-guest interaction or *biomimetic recognition elements.*

Firstly, it has led to the computer-aided design of models of the active site and to the design of its synthesis using an analytical approach where the synthetic analogues or synthons are screened. Secondly, it has stimulated the mimicking of the chemical reactivity of the active site with a host-guest interaction without truely mimicking the turnover and generating the typical products. The first approach to designing biomimetic recognition elements is the objective of an ongoing National Foundation project in our Centre. A combination of the two dichotomous approaches, which would allow functional and structural analogy to be combined, would be the target to aim for.

One of the objectives of designing chemical sensors is to incorporate the chemical recognition process into a typical layer which is part of a two-phase system in contact with the specimen. In view of the significant contribution of the environment to the host-guest interaction within the enzyme's cavity, the physico-chemical features of the bulk of the sensing layer are decisive. This design provides two advantages: Firstly, the selectivity of the recognition process is supported by a phase-transfer reaction and, thus, by an extraction and separation step. Secondly, the bulky constitution of the membrane phase allows additives to be incorporated which can act as catalysts of the chemical reaction or as phase-transfer catalysts. Alternatively, the transfer enthalpy of the analyte contributes considerably to the total free energy of the sensing process and might constitute a crucial factor, as will be demonstrated with the creatinine sensor.

A fundamental problem to be solved in new designs of substrate sensing probes is that of converting the specific chemical information inherent in a recognition system into a signal which can be read electronically. In order to create the prerequisites for such a conversion, a chemical recognition process must be coupled to a physical transducer. Coenzymes, such as $NADH/NAD^+$ and $FADH_2/FAD$, change their optical properties when reduced and oxidized. As in enzymatic assays where, in addition to its cosubstrate activity, the coenzyme is analytically used as the chemical transducer of the enzymatic turnover, a second chemical transducer is necessary in many synthetic approaches. While ions create a boundary potential upon partition between two phases, neutral compounds such as saccharides, urea, creatinine and many other biological substrates generate no easily detectable physical quantity upon extraction. Currently, novel transducers and novel chemical recognition models are being developed (Proceedings EUROPT(R)ODE '96: Freiner et al., 1995; Rapp et al., 1995; Scheggi, 1995; Gauglitz et al., 1995; Forschungszentrum Karlsruhe, 1995). This means that a wide range of designs for tackling specific analytical problems will become available in the future.

In the following, some preliminary examples of structural and functional analogues, i.e. of *biomimetic recognition elements for sensor applications,* will be presented. The first example mimicks the chemical reactivity at the active site. This sensing principle has been applied to optical monitoring of ethanol production along with the proceeding bioreactor process. The second example shows two very preliminary approaches to glucose recognition using boronic acid and octahydroxycalix [4] arene derivatives. Optode membranes have been developed for boronic acid esterification, coupling a second pH-dependent chemical recognition process to the basic host-guest interaction. The third section presents the development of biomimetic elements for creatinine recognition involving a chromogenic host compound. This section will elucidate the problems associated with implementing a basically working recognition principle into a working sensing element. Other recent approaches have been discussed earlier (Spichiger, 1993a; Spichiger et al., 1993b).

Biomimetic optode membranes for ethanol and humidity

The alcohol dehydrogenases (ADH), alcohol: NAD^+-oxidoreductases (EC.1.1.1.1.) are enzymes of broad specificity which are frequently used in biosensors. They reversibly oxidize a wide range of aliphatic and aromatic alcohols to their corresponding aldehyde and ketones using NAD^+ as a coenzyme. The crystal structure of the horse liver apo- and holoenzymes of ADH has been elucidated before 1983 (Spiro, 1983; Fersht, 1985). ADH belongs to a group of dehydrogenases where the redox reaction is activated by *metal-ion catalysis.* A zinc ion sits within a hydrophobic pocket and is coordinated to 2 cysteine moieties (Cys-46 and Cys-174) and to a histidine nitrogen (His-67) (see Fig. 1). In the hydrophobic pocket of the enzyme, the zinc atom acts as a Lewis acid and polarizes the hydroxyl group of the substrate, e.g. ethanol, in order to stabilize the anionic transition state and to prepare the transition of the leaving hydride ion. The zinc ion is known to possess an extremely weak Lewis acidity, and thus has an extremely low and reversible anion-binding capacity as well as reduced polarizing properties. Due to the coordination of the hydroxy-group to the metal center, the acidity of the alcohol is drastically increased and the oxidation and hydride transfer to NAD^+ is facilitated. In the NAD^+ binding region, the

THE SUBSTRATE (GUEST) BINDING POCKET

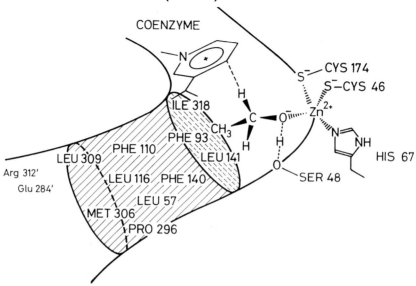

Figure 1. Model of the hydrophobic substrate binding pocket and the active site of ADH according to the data by Spiro et al. (1983) involving NAD⁺. Illustration of the position of the nicotinamide moiety of NAD⁺ and ethanol (substrate) in the active center of the LADH (human liver ADH) during oxidation and hydride transfer.

adenine moiety is inserted into a hydrophobic crevice, whereas the reactive nicotinamide unit makes contact to the polar active site. This principle is common to various enzymes and looks like a fundamental structural motif of NAD⁺-linked dehydrogenases (Stryer, 1995).

Enzyme catalyzed reaction (Bonnichsen and Theorell, 1951):

$$CH_3\text{-}CH_2OH + NAD^+ \underset{}{\overset{\text{ADH}}{\rightleftharpoons}} CH_3\text{-}CHO + NADH + H^+ \tag{1}$$

According to Holzer (1962) and Bonnichsen (1951) the equilibrium of the enzymatic reaction is on the left side. The pH of the in vitro reaction was optimized to pH 8.2–9.5 and semicarbazide was added to remove the aldehyde from the equilibrium reaction. LADH does not discriminate higher alcohols like amyl and allyl alcohol, propanol, butanol; also it is selective for methanol. Ethanol, which is coordinated to the metal center, is mainly present as an alcoholate at a local pH which must be < 7.4. The pK_a of ethanol is thus decreased from 16 to values around 7 within the lipophilic cavity of the enzyme and the vicinity of the zinc ion.

The principle of this reaction is especially effective where there is no water present to act as a competitive strong dipole. The fourth ligand of the zinc atom in the crystal structure is a water molecule. However, the lipophilic channel excludes water from the active cavity of the enzyme and enhances electrostatic and ion-dipole interactions. The hydrophobic environment of the pocket is also relevant since the free energy of electrostatic interactions is increased typically by a factor of 2 to 20 in apolar solvents with a lower relative dielectic permittivety, e_r, than water (Born, 1920; Landau und Lifshitz, 1963, 1984). Carbonyl oxygens of an amide, an aldehyde, a ketone or an alcohol as substrates can coordinate to the Zn^{2+} at the active site of the enzyme.

The analogy between the natural biological mechanism of alcohol recognition and a synthetic recognition process based on nucleophilic addition of ethanol to substituted trifluoroacetanilides is striking. In both cases an electrophilic center interacts with the alcoholic hydroxy group inducing the chemical transformation of ethanol.

The synthetic mechanism relies on the base-catalyzed formation of a hemiacetal with alcohol. The carbonyl carbon is polarized by the fluorine atoms and activated by the nucleophilic attack by OH^- (see diagram of the reaction principle in Fig. 2). Within the hydrophobic membrane phase, OH^- is provided by methyl-tridodecyl ammonium hydroxide, MTDDAOH, which is generated during the conditioning period of the optode membrane in water. OH^- is in thermodynamic equilibrium with the nucleophilic analyte and is quantitatively replaced by the alcoholate due to the electrostatic interaction with the lipophilic ammonium cation. The function of

MECHANISM OF ETHANOL RECOGNITION BY ETH 6022

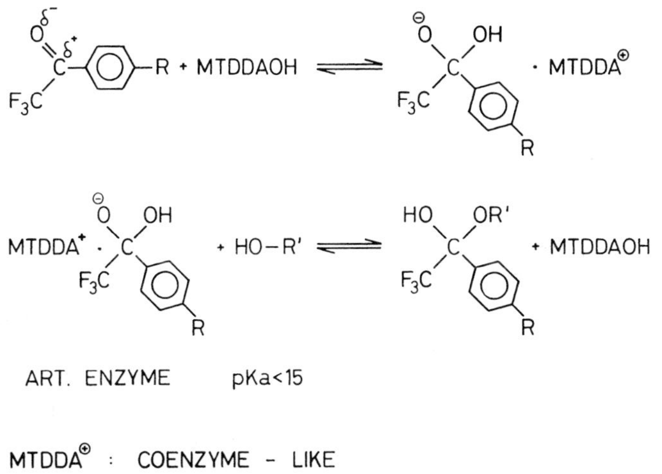

Figure 2. Suggested mechanism for ethanol recognition by ETH 6022 (*N*-dodecyl-*N*-(4-trifluoroacetylphenyl)-acetamide) trapped in a plasticized PVC bulk membrane. The interaction between the ligand and analyte is catalyzed by the lipophilic cation MTDDA$^+$ and OH$^-$. The pK$_a$ of ethanol within the membrane medium is considerably decreased (Spichiger et al., 1992).

the ammonium cation is analogous to that of the coenzyme, NAD$^+$, in this step, which induces the hydride transfer from the alcohol to the coenzyme. In this case no second chemical recognition step is needed. The optical absorption spectrum of the hemiacetal is shifted to higher frequencies as compared to the spectrum of the free ligand at $\lambda_{max} = 305$ nm.

Within the hydrophobic environment of a plasticized, polymeric bulk membrane, the pK$_a$ of the alcohol's hydroxy group is significantly decreased. The same is true for water dissolved within the membrane medium (max. 5% for a membrane with a low dielectric constant of approx. 4). Hence the water is also in a thermodynamic equilibrium with the ligand and ethanol. The selectivity coefficient, $K_{ROH.H2O}^{opt}$, of the ligand, L, for alcohols, ROH, compared with that of water has to be taken into account when the water concentration or activity, a_{H2O}, of the specimen changes significantly. The mixed thermodynamic equilibrium between water with the overall equilibrium constant K_{H2O} and alcohol (K_{ROH}) can be described by eq. 2 and 3. The square brackets indicate the concentration of the ligand and the specific hemiacetale within the membrane. a_{ROH} denotes the alcohol activity of the specimen.

$$K_{ROH.H2O}^{opt} = K_{H_2O}/K_{ROH} = \frac{[L \cdot H_2O]}{[L] \cdot a_{H_2O}} \cdot \frac{[L] \cdot a_{ROH}}{[L \cdot ROH]} \tag{2}$$

$$= k_{H_2O} \cdot \beta_{H_2O}/k_{ROH} \cdot \beta_{ROH} \tag{3}$$

k denotes the partition coefficient of water or alcohol between the membrane and the aqueous phase, and β specifies the binding constant of each species with the ligand. The selectivity coefficients of ETH 6022 for ethanol against methanol, n-propanol and 2-propanol were determined in the aqueous and in the gas phases (Kuratli, 1993). These have been confirmed by recent investigations (Wild et al., 1996). Water contributes with approx. 11% to the signal of ethanol in aqueous solution and with approx. 47% to the signal in the gas phase. In most cases, the water concentration of a specific type of specimen can be assumed to be constant.

The selectivity of the host molecule is influenced by the p-substituents of the anilinamide which define the electrophilicity of the carbonyl carbon by the σ-effect (Spichiger et al., 1992). Due to the increased transfer enthalpy, analytes with higher polarity and lower pK$_a$, such as anions or water, need membranes that incorporate ligands where the carbonyl carbon is more electrophilic and the p-substituents are strong electron acceptors (pos. σ-effect). The increased electrophilicity of the carbonyl carbon compensates for the lower solubility (k in eq. 1) of these analytes in the apolar membrane phase. This finding was decisive for using ETH 6010 (heptyl-4-trifluoro-acetyl benzoate) and ETH 6019 (1-(n-dodecylsulfonyl)-4-trifluoracetyl benzene) to prepare two humidity sensing optodes with different sensitivi-

Table 1. Constitutions and Hammett constants of 4-trifluoroacetyl benzene derivatives used as synthetic host molecules as discussed in the text.

Synonym	Derivatives R $F_3CCO-\langle\bigcirc\rangle-R$	Hammett constants* σ
ETH 6004	$-N^{\underset{\sim}{N}}N-\langle\bigcirc\rangle-N\begin{smallmatrix}C_2H_4OCOC_3H_7\\C_2H_4OCOC_3H_7\end{smallmatrix}$	aryl azo: 0.13
ETH 6010	$-COOC_7H_{15}$	0.52
ETH 6019	$-SO_2C_{12}H_{25}$	0.73
ETH 6022	$-N\begin{smallmatrix}C_{12}H_{25}\\COCH_3\end{smallmatrix}$	-0.01

* Jaffé, H.H. (1953) A Reexamination of the Hammett Equation. Chemical Reviews 53, 191–261.

ties (Wang et al., 1991) and an optode which was carbonate sensitive in aqueous solutions based on ETH 6004 (azobenzene derivative) (Behringer et al., 1990) in the absence of alcohols. A strong competition between chemical selectivity and solubility or partition of the analyte between the aqueous phase and sensing layer obviously characterizes the behavior of these sensing membranes.

The ligand, ETH 6022 (N-dodecyl-N-(4-trifluoroacetylphenyl)-acetamide), was used recently as a biomimetic recognition molecule in optode membranes in order to monitor the increasing concentration of ethanol in the vapor phase of a bioreactor (Wild et al., 1996). This may be the first report on continuous process monitoring using a fully synthetic chemical sensing membrane. The analysis is performed without any sample pretreatment. The same sensing membrane was used over more than 6 days.

Tests on a number of different critical interfering analytes in aqueous solutions showed no cross-reactivity (Seiler et al., 1991). Recently, a comparison of the reactivity of different alcohols was made by Wild et al. (1996). Due to the high activity of n-butanol in water and its solubility within the membrane phase, even the low vapor pressure was found to be

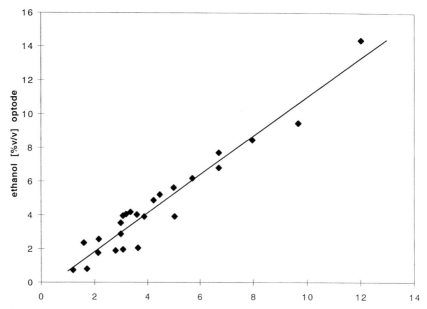

Figure 3. Summarized results of a method comparison between the optical sensing layer and the reference method for 4 different batches of the reactor filling using the same pair of active and reference membranes. Slope 1.14 ± 0.06; $R^2 = 0.9354$; intercept -0.48 ± 0.32; $s_{y.x} = 0.80\%$ v/v. The comparison is based on the first order linear regression model. Data of the reference method (x-axis) were assumed to be error-free.

more than compensated for and n-butanol was preferred over ethanol, methanol and propanol in gas-phase monitoring (Wild et al., 1996). In most cases, the lack of discrimination of n-butanol will not be relevant. Characteristic curves of ethanol production, e.g. by yeast cells, were monitored over 4 runs on 4 different days. After calibration of the optode membrane, a determination coefficient of $R^2 = 0.9354$ was evaluated from the 1st order linear regression of the optode results vs. the results of the reference method on the x-axis using distillation and measurement of the refractive index (Helrich, 1990) ($Y_{optode} = -0.48 + 1.14\ X_{reference}$; $s_{y.x} = 0.8\%$ v/v ethanol). The results with the optode generally fitted those predicted by the reference method well. The short lifetime of the solvent polymeric membrane was no problem for measurements between 1 and 10% ethanol in the gas phase. Nevertheless, it is a problem for solutions with ethanol concentrations of $\geq 12\%$, although these membranes with a thickness of $2-4\ \mu m$ have never been adapted to these specific applications. For optical membranes generally, it was shown that the viscosity of the sensing layers might be increased, thus making them mechanically much more resistant and less sensible to leaching of components (Spichiger et al., 1995). The activity of water was taken into account for calculations of the alcohol concentrations in various beverages (Seiler et al., 1991).

Molecular recognition of monosaccharides and catechol

In recent years, considerable attention has been focused on the development of electroanalytical biosensors based on glucose oxidase (GOD, EC 1.1.3.4) and glucose dehydrogenase (GDH, EC 1.1.1.47), which have been the most successful biorecognition elements (for reviews, see Gorton et al., 1991; Gilmartin et al., 1995). Developments have concentrated mostly on the medical market. Although the surface modification and immobilization methods, as well as the applied mediators and electrode materials, have become increasingly sophisticated, there are no well-established general strategies for preparing a glucose sensor at present. There is an enormous number of publications on biosensing systems produced every year. Nevertheless, the actual number of commercialized biosensors is comparatively small. The reasons are discussed elsewhere in this volume (e. g. Pfeiffer, Lüdi).

At first glance, enzymatic sensors offer a solution to the selectivity problem and allow a generally working sensor to be produced. The chemical host, on the other hand, needs a relatively long development period in order to yield the required selectivity. Additionally, there are problems implementing the host into a sensing layer and coupling it to a transducer. The following examples show that host-guest interactions do not obey the same rules in a pure organic solvent as compared to a separated layer with a phase transfer involved. The phasetransfer problems might be underestimated in some cases. Along with the development of biosensors, considerable effort has been made to develop synthetic glucose- or carbohydrate-selective compounds. These efforts have resulted in sensing devices which favor the more lipophilic catechol derivatives. In the following section two examples of these will be discussed.

Molecular recognition by lipophilic boronic acids, optical oligool-selective bulk membranes

When titrating boric acid, Mellon and Morris (1924) found that the pK_a of boric acid was significantly decreased (pK_a ca. 5) upon adding fructose and hexitols (mannitol, sorbitol) to the solution. In comparison, glucose induced a pK_a of about 7.2, which differed from the equilibrium constant of the boric acid acidity by a factor > 100. It has been established that the basis of the acidification effect is the formation of cyclic anionic boric and boronic acid diesters of the type shown in eq. 4.

erythritan

catechol

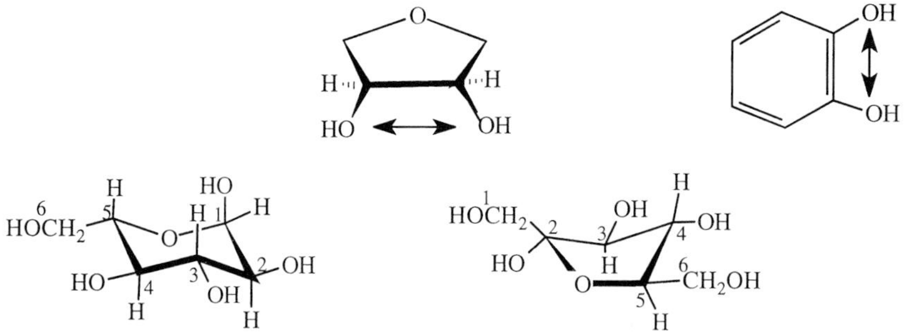

α-D-glucopyranose   β-D-fructofuranose

Figure 4. Structures of diol substrates/analytes.

Further, acidification has been shown to be accompanied by an increase in electrical conductivity which can be used for analytical purposes. The enhancement of the conductivity is a measure of the extent of formation of the esters and their stability at low pH values. The steric conditions (distance) between the neighbored diol groups clearly played an important role enabling the esterification with the boric and boronic acids. The lowest pK_a of boric acid was determined with erythritan (3,4-cis-dihydroxy-tetrahydrofuran) resulting in a pK_a of 4.8. The intramolecular distance between the two exocyclic oxygen atoms of an erythritan as derived from crystal structures is 254 pm. Comparatively, that between exocyclic oxygen atoms in catechol is 279 pm (Sybyl software package, 1995).

In diluted solutions, mononuclear species of boric acid are formed opposed to the crystalline borates. The boric acid molecule is trigonal and tetrahedral as the hydroxide anion. Boric acid acts not as a proton donor, but as a Lewis acid, accepting hydroxyl ions. The borate anion is tetrahedral, with bond angles of ~109°. Because of the small size of the atom, boron has a maximum coordination number of 4. In pure water, boric acid is a weak Lewis acid with a pK_{a1} of 9.24 and of pK_{a2} 12.74 (Dean, 1979).

Boronic acid derivatives can reversibly be converted to alkyl or aryl orthoborates. Nevertheless, steric considerations and the conformation of the oligohydroxy compound or *oligool* are very critical. Mazurek and Perlin (1963) found that the most stable diesters were formed with cisdiols attached to a furanoid ring. ^{11}B and ^{13}C NMR studies elucidated the structures of the diesters and showed that, in the case of acyclic hexitols, the diester formation occurs between trans-OH. However, the interaction with the furanoid form yields the lowest free energy of conformation since the position of the OH-groups, and thus the distance between the oxygens of the diol is fixed. Additionally, the ring oxygen activates, as an electron acceptor, the OH groups. According to Makkee et al. (1985), a dihedral angle of 0–40° for the O-C-C-O group favors the formation of stable anionic boronic acid diesters corresponding to a predicted distance between O-O centers of 249–263 pm. This distance corresponds best to the furanoid cis-diols, such as ribose and α-D-fructofuranosid where the distance between exocyclic oxygen atoms is 260 pm between C_1C_3 oxygen atoms for the α-D-fructofuranose and 261 pm for the C_2C_3 oxygens of β-D-fructofuranose (see Fig. 4) as determined from crystal structures. In contrast, the hydroxy groups of α- or β-D-glucopyranose do not fit the boronic acid diols without conformational changes. The distances between the exocyclic oxygens C_1C_2, C_2C_3 and C_3C_4 are in the range from 283 to 289 pm, whereas the C_4C_6 distance amounts to 442 pm. In aqueous solutions of fructose, the furanosid cis-diol groups are naturally occurring; fructose equilibrates between 26.6% of β-D-fructofuranose and 73.4% β-D-fructopyranose due to mutarotation whereas glucose equilibrates between 65% of β-D-glucopyranose and 35% α-D-glucopyranose. Based on these considerations, a higher stoichiometry than 1:1 of the fructose- and also glucose-boronic acid complexes are rather unlikely. For catechols with a distance of 279 pm between the exocyclic oxygens atoms, the acidity of the HO-groups rather than the conformation might contribute to the reactivity.

Investigations of optical diol-selective bulk membranes were focused on the three different approaches described in the following:

1. ETH 1999 (octadecanoic acid 3-(4-(dihydroxyborylphenoxy)-propyl-ester) (see Fig. 5) was incorporated into a solvent polymeric membrane phase using a non-aromatic plasticizer with low permittivity (bis(ethyl-hexylsebacate), DOS) and PVC (poly(vinylchloride)) as bulk constituents. The cyclic triesterboroxin shows aromatic character and a UV absorption maximum at 236 nm. The decreasing absorbance of the boroxin at $\lambda = 235$ nm upon complexation of diols was determined at pH 7.16, 8.15 and 9.19. In order to fulfill the electroneutrality conditions, an equivalent amount of lipophilic ammonium ions (tridodecylmethyl ammonium chloride), R^+, with their counterions, X^{n-}, was added to the membrane bulk.

The optode reaction can be described by the following equilibrium between aqueous sample phase (aq) and bulk membrane (m):

$$\text{oligool (aq)} + nR^+(m) + \frac{n}{v} X^{v-}(m) + n/3 \text{ (boroxin) (m)} + 3H_2O$$
$$\rightleftharpoons \text{(oligool. boronate)}^{n-}(m) + nR^+(m) + \frac{n}{v} X^{v-}(aq)$$
$$+ H_3O^+(aq) + (3-n) \text{ boronic acid} \tag{5}$$

where the stoichiometry, n, of the diol-boronic acid ester is assumed to be 1:1 or 1:2; and v-denotes the charge number of water soluble anions within the membrane equilibrating the charge of R^+. Based on measurements at $\lambda = 235$ nm, the cyclic trimeric anhydride of a boronic acid derivative was shown to be in equilibrium with the free compound and the ester when adding a diol. The interaction between hydrogen ions, boroxin and the

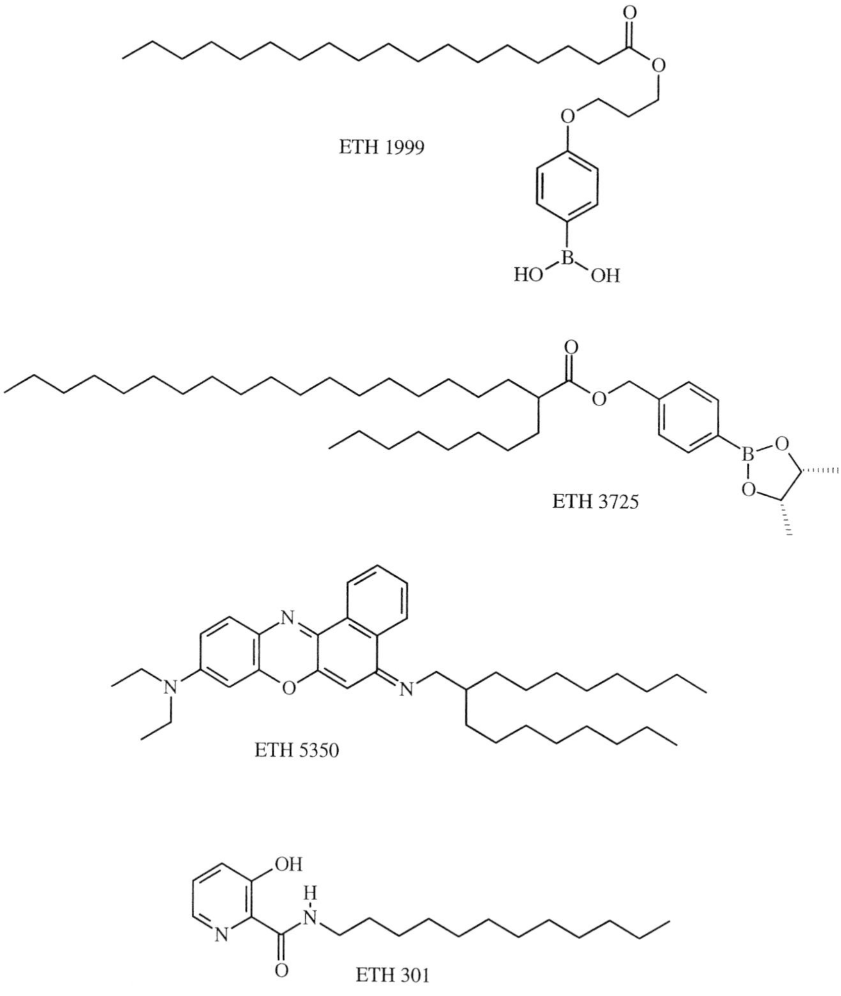

ETH 1999

ETH 3725

ETH 5350

ETH 301

Figure 5. Constitutions of two host molecules and two pH-sensitive optochemical transducers as used and cited for glucose recognition.

anions, X^{v-}, as well as the release of H_nX to the aqueous phase, was shown to induce a pH-dependence of the whole system and an anion sensitivity.

The detection limit for glucose was $2.5 \cdot 10^{-2}$ M (at pH 9.2 in Tris-sulfate buffer). Fructose was preferred (detection limit: $5.8 \cdot 10^{-3}$ M). The reaction principle of this optode is based on the release of an equivalent amount of hydrogen ions and anions to the sample phase.

2. In this approach another prototype membrane, based on the boronic acid derivative, ETH 3725 (2-octyl eicosanoic acid 4-[4,5-dimethyl-(1,3,2)-dioxaborolan-2-yl]-benzyl ester) (see Fig. 5) was used. By adding a basic indicator, ETH 5350 (9-(diethylamine)-5-[(2-octadecyl)imine]-5H-

benzene[a]phenoxazine) (see Fig. 5) to the membrane phase as a hydrogen ion acceptor, the sensitivity and other characteristics of the optode membrane were improved. The esterification is coupled to the protonation of the lipophilic indicator resulting in a bathochromic shift in its absorbance spectrum from $\lambda = 498$ nm of the unprotonated, C, to $\lambda = 655$ nm of the protonated pH-sensitive chromoionophore, CH^+ (see Fig. 6). The indicator is strongly basic (pK$_a$ 12.0 in methanol; pK$_a$ 13.4 within a DOS/PVC membrane (Brakker, 1993) relative to the weak acidity of the boronic acid host and catalyzes the esterification. The analytical wavelength is shifted to the visible range. At 235 nm, the generation of boroxin was no longer observed. The fraction of boroxin depends largely on the degree of hydration of the membrane bulk, and on the pH and the pK$_a$ of the boronic acid derivative used as a host molecule. Nevertheless, a high working pH ist required in the sample solution buffered to pH > 9, in order for this type of membrane to work. The extraction of the oligool can be denoted by the following thermodynamic equilibrium between aqueous sample phase (aq) and bulk membrane (m):

oligool (aq) + nC (m) + nboronic acid (m)

$$\rightleftharpoons ester^{n-}(m) + nCH^+(m) + nH_2O \text{ (aq)} \qquad (6)$$

With ETH 3725, the sensitivity is enhanced compared to that of ETH 1999 due to the slightly decreased pK$_a$ of the boronic acid derivative. The p-substituent in ETH 1999, -OCH$_2$R, exhibits a more negative σ-effect and, thus,

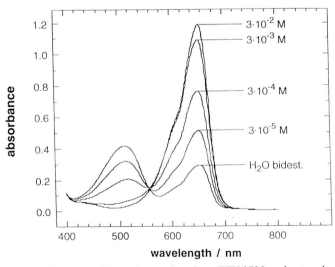

Figure 6. Spectrum of two optode membranes based on ETH 3725 as host molecule and ETH 5350 as indicator equilibrated with unbuffered aqueous catechol solutions (concentrations $3 \cdot 10^{-5}$ to $3 \cdot 10^{-2}$ M). For membrane composition, see text. Membrane thickness and total path length are approx. 6 µm (2 optodes).

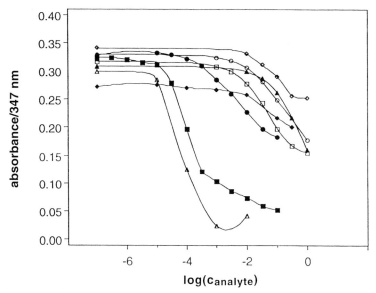

Figure 7. Optical absorbance at 347 nm as a function of the analyte concentration (for the path length, compare legend of Fig. 6). The aqueous solutions were buffered to pH 9.2 with Tris/Cl⁻. The analytes were △ catechol, ■ salicylate, ● oxalate, ◆ 1,2-hexanediol, □ fructose, ○ glucose, ▲ galactose and ◇ xylose. For membrane composition, see text. The incorporated host compound was ETH 1999.

increases the pK_a relative to the p-substituent, -CH$_2$OR, in ETH 3725. The ligand was isolated as the ester which is supposed to equilibrate with the free boronic acid within the hydrated membrane bulk. The membrane was prepared with the following components: ETH 3725 (2.5 wt%), ETH 5350 (2.3 wt%), o-NPOE as a plasticizer (78.2 wt%) (DK ≈ 23) and PVC (28.8 wt%). The detection limit for catechol was 10^{-5} M, that for glucose approx. $2.0 \cdot 10^{-2}$ M.

3. In the third approach, the acid indicator, ETH 301 (N-dodecyl-3-hydroxypicolinic acid amide), was used. Here the lowest detection limits were observed ($2.0 \cdot 10^{-6}$ M), especially for catechols, whereas the detection limit for glucose did not change significantly ($4.0 \cdot 10^{-2}$ M). The membrane showed a considerable drift in adsorbance, probably due to association between R⁺ and the negatively charged ester or due to an interaction between the indicator and boronic acid. The drift was influenced by the permittivity of the plasticizer. Deprotonation of the indicator was not fully reversible.

$$\text{oligool (aq)} + nR^+(m) + nC^-(m) + n\text{boronic acid (m)}$$
$$\rightleftharpoons \text{ester}^{n-}(m) + nR^+(m) + CH(m) + nH_2O \text{ (aq)} \qquad (7)$$

With ETH 301, the absorbance was evaluated at $\lambda = 347$ nm (see Fig. 7); however, the quenching of the fluorescence emission might have been evaluated as an alternative. Fructose was preferred by a factor of about 10. With ETH 3725 or ETH 1999 (2.5 wt%), ETH 301 (2.3 wt%), MTDDACl (2.5 wt%) and o-NPOE (57.7 wt%) as a plasticizer, the detection limits were as follows: $2.0 \cdot 10^{-6}$ M catechol, $6.3 \cdot 10^{-3}$ M fructose and $4.0 \cdot 10^{-2}$ M glucose.

Conclusion: The structures of the boronic acid derivatives used as synthetic host molecules do not ideally fit the vicinal exocyclic diol groups in glucose. From the conformational point of view, fructose and also some pentoses must be preferred over glucose. For catechol, the solubility in the apolar phase and the acidity of the diol-groups are likely to contribute to the discrimination pattern. Nevertheless, there is a real possibility that the esterification and solubilization of glucose can be favored by multipodal ligands. Progress can also be made by using boronic acid derivatives substituted by electron acceptors.

Lipophilic resorcinol-aldehyde cyclotetramers

Multidentate alternative hosts have been described by Aoyama et al. (1989) and Aoyama (1993). The host molecule consists of a lipophilic resorcinol cyclic tetramer incorporating four pairs of hydrogen-bridged hydroxyl groups on adjacent benzene moieties. Some monosaccharides, especially ribose, are selectively solubilized in chloroform or carbon tetrachloride that both contain the synthetic host molecule. Complexation was studied by ^{1}H NMR and circular dichroism (CD) spectroscopy. These ligands were not incorporated in sensing layers or devices.

Molecular recognition of creatinine

Creatinine and urea are diagnostic target substrates which allow the overal renal function to be examined by so-called renal clearance studies. Creatinine is the most commonly used endogenous substance for clearance studies since the venous blood level is normally independent of nutrition. Creatinine is a degradation product of creatine and consequently of creatine phosphate; production and excretion relate directly to the muscle mass and are individual characteristics. Creatinine belongs to the most frequently measured analytical parameters in the clinical laboratory.

Creatinine can basically form 5 tautomers. However only the two tautomers, a and b, were traced in the solid state or in solution (see Fig. 8). Theoretically, the tautomer a is the most stable, and certainly more stable than b (Butler, 1985). This was confirmed by Bell et al., (1995) and by Bühlmann (1993).

Bell et al., (1995) identified a reversed order of the relation between the bond length of the intracyclic and the exocyclic nitrogen to the C_2 carbon

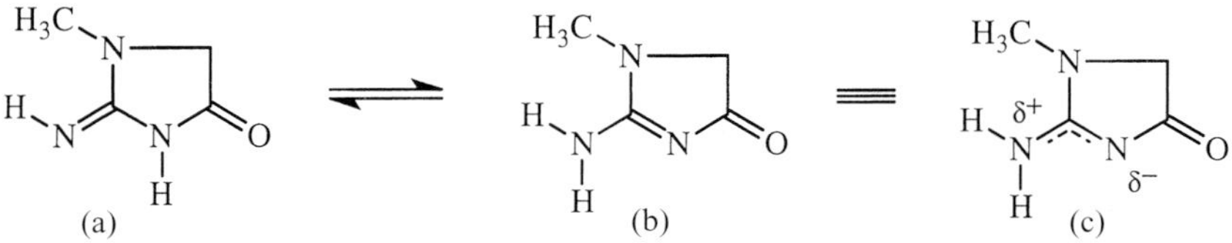

Figure 8. The tautomers a and b as well as the transition state c discussed by Bell et al. (1995).

atom due to electron delocalization. Their investigations relied on crystal structure analysis which enabled them to derive structure c as the transition state. Other groups found that only the tautomer b exists in aqueous solutions (Butler and Glidewell, 1985; Kenyon and Rawley, 1971). Creatinine is characterized by two pK_a values of 4.88 (20 °C) for protonation and 12.7–13.4 for deprotonation of the exocyclic amine group (discussed by Bühlmann, 1993). This suggests that creatinine predominantly occurs in aqueous solution and in blood plasma at pH 7.4 in neutral form.

The classical method for determining creatinine in the daily routine is the Jaffé reaction. The adduct or complex between picrate and creatinine in alkaline solution is evaluated in an optical assay. In view of the large commercial potential for a creatinine sensor, descriptions of a number of potentiometric and amperometric creatinine biosensors have been published (for an overview see Bühlmann, 1993), but these do not fulfill all the requirements. Further, the low physiological concentrations in human blood are crucial: 57–90 µmol L^{-1} for females and 72–105 µmol L^{-1} for males (Spichiger, 1989). Since even the routine optical assays have certain typical drawbacks (Henry, 1991), a selective synthetic recognition process can still contribute significantly to more reliable analysis and provide an improved device for direct analysis in blood plasma or whole blood.

Two different preliminary approaches to developing optical sensors will be described in the following. In both cases, the ligand is at the same time the opto-chemical transducer.

Host-guest interaction by cooperative hydrogen bonds

Hydrogen bonds constitute a biologically adequate compromise for receiving relatively stable and efficient interactions which are also reversible in an aqueous environment. The exchange of hydrogen ions is a basic principle in enzyme-catalyzed reactions; hydrogen bonds are basic in fixing the conformation of polypeptides and nucleotides as postulated by Frey-Wyssling (1938) for plants. In 1951, Pauling suggested the two protein structures, the α-helix and the β-strand, based on x-ray data (Pauling and Coray, 1954). In 1953, Watson and Crick published the x-ray diffraction pattern of a hydrated DNA string and derived a model of its structure (Watson and Crick, 1953).

The first attempt at molecular recognition of creatinine made use of these models (Bühlmann, 1993). First investigations were based on isocytosine isologues as ligands in order to solubilize creatinine within an apolar sensing layer.

The ability of pyrimidone derivatives to complex creatinine has been investigated by spectroscopic methods (Bühlmann, 1993; Bühlmann and Simon, 1993). In organic solvents containing 4-(1-heptyloctylamino)-4($3H$)-pyridone (ETH 1417) and 2-amino-6-(1-octylnonyl)-4($3H$)-pyrimidone (ETH 1413 see Fig. 9), creatinine proved to be strongly solubilized. The latter showed an absorption maximum at $\lambda = 284$ nm with a molar absorption coefficient, $\varepsilon_{\lambda max,d} = 7.45 \cdot 10^3$ M^{-1}cm^{-1}. Adding an alkylated creatinine isologue (ETH 1414, 2-amino-1,5-dihydro-1-(1-heptyloctyl)-4H-imidazol-4-one) to a 0.51 mM solution of the ligand in methylene chloride and increasing the concentration from 0 to 10 mM, raised the absorbance by up to 14% (1 mm optical pathlength). The tautomers thought to participate in the reaction were discussed based on spectroscopic data and considerable attention was given to their dimerization. The latter is a considerable drawback when using allophanate derivatives.

The ligand ETH 1418 (2-(2-naphtylamino)-6-propyl-4($3H$)-pyrimidone) (see Fig. 10) proved to have very interesting properties. The UV-spectrum of this compound in methylene chloride at $\lambda = 308$ nm shows an absorption coefficient of $1.56.10^4$ M^{-1}cm^{-1}, which is 2.1 times larger than that of ETH 1413 at 284 nm. The addition of one mole equivalent of the alkylated creatinine analogue ETH 1414 to a 0.5 mM solution of ETH 1418 resulted in an increase by 0.418 absorption units, i.e. a 54% increase compared with the absorbance of a solution containing pure ETH 1418 (see Fig. 10). Based on the results, bulk optode membranes were prepared from 2 wt% of the ligand, using PVC as the polymer and various plasticizers as solvents. A 200 mM creatinine solution in phosphate buffer pH 7 did not

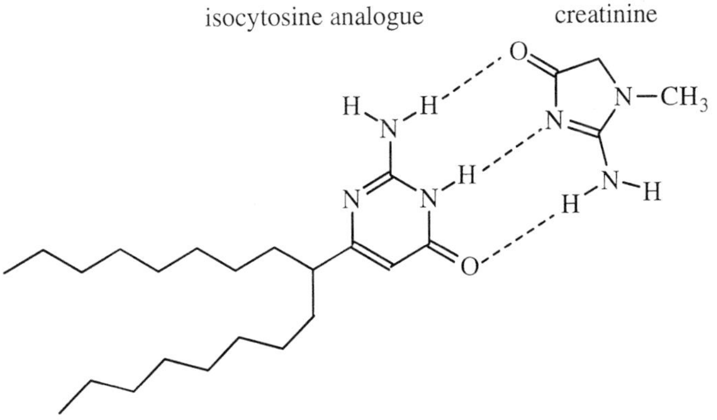

Figure 9. The Watson-Crick like model of creatinine with ETH 1413.

ABSORBANCE

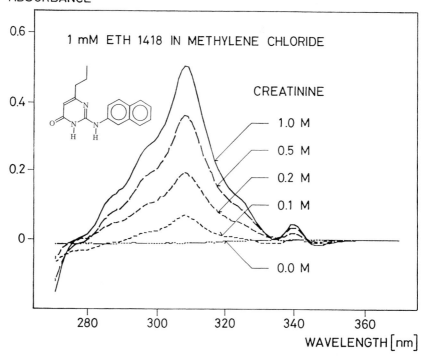

Figure 10. Differential UV spectrum between a cell containing a 1 mM chloroform solution of ETH 1418 equilibrated with aqueous creatinine solutions vs. a reference cell containing 1 mM ETH 1418 in chloroform equilibrated with dist. water. The two absorption maxima are at $\lambda = 308$ and 340 nm.

alter the spectrum of the ligand within the membrane. It was, however, observed that any change in the concentration of the sample solution induced a loss in absorbance of the ligand. This was interpreted as a loss of ligand to the aqueous sample solution. A more lipophilic isologue ligand, ETH 1421 (2-(2-naphtylamino)-6-(1-octylnonadecyl)-4(3H)-pyrimidone), was evaluated. Even if the extraction of creatinine into methylene chloride was as successful as with ETH 1418, the extraction into bulk optode membranes made of 2.6 wt% ligand, 64.6 wt% DOS and 32.8 wt% PVC was not sensitive enough for practical use. It was however, reversible and concentration-dependent at the level of 100 to 500 mM creatinine in the aqueous phase (1.5% absorbance change upon exposure to a 500 mM creatinine solution). Clearly, the phase-transfer enthalpy of an analyte into organic solvents varies considerably from that of the extraction into a solvent polymeric membrane.

From these experiments it could be shown that two and three hydrogen bonds correspondig to approx. 7.5 kJ mol^{-1} are not sufficient to extract creatinine into an organic solvent with the relevant concentrations. The solubility of creatinine in water is relatively high (87 g L^{-1} (16 °C). Molecular models for multitopic recognition were proposed.

A chromogenic creatinine ligand

The second approach relies on the chromogenic and charge-transfer properties of picrate. Bell et al. (1995) synthesized a tripodal receptor incorporating a nitroaromatic chromophore (9-*n*-butyl-13-hydroxy-10,12-dinitro-7,8-dihydro-naphthyridino[2,3-c]acridine). Upon complexation of creatinine by three hydrogen bonds, an anionic charge-transfer complex results. The delocalization of a hydrogen ion and of the charge produces a spectral shift and a broad absorption band in a water-saturated chloroform solution at $\lambda = 444$ nm. Creatinine was extracted from an aqueous solution buffered to pH 6.0 (0.1 M MES-buffer) into a 0.14 mM chloroform solution of the ligand. A titration curve, absorbance = $f(c_{creatinine})$, where the creatinine concentration, c, ranged from 0 to 16 mM, resulting in a change of absorbance between > 0.2 to 1.4, was reported. The selectivity against electrolyte ions was determined for chloroform solutions and resulted in recommending a correction of the physiological Na^+ background between 130 to 150 mM. However, the selectivity factor of 80 might cause problems in real samples.

The paper of Bell et al. (1995) focuses rather on structure elucidation than on the preparation of a sensing structure. The dissociation constant of the creatinine-ligand complex of 0.5 µM in water-saturated chloroform might allow for a reversible operation. The use of the host-guest interaction for preparing a sensing device appears attractive. A higher alkylation of the ligand to improve its solubility in an apolar environment is necessary.

For measurements in blood directly at pH 7.4, the low pK_a of the dinitrophenol moiety ($pK_a \approx 4$) along with the low basicity of the aromatic nitrogens ($pK_a \approx 5$) of the polycyclic ligand may constitute a problem. The complexation and the optical absorbance spectrum will show a relevant pH and solvent sensitivity. In an apolar membrane environment, the pK_a of the ligand can be shifted by 3–4 orders of magnitudes. Under these conditions, the ligand will mainly be present in the zwitterionic or neutral form. An intelligent design of a sensing layer is necessary in this case and could also improve the discrimination of sodium ions.

The structure of the free ligand and the complex was confirmed by x-ray structure analysis. The molecular recognition process can be compared to the electrophilic reaction and acid catalysis as exhibited by proteases (Fersht, 1984). The ligand presented by Bell et al. (1995) acts on the basis of the electronic complementarity and the geometrical, conformational fit between ligand and substrate and is not intended to hydrolyze the compound. Also, it does not need the extended area of the active site of an enzyme, e. g. pepsin.

Conclusions and outlook

Taking into account the relative success of the investigations of synthetic recognition elements for sensors reported here and their potential for chemically modifying the characteristic features of a ligand and for influencing the selectivity assisted by the solubility of the participating compounds, the question arises why this field has been neglected in comparison with the large number of biosensors studied. Many examples of chemical ion-selective sensors exist which demonstrate the practicability of the synthetic approach. A large number of other models, such as the research on urea ligands, the chemical discrimination between enantiomers of 1-phenylethylamine cations, the determination of its enantiomer excess, and the optical sensing of ammonia in the gas phase are further examples of the range of possible synthetic recognition elements.

Acknowledgements
Part of this work was supported by the Swiss National Science Foundation, by the Swiss Priority Program "OPTIK" and by the EUREKA-project "ASREM". This financial support is highly acknowledged.

References

Aoyama, Y. (1993) Molecular recognition of sugars. *Trends Anal. Chem.* 12:23–28.

Aoyama, Y., Tanaka, Y. and Sugahara, S. (1989) Molecular Recognition. 5. Molecular Recognition of Sugars via Hydrogen-Bonding Interaction with a Synthetic Polyhydroxy Macrocycle. *J. Am. Chem. Soc.* 111:5397–5404.

Auterhoff, H., Knabe, J. and Höltje, H.-D. (1991) *Lehrbuch der Pharmazeutischen Chemie.* Wiss. Verlagsgesellschaft mbH, Stuttgart, pp 320–323.

Bakker, E., (1993) *Die Bedeutung von Phasentransfergleichgewichten für die Funktionsweise von ionenselektiven Flüssigmembranoptoden* und *-elektroden.* PhD Thesis Swiss Federal Institute of Technology (ETHZ), Zürich, Nr. 10229.

Bartels, H. and Cikes, M. (1969) Ueber Chromogene der Kreatininbestimmung nach Jaffé. *Clin. Chim. Acta* 26:1–10

Behringer, C., Lehmann, B., Haug, J.-P., Seiler, K., Morf, W.E., Hartmann, K. and Simon, W. (1990) Anion selectivities of trifluoroacetophenone derivatives as neutral ionophores in solvent-polymeric membranes. *Anal. Chim. Acta* 233:41–47.

Bell, T.W. and Liu, J.H. (1988) Hexagonal Lattice Hosts for Urea. A New Series of Designed Heterocyclic Receptors. *J. Am. Chem. Soc.* 110:3673–3674.

Bell, T.W., Hou, Z., Luo, Y., Drew, M.G.B., Chapoteau, E., Czech, B.P. and Kumar, A. (1995) Detection of Creatinine by a Designer Receptor. *Science* 269:671–674.

Böeseken, J. (1949) The Use of Boric Acid for the Determination of the Configuration of Carbohydrates. *Adv. Carbohydr. Chem.* 4:189–210.

Bonnichsen, R.K. and Theorell, H. (1951) An enzymatic method for the microdetermination of ethanol. *Scand. J. Lab. Invest.* 3:58–62.

Born, M. (1920) Volumen und Hydrationswärme der Ionen. *Z. Physik* 1:45–48.

Brookhaven National Laboratory, Department of Energy, Upton, N.Y., USA.

Bühlmann, P. (1993) *Molecular Recognition of Creatinine.* Thesis Swiss Federal Institute of Technology (ETHZ), Zürich, Nr. 10066.

Bühlmann, P. and Simon, W. (1993) Neutral Hosts for the Complexation of Creatinine. *Tetrahedron* 49:7627–7636.

Butler, A.R. and Glidewell, C.J. (1985) Creatinine: an Examination of its Structure and Some of its Reactions by Synergistic Use of MNDO Calculations and Nuclear Magnetic Resonance Spectroscopy. *J. Chem. Soc. Perkin Trans.* II:1465–1467.

Dean, J.A. (1979 *Lange's Handbuch of Chemistry*, 12th edn. McGraw-Hill, New York.

Fersht, A. (1984) *Enzyme Structure and Mechanism.* W.H. Freeman and Comp., New York.

Forschungszentrum Karlsruhe (1995), Technik und Umwelt, Nachrichten 1.

Freiner, D., Kunz, R., Citterio, D., Spichiger, U.E. and Gale, M. (1995) Integrated optical sensors based on refractometry of ion-selective membranes. *Sensor. Actuator. B* 29:277–285.

Frey-Wyssling, A. (1938) *Submikroskopische Morphologie des Protoplasmas and seiner Derivate*. Burnträger, Berlin, pp 112–119.

Gauglitz, G., Brecht, A. and Kraus, G. (1995) Interferometric biochemical and chemical sensors. *In*: A.V. Scheggi (ed.) *Chemical, Biochemical, and Environmental Fiber Sensors VII.*, Proc. SPIE 2508, 1995, pp 41–48.

Gilmartin, M.A.T. and Hart. J.P. (1995) Sensing With Chemically and Biologically Modified Carbon Electrodes, A Review. *Analyst* 120:1029–1045.

Gorton, L., Csöregi, E., Dominguez, E., Emneus, J., Jönsson-Pettersson, G., Marko-Varga, G. and Persson, B. (1991) Selective detection in flow analysis based on the combination of immobilized enzymes and chemically modified electrodes. *Anal. Chim. Acta* 250: 203–248.

Haug, J.-P. (1993) *Einsatz von lipophilen Boronsäuren in Optoden: Ein Ansatz zur Realisierung eines nicht enzymatischen Glucose-Sensors.* PhD Thesis Swiss Federal Institute of Technology (ETHZ), Zürich, Nr. 10230.

Helrich, K. (1990) *Official Methods of Analysis of the Association of Official Analytical Chemists*, 15th ed, vol. 2. Association of Official Analytical Chemists, Arlington, Virginia, p 739.

Henry, J.B. (1991) *Clinical Diagnosis & Management by Laboratory Methods.* W.B. Saunders Comp., Philadelphia.

Holy, P., Morf, W.E., Seiler, K., Simon, W. and Vigneron J.-P. (1990) Enantioselective Optode Membranes with Enantiomer Selectivity for (*R*)- and (*S*)-1-Phenylethylammonium ions. *Helv. Chim. Acta* 73:1171–1181.

Holzer, H. and Sölig, H.D. (1962) Bestimmung von L-Lactat, L-Malat, L-Glutamat and Aethylalkohol im enzymatisch-optischen Test mit Hilfe des DPN-Analogen 3-Acetyl-Pyridin-DPN. *Biochem. Z.* 336:201–214.

Israelachvili, J. (1992) *Intermolecular & Surface Forces.* Academic Press, London.

Kenyon, G.L. and Rowley, G.L. (1971) Tautomeric Preferences among Glycocyamidines. *J. Am. Chem. Soc.* 93:5552.

Kovar, K.-A. (1972) Meisenheimer-Komplexe. *Pharmazie in unserer Zeit* 1:17–20.

Krantz, J.C. Jr., Carr, C.J. and Beck, F.F. (1936) A Further Study of the Effect of Sugar Alcohols and Their Anhydrides on the Dissociation of Boric Acid. *J. Phys. Chem.* 40:927–931.

Kuratli, M. (1993) *Beitrag zur Entwicklung von optischen chemischen Sensoren. Organischchemische Reaktionen als Erkennungsprozesse.* PhD Thesis Swiss Federal Institute of Technology (ETHZ), Zürich, No 10380.

Kvassman, J., Laisson, A. and Petterson, G. (1981) *Env. J. Biochem.* 114:555–563.

Landau, L.D. and Lifshitz, E.M. (1963, 1984) *Electrodynamics of Continuous Media*, vol. 8, 2nd edn. Pergamon, Oxford.

Lindenmann, B.A. (1989) *Beitrag zur Entwicklung eines neuartigen Glucosesensors auf der Basis von PVC-Flüssigmembranen mit lipophilen Boronsäuren als substratselektiven Carriern.* PhD Thesis Swiss Federal Institute of Technology (ETHZ), Zürich, Nr. 8957.

Makkee, M., Kieboom, A.P.G. and van Bekkum, H. (1985) Studies on borate esters III. Borate esters of D-mannitol, D-glucitol, D-fructose and D-glucose in water. *Recl. Trav. Chim. Pays-Bas* 104:230–235.

Mazurek, M. and Perlin, A.S. (1963) Borate Complexing by Five-Membered-Ring *vic*-Diols. *Can. J. Chem.* 41:2403–2411.

Meyerhoff, M.E., Pretsch, E., Welti, D.H., and Simon, W. (1987) Role of Trifluoroacetophenone Solvents and Quarternary Ammonium Salts in Carbonate-Selective Liquid Membrane Electrodes. *Anal. Chem.* 59:144–150.

Pauling, L. and Coray, R.B. (1954) The Configuration of Polypeptide Chains in Proteins. *Fortschr. Chem. org. Naturstoffe* 11:180–239.

Proceedings of 3rd EUROPT(R)ODE`96 in Zürich, Switzerland (1996). Sensor. Actuator. B, Elsevier, Amsterdam.

Rapp, M., Moss, D.A., Reichert, J. and Ache, H.J. (1995) Acoustoelectric Immunosensor Based on Surface Transverse Waves for in Situ Measurements in Water. Proceedings of the 7th International Conference on Solid-State Sensor. Actuator. pp 538–540.

Santucci, L. (1983) Hydrolytic cleavage of triphenylboroxin in cyclohexane medium. *Gazz. Chim. Ital.* 113:515.

Seiler, K., Wang, K., Kuratli, M. and Simon, W. (1991) Development of an ethanol-selective optode membrane based on a reversible chemical recognition process. *Anal. Chim. Acta.* 244:151–160.

Spichiger, U.E. (1989) *A Selfconsistent Set of Reference Values.* PhD Thesis Swiss Federal Institute of Technology (ETHZ), Zürich, Nr. 8830.

Spichiger, U.E. (1994) *Chemical Sensors and Biosensors for Medical and Biological Applications: An area of Analytical Chemistry.* Habilitation Thesis, Swiss Federal Institute of Technology (ETH), Zürich, Switzerland.

Spichiger, U.E., Kuratli, M. and Simon, W. (1992) ETH 6022: An Artificial Enzyme? A Comparison Between Enzymatic and Chemical Recognition for Sensing Ethanol. *Biosensors Bioelectron.* 7:715–723.

Spichiger, U.E., Freiner, D., Bakker, E., Rosatzin, T. and Simon, W. (1993a) Optodes in Clinical Chemistry: Potential and Limitations. *Sensor. Actuator. B* 11:263–271.

Spichiger, U.E., Simon, W., Bakker, E., Lerchi, M., Bühlmann, P., Haug, J.-P., Kuratli, M., Ozawa, S. and West, S. (1993b) Optical Sensors Based on Neutral Carriers. *Sensor. Actuator.* 11:1–8.

Spichiger, U.E., Citterio, D. and Bott, M. (1995) Analyte-selective optodes membranes and optical evaluation techniques: characterization of response behaviour by ATR measurements. *In:* A.V. Scheggi (ed.) *Chemical, Biochemical, and Environmental Fiber Sensors VII.,* Proc. SPIE 2508, 1995, pp 179–189.

Spiro, T.G. (1983) *Zinc enzymes.* John Wiley & Sons, New York, pp 123–152.

Stryer, L. (1995) *Biochemistry.* W.H. Freeman and Comp., New York.

Sybyl, Software package Version 6.1 (August 1994). Tripos Inc., St. Louis, Missouri 63: 144–2913.

Wang, K., Seiler, K., Haug, J.-P., Lehmann, B., West, S., Hartmann, K. and Simon, W. (1991) Hydration of Trifluoroacetophenones as the Basis for an Optical Humidity Sensor. *Anal. Chem.* 63:970–974.

Watson, J.D. and Crick, F.H.C. (1953) Molecular Structure of Nucleic Acid. *Nature* 171:737f.

West, S.J., Ozawa, S., Seiler, K., Tan, S.S.S. and Simon, W. (1992) Selective Ionophore-Based Optical Sensors for Ammonia Measurments in Air. *Anal. Chem.* 64:533–540.

Wild, R., Critterio, D., Spichiger, J. and Spichiger, U.E. (1996) Continuous monitoring of ethanol for bioprocess control by a chemical sensor. *J. Biotech.* 50:37–46.

Frontiers in Biosensorics I
Fundamental Aspects
ed. by F.W. Scheller, F. Schubert and J. Fedrowitz
© 1997 Birkhäuser Verlag Basel/Switzerland

Screening and characterization of new enzymes for biosensing and analytics

W. Hummel

Institut für Enzymtechnologie der Heinrich-Heine-Universität Düsseldorf, Forschungszentrum Jülich, D-52404 Jülich, Germany

Summary. The development of new or improved analytical methods requires new enzymes. Screening techniques utilizing enrichment cultures and rapid assay methods supported by automated or miniaturized methods are useful tools to detect new enzyme producers. Notably, oxidoreductases are well suited for analytical purposes. The NAD(P)- and oxygen-independent quinoprotein dehydrogenases with a covalently bound redox cofactor can be used advantageously for the development of biosensors. Examples are given of selective enrichment methods used in screening for useful enzyme-producing microorganisms. Enrichment under chemostatic conditions proved to be successful because enzymes with a remarkably high affinity against the analyte could be obtained. This is demonstrated by the screening of a trimethylamine-converting enzyme. The frequently observed high selectivity of these enzymes against the substrate is demonstrated in a few examples. In exploitation of these new oxidoreductases, new analytical methods were developed which are useful for the detection and during monitoring of phenylketonuria (PKU) or maple syrup urine disease (MSUD).

Introduction

Enzymes as the biological compound of biosensors

Biosensors represent a highly sophisticated analytical tool consisting of the integrated arrangement of a biological signal producing component and a transducing measuring element. In principle, any system that reveals biological or biochemical reactivity against the substrate can be applied as the biological part, and indeed many kinds of biosensors have been developed using enzymes, multifunctional enzyme complexes, tissue slices, isolated cell organelles (e.g. chloroplasts), whole cells or antibodies. Animal or plant cells, cell cultures or microorganisms represent the sources of these compounds. In particular, isolated purified enzymes show some advantage for the development of biosensors; they are free of side reactions and they can be applied with a high specific activity which results in a dense coating of the transducing component with catalytic sites.

Requirement for enzyme properties

Enzymes as biological compounds of biosensors should meet several requirements that apply to other analytical devices as well, e.g. they should be absolutely specific for the analyte, or show a high stability. In contrast to other analytical devices such as cuvette assays, flow injection analysis, or test strips the sensing area element of biosensors is remarkably small and thus requires enzymes with a high specific activity. Moreover, coupling with additional reactions catalyzed by a second or even third enzyme limits the available area. Concerning the kind of reaction, it is to be seen from the literature that very often enzymes catalyzing redox reactions (oxidoreductases; E.C. group 1) seem to be well suited for biosensors. These reactions can be monitored with the aid of an electrode or redox dyes leading to amperometric or optical sensors, respectively. In addition to NAD(P)-dependent dehdrogenases and perioxide-producing oxidases, the group of quinoprotein dehydrogenases (Duine, 1991; Duine and Jongejan, 1989; Duine et al., 1980; D'Costa et al., 1986) is of particular interest because the cofactor is covalently bound to the enzyme and the reaction does not depend on the oxygen supply.

Enzyme sources

Enzymes can be obtained commercially or by isolation of the user himself. Although there is an increasing number of enzymes commercially available, new analytical problems or improvements of existing analytical methods still require new enzymes. This refers to the development of new enzymes for reactions that are so far unknown to be catalyzed by enzymes as well as to the development of altered properties of enzymes that are known in principle. Many enzymes were found to occur in varying structures with different molecular masses and differences concerning the amino acids in the active site. For example, there are about 20 variants of NAD-dependent alcohol dehydrogenase (E.C. 1.1.1.1) commercially available.

Results and discussion

Screening for new enzymes

The development of a new or improved enzyme generally starts with a search in literature data bases to find sources and isolation methods if there is no commercial source available. If such data are lacking, screening of organisms and cultures follows. Suitable methods for a screening depend in detail on the problem to be solved; only a few methods and techniques generally applicable are described and only a few reviews are available on this subject (Cheetham, 1987; Elander, 1987; Goodhue, 1982).

Selective methods for the enrichment of enzyme producers

Screening requires the development of an assay procedure to detect the desired activity as well as the development of enrichment strategies if microorganisms from soil samples are used. Microorganisms represent an inexhaustible source for new enzymes and enzyme variants (see "hydroxy-isocaproate dehydrogenase"-screening in this contribution), and on the other hand selective enrichment strategies are valuable tools to obtain the required enzymes.

Every kind of isolation of a microorganism can be considered as an enrichment procedure. For example, isolation under normal atmospheric aeration conditions leads to the enrichment of aerobic organisms while anaerobes are simultaneously suppressed; isolation at pH 7 enriches organisms growing at this pH and prevents at the same time the growth of acido- and alcalophiles (see "hydantoinase"-screening in this contribution). Efficient enrichment procedures however require detailed knowledge of the physiology involved. Proceeding on this assumption strategies may be developed to impose a selection pressure on the organisms containing the desired enzyme activity. The choice of the carbon and nitrogen sources or defined growth conditions can be decisive for the successful enrichment of organisms producing new enzymes or of similar organisms producing enzyme variants. In particular continuous cultivation techniques represent efficient methods of producing a selection pressure on cultures containing mixed populations. Enrichment under chemostatic conditions for example favors the growth of organisms which can survive under low limiting substrate levels (see "trimethylamine dehydrogenase" in this contribution).

Rapid detection of enzyme producers

Selective enrichment methods can only help to reduce the amount of potential organisms to be tested. In practice, several hundred strains or more must be assayed for the desired enzyme activity. This time-consuming part of the screening can be facilitated by rapid assays or techniques which support a high frequency of samples. In some cases, assays can be carried out directly on agar plates. For example a simple spot test for the detection of hydantoinase activity was developed (Morin et al., 1986) where enzyme activity can be detected by the yellow color around the active colonies which develops after application of a drop of p-dimethylaminobenzaldehyde. For the detection of acylase (Yamazaki et al., 1987) or esterase activity (Yamazaki and Kula, 1987) a layer technique was developed that can be carried out directly on agar plates. Because of the occurrence of interfering material from the cells methods are often helpful in excluding a large amount of negative enzyme producers.

Usually, the enzymes of interest for analytical application are located intracellular. In the progress of a screening, only a small amount of the cell-free extract sufficient for one or two assays is needed. For the disintegration of cell suspensions in a microliter scale a simple method is applicable

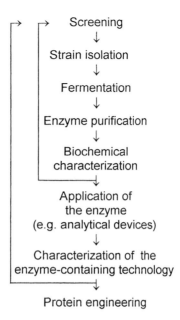

Figure 1. Schematic presentation of the different steps nessecary for the development of new enzyme.

where up to ten samples can be treated simultaneously. The cells are disrupted when ground with glass beads in disposable plastic reaction vials (Hummel and Kula, 1989). A high sample frequency can also be reached using automation of operation units such as automated chromatographic plants (Schütte et al., 1985), robots or computer-assisted solutions.

Biochemical characterization of new enzymes
For any analytical application, enzymes are needed with a high specificity and selectivity against the analyte. In order to estimate the usefulness of a new enzyme activity a comprehensive biochemical characterization of the purified enzyme is required.

Figure 1 demonstrates that at this step these data help to decide whether to continue the screening or to apply this enzyme to an analytical device. The application of the enzyme-supported analytical method on real samples represents another critical step. Interfering substances from the matrix or a reduced stability under assay conditions might be reasons to look for alternate enzymes. Problems like these can arise when several organisms producing the desired enzyme or variants of the desired enzyme resp. are obtained during the same screening procedure.

The use of different screening and enrichment methods for the detection of new enzymes as well as the most important biochemical data of some new enzymes will be demonstrated in the following examples.

Examples

Phenylalanine dehydrogenase

L-Phenylalanine is an important clinical parameter in diagnosis and therapy of disorders of phenylalanine catabolism. The Guthrie test (Guthrie and Susi, 1963), which is used worldwide for newborn screening only gives semiquantitative results with low precision, whereas column chromatography is more precise but time consuming and not applicable for the screening of newborns. We carried out several enzyme screening procedures in order to find an NAD-dependent phenylalanine dehydrogenase catalyzing the following reaction:

$$\text{L-phenylalanine} + NAD^+ \rightarrow \text{phenylpyruvate} + NADH + NH_4^+$$

In principle, such reactions are catalyzed by amino acid dehydrogenases such as leucine dehydrogenase or alanine dehydrogenase but none of the known enzyme converted any aromatic amino acid. The screening was carried out by direct plating of several soil samples on agar dishes with a medium containing L-phenylalanine as carbon and nitrogen source at 1% concentration (method A in Tab. 1). Only one active strain could be found belonging to the genus *Brevibacterium* (Hummel et al., 1984). Repeated attempts with further soil samples to obtain further active stains were absolutely unsuccessful. Only after coupling this enrichment method with a selective preculture step (method C), could significantly higher amounts of phenylalanine dehydrogenase-possessing microorganisms be obtained (Tab. 1). Microbiological characterization of the isolated strains (Hummel et al., 1987a) proved that they all belong either to the *Brevibacterium* or *Rhodococcus* group (Tab. 2).

Purification and characterization of phenylalanine dehydrogenases of *Brevibacterium* (Hummel et al., 1986; Hummel et al., 1988b) and *Rhodococcus* (Hummel et al., 1987b) revealed that the enzyme from *Brevibacterium* is quite unstable, whereas the enzyme from *Rhodococcus* can easily be purified (Tab. 3).

Table 1. Comparison of different methods for the isolation of microorganisms with phenylalanine dehydrogenase. Method A: L-Phenylalanine (1%) as carbon and nitrogen source. Method B: Growth on propionic acid. Method C: Enrichment of soil bacteria by growth on propionic acid, transfer on plates with 1% phenylalanine as carbon and nitrogen source

Isolation method	Number of soil samples	Number of isolated organisms	Number of phenylalanine dehydrogenase positive strains
A	106	1625	1
B	23	395	0
C	40	220	21

Table 2. Results of the screening of microorganisms containing phenylalanine dehydrogenase.
Method A and C as described in the legend of Table 1

Isolation method	Genus	Strain	Phenylalanine dehydrogenase	
			Specific act. [U/mg]	Volume act. [U/L medium]
A	*Brevibacterium*	VII/1	0.9	710
C	*Rhodococcus*	M4	16.0	13 000
C	*Rhodococcus*	F6	14.5	12 800
C	*Rhodococcus*	H8	13.3	11 800
C	*Rhodococcus*	I/3	11.0	8 600
C	*Brevibacterium*	I/2	1.5	640
C	*Brevibacterium*	V/II	0.5	205

Table 3. Biochemical data of phenylalanine dehydrogenase

Parameter	Properties
Enzyme source	*Rhodococcus spec.* M4
Specific activity[a]	technical product: 20 U/mg purified: 270 U/mg
Electron acceptor	only NAD^+
K_m-value (L-phe)	$7.5 \cdot 10^{-4}$ M
K_m-value (NAD^+)	$1.27 \cdot 10^{-3}$ M
pH optimum	9 – 10
Substrate specificity	L-Phenylalanine

[a] Assayed with L-phenylalanine and NAD^+.

For the intended analytical applications it was found that the enzyme did not react significantly with any compound occurring in physiological fluids, only L-tyrosine gave a slight signal in the range of < 5 %, compared to the signal of equimolar amounts of L-phenylalanine (Fig. 2).

This lack of interfering reactions makes the enzyme highly suitable for tests in physiological environments. By coupling with diaphorase/iodonitro tetrazolium chloride (INT) a fast and precise photometric assay could be developed (Wendel et al., 1989). A linear relationship between the dA (formed formazan) and the phenylalanine concentration was obtained with standard solutions in the range from 30 to 1200 µmol/liter. The applicability of this method is demonstrated by comparative determinations of plasma L-phenylalanine of treated phenylketonuria (PKU) patients by enzyme-catalyzed photometric assay and automated amino acid analysis (Fig. 3). Further studies demonstrated the applicability of this enzyme for the screening of PKU and its monitoring during dietary treatment (Hummel et al., 1989; Wendel et al., 1990a, b; Wendel et al., 1991).

Hydroxyisocaproate dehydrogenase
Ketoisocaproate (2-oxoisocaproate) is the toxic compound in MSUD (maple syrup urinary disease), an inherited deficiency of the branched-

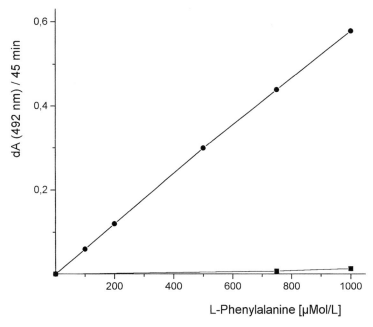

Figure 2. Calibration curves for L-phenylalanine (● – ●) and L-tyrosine (■ – ■) in a photometric assay with phenylalanine dehydrogenase (0.1 U/ml assay).

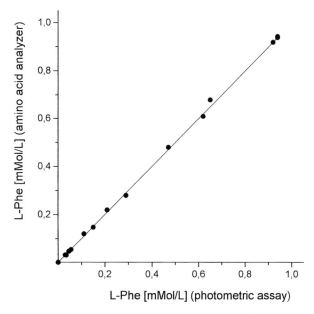

Figure 3. Plasma phenylalanine concentration in treated PKU patients measured by the enzyme-catalyzed photometric assay and on an amino acid analyzer.

chain 2-oxo acid dehydrogenase complex. This metabolic block results in a marked increase in plasma and tissue concentrations of leucine, valine, and isoleucine and their corresponding 2-oxo acids (2-oxoisocaproate = KIC, 2-oxoisovalerate = KIV, 2-oxo-3-methylvalerate = KMV) but KIC is the main neurotoxic substance among the accumulated branched-chain compounds if it exceeds 1 mM. A selective chromatographic determination of KIC is difficult to achieve, thus we tried to develop an enzymatic assay using hydroxyisocaproate dehydrogenase. This enzyme catalyzes the NADH-dependent reduction of keto acids:

$$\text{2-oxo acid} + \text{NADH} \rightarrow \text{2-hydroxy acid} + \text{NAD}$$

Such kinds of enzymes occur in strains of the genus *Lactobacillus*. We isolated and characterized several hydroxyisocaproate dehydrogenases in order to find a variant that converts KIC selectively (Schütte et al., 1974; Hummel et al., 1985, 1988a). The K_m-values of several *Lactobacillus*-enzymes for the three keto acids (Tab. 4) demonstrate, that the three structurally related compounds are all converted by most of the enzymes with nearly the same affinity. Solely the enzyme from *Lactobacillus casei* (DSM 20008) showed a pronounced specificity against KIC. This was remarkable because the enzyme from another *L. casei* strain (DSM 20244) was absolutely unspecific. The KIC-specificity of the *L. casei* DSM 20008 enzyme is confirmed by measurements with standard solutions of the three keto acids (Fig. 4). Comparative determinations of plasma samples using the enzymatic assay (photometric measurements) and HPLC resulted in a good correlation (Schadewaldt et al., 1989). Hydroxyisocaproate dehydrogenases can also be used for the determination of branched-chain amino acid aminotransferase activity in plasma (Schadewaldt et al., 1995). It is evident from these studies on the biochemical properties of several enzyme variants that for a successful development of an enzyme-supported method it is profitable to have enzyme variants from different sources available.

Table 4. Biochemical characterization of hydroxyisocaproate dehydrogenases from different sources

Strain	K_m-value [mM]			Relation K_m-values KIC : KIV : KMV
	Ketoiso-caproate (KIC)	Ketoiso-valerate (KIV)	Keto-methylvalerate (KMV)	
Lactobacillus confusus	0.06	0.07	0.50	1 : 1 : 8
Leuconostoc oenos	0.07	0.09	0.09	1 : 1 : 1
Lactobacillus curvatus	0.09	0.10	0.18	1 : 1 : 2
Lactobacillus casei (DSM 20244)	0.05	0.09	0.06	1 : 2 : 1
Lactobacillus casei (DSM 20008)	0.06	4.80	2.20	1 : 80 : 37

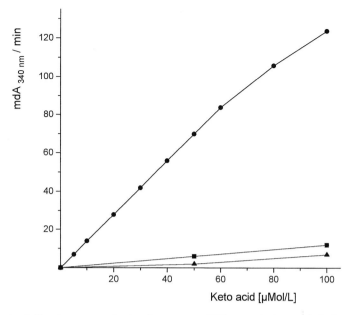

Figure 4. Calibration curves for ketoisocaproate (KIC, ● – ●), ketomethylvalerate (KMV; ■ – ■) and ketoisovalerate (KIV; ▲ – ▲).

Table 5. Results of the isolation of microorganisms with hydantoinase activity from a soil sample. Medium M2 is a mineral salt medium (phosphate buffer + trace elements) containing 2 g/L hydantoin, 0.1 g/l sodium citrate and 0.5 g/L yeast extract; pH 7.0

Medium/Isolation method	Enrichment of	No. of strains isolated	No. of hydantoinase-positive strains
M2	Mesophiles	9	1
M2 + 10 g glycerol	Spore-formers	6	0
Dried soil sample; M2	Drought resistants	9	1
M2; pH 4	Acidophiles	3	0
M2; pH 10	Alkalophiles	4	4
M2; 10 °C	Psychrophiles	6	1
M2; 55 °C	Thermophiles	5	2
M2 + penicillin	Gram-negatives	5	2
M2 + polymyxin	Gram-positives	5	2

Hydantoinase

The screening for hydantoinase activity demonstrates different simple enrichment methods for microorganisms from soil samples (Morin et al., 1987). Depending on the growth conditions different kinds of organisms can obtained (Tab. 5).

After purification and biochemical characterization several variants of the enzyme hydantoinase are available by this procedure.

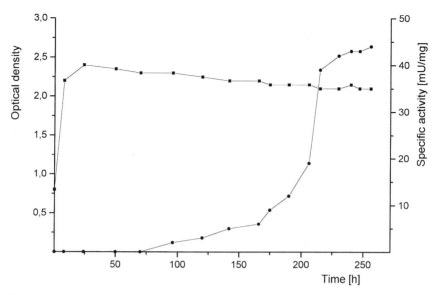

Figure 5. Enrichment of trimethylamine degrading microorganism using continuous cultivation technique (chemostat). Dilution rate = 0.1 h⁻¹ with trimethylamine as the carbon source. During cultivation the optical density (■ – ■) at 554 nm as the growth parameter and the enzyme activity (NAD(P)-independent trimethylamine dehydrogenase (● – ●)) were measured.

Trimethylamine dehydrogenase

Determination of trimethylamine (TMA) is of importance for the food industry as a parameter of fish freshness and in the clinical chemistry for the diagnosis of trimethylaminuria, a disorder of the metabolism of dietary choline. A TMA assay can also be used for the determination of TMA-containing compounds such as choline, betaine or carnitine, when they are cleaved by a preceding step liberating TMA. In order to develop an enzymatic method for the measurement of TMA with a biosensor we screened for a quinoprotein dehydrogenase. In the literature, some quinoprotein amine dehydrogenases are described which convert methylamine or diamines. Quinoprotein dehydrogenases are well suited for biosensors because they are NAD(P)- and O_2-independent and contain a tightly bond cofactor.

Isolation of TMA-dehydrogenase containing organisms by direct plating was found to be absolutely unsuccessful, but by using an enrichment technique in a chemostat one strain of *Paracoccus* with a NAD(P)-independent dehydrogenase could be isolated (Fig. 5).

The enzyme can be assayed with the aid of the artificial electron acceptor phenazine ethosulfate (PES), according to the following reaction:

$$\text{trimethylamine} + \text{PES}_{ox.} \rightarrow \text{dimethylamine} + \text{HCHO} + \text{PES}_{red.}$$

The enrichment of microorganisms in a chemostat is characterized by a rather low substrate level in the vessel which can be maintained by feeding

the substrate continuously at a low rate (dilution rate (D) in the range of
< 0.1 h⁻¹). For example, this technique can be applied profitably to isolate
organisms degrading inhibiting or toxic substances or for the selection of
strains after mutations in order to obtain constitutive enzyme producers.
Other applications derived from the fact that only organisms with a high
affinity towards the substrate are able to grow under these conditions. This
high affinity must be correlated with a high substrate affinity of the enzymes themselves or the transport processes involved, in particular the primary
transport or metabolizing step. After purification and characterization of

Table 6. Biochemical data of trimethylamine dehydrogenase

Parameter	Properties
Enzyme source	*Paracoccus* spec. TMA-1
Specific activity	4.2 U/mg
Electron acceptor	phenazine metho- or ethosulfate
	NAD(P)-independent; O_2-independent
K_m-value (TMA)	$1 \cdot 10^{-7}$ M
pH optimum	9.0
Substrate specificity	only trimethylamine

Table 7. Summary of new enzymes available for the development of new analytical applications and biosensors

Analyte	Enzyme	Organism
Amino acids		
L-Phenylalanine	NAD dehydrogenase	*Rhodococcus* spec.
L-Glutamate	Oxidase	*Streptomyces* spec.
Branched-chain amino acids	NAD dehydrogenase	*Bacillus* spec.
Hydroxy and Keto acids		
Ketoisocaproate	NAD-dependent dehydrogenase	*Lactobacillus casei*
Keto acids	NAD-dependent dehydrogenase	*Lactobacillus* spec.
Hydroxy acids	NAD-dependent dehydrogenase	*Lactobacillus* spec.
Amines		
Trimethylamine	NAD(P)-independent dehydrogenase	*Paracoccus* spec.
Dimethylamine	NAD(P)-independent dehydrogenase	*Paracoccus* spec.
Histamine	NAD(P)-independent dehydrogenase	
Alcohols/Ketones		
Aceton	NAD-dependent dehydrogenase	*Lactobacillus* spec.
Diacetyl	NAD-dependent dehydrogenase	*Lactobacillus* spec.
Ethanol	NAD (P)-independent dehydrogenase	*Pseudomonas* spec.
Ions		
Potassium	NAD-dependent dehydrogenase	*Pseudomonas* spec.
Magnesium	NAD-dependent dehydrogenase	*Pseudomonas* spec.
		Lactobacillus spec.

the TMA-oxidizing enzyme from *Paracoccus* spec. (Hummel et al., 1992) which was isolated by enrichment in the chemostat, a quite low K_m-value for TMA was observed (Tab. 6). This is in good correlation with the expected high affinity of the first metabolizing step (oxidation of TMA) in the organism.

These examples should demonstrate some possibilities in developing successful screening methods. A summary of some new enzymes available for analytic application and the construction of biosensors given in Table 7.

Acknowledgements
The author thanks Bea Weitz and Vera Ophoven for expert technical assistance.

Reference

Cheetham, P.S.J. (1987) Screening for novel biocatalysts. *Enz. Microb. Technol.* 9:194–213.

D'Costa, E.J., Higgins, I.J. and Turner, A.P.F. (1986) Qunioprotein glucose dehydrogenase and ist application in an amperometric glucose sens. *Biosensors* 2:71–87.

Duine, J.A. (1991) Quinoproteins: enzymes containing the quinoid cofactor pyrroloquinoline quinone, topaquinone or tryptophan-tryptophan quinone. *Eur. J. Biochem.* 200:271–284.

Duine, J.A. and Jongejan, J.A. (1989) Quinoproteins, enzymes with pyrrolo-quinoline quinone as cofactor. *Ann. Rev. Biochem.* 58:403–426.

Duine, J.A., Frank, Jzn.J. and Verwiel, P.E.J. (1980) Structure and activity of the prosthetic group of methanol dehydrogenase. *Eur. J. Biochem.* 108:187–192.

Elander, R.P. (1987) Microbial screening, selection and strian improvement. *In*: J. Bu'Lock and B. Kristiansen (eds): *Basic Biotechnology*. Academic Press, London, pp 217–251.

Goodhue, C.T. (1982) The methodoly of microbial transformation of organic compounds. *In*: J.P. Rosazza (ed): *Microbial transformations of bioactive compounds*. Vol I. CRC Press, Boca Raton, Florida, pp 9–44.

Guthrie, R. and Susi, A. (1963) *Pediatrics* 32:338–343.

Hummel, W. and Kula, M.-R. (1989) Simple method for small-scale disruption of bacteria and yeasts. *J. Microbiol. Meth.* 9:201–209.

Hummel, W., Weiss, N. and Kula, M.-R. (1984) Isolation and characterization of a bacterium possessing L-phenylalanine dehydrogenase activity. *Arch. Microbiol.* 137:47–52.

Hummel, W., Schütte, H. and Kula, M.-R. (1985) D-2-Hydroxyisocaproate dehydrogenase from *Lactobacillus casei* – A new enzyme suitable for the stereospecific reduction of 2-ketocarboxylic acids. *Appl. Microbiol. Biotechnol.* 21:7–15.

Hummel, W., Schmidt, E., Wandrey, C. and Kula, M.-R. (1986) L-Phenylalanine dehydrogenase from *Brevibacterium* sp. for production of L-phenylalanine by reductive amination of phenylpyruvate. *Appl. Microbiol. Biotechnol.* 25:175–185.

Hummel, W., Schmidt, E., Schütte, H. and Kula, M.-R. (1987a) Isolation of microorganisms containing high levels of phenylalanine dehydrogenase. *Proc. Biochemical Engineering*, Fischer Verlag, Stuttgart, pp 392–395.

Hummel, W., Schütte, H., Schmidt, E., Wandrey, C. and Kula, M.-R. (1987b) Isolation of L-phenylalanine dehydrogenase from *Rhodococcus* sp. M4 and its application for the production of L-phenylalanine. *Appl. Microbiol. Biotechnol.* 26:409–416.

Hummel, W., Schütte, H. and Kula, M.-R. (1988a) D-(-)-Mandelic acid dehydrogenase from *Lactobacillus curvatus. Appl. Microbiol. Biotechnol.* 28:433–439.

Hummel, W., Schütte, H. and Kula, M.-R. (1988b) Enzymatic determination of L-phenylalanine and phenylpyruvate with L-phenylalanine dehydrogenase. *Anal. Biochem.* 170:397–401.

Hummel, W., Tauschensky, S., Spohn, U., Wendel, U. and Langenbeck, U. (1989) Towards Home-Monitoring and screening of phenylketonuria by biosensors. Studies on flow-injection analysis. *In*: R.D. Schmid and F. Scheller (eds): *Biosensors: Applications in Medicine, Environmental protection and process control*. GBF-Monographs 13. VCH Verlagsgesellschaft, Weinheim, pp 313–318.

Hummel, W., Wendel, U. and Sting, S. (1992) Biochemical characterization of a highly specific trimethylamine dehydrogenase suited for the application in biosensors. *In*: F. Scheller and R.D. Schmid (eds) *Biosensors: Fundamentals, Technologies and Applications*. GBF Monographs 17. VCH Publishers, Weinheim, pp 381–384.

Morin, A., Hummel, W. and Kula, M.-R. (1986) Rapid detection of microbial hydantoinase on solid medium. *Biotechnol. Lett.* 8:571–576.

Morin, A., Hummel, W. and Kula, M.-R. (1987) Enrichment and selection of hydantoinase producing microorganism. *J. Gen. Microbiol.* 133:1201–1207.

Schadewaldt, P., Hummel, W., Trautvetter, U. and Wendel, U. (1989) A convenient enzymatic method for the determination of 4-methyl-2-oxopentanoate in plasma: Comparison with High Performance Liquid Chromatographic analysis. *Clin. Chim. Acta* 183:171–182.

Schadewaldt, P., Hummel, W., Wendel, U. and Adelmeyer, F. (1995) Enzymatic method for determination of branched-chain amino acid aminotransferase activity. *Anal. Biochem.* 230:199–204.

Schütte, H., Hummel, W. and Kula, M.-R. (1984) L-Hydroxyisocaproate dehydrogenase – a new enzyme from *Lactobacillus confusus* for the stereospecific reduction of 2-ketocarboxylic acids. *Appl. Microbiol. Biotechnol.* 19:167–176.

Schütte, H., Hummel, W. and Kula, M.-R. (1985) Improved enzyme screening by automated fast protein liquid chromatography. *Anal. Biochem.* 151:547–553.

Wendel, U., Hummel, W. and Langenbeck, U. (1989) Monitoring of Phenylketonuria: A colorimetric method for the determination of plasma phenylalanine using L-phenylalanine dehydrogenase. *Anal. Biochem.* 180:91–94.

Wendel, U., Koppelkamm, M, Hummel, W., Sander, J. and Langenbeck, U. (1990a) A new approch to the newborn screening for hyperphenylalaninemias: Use of L-phenylalanine dehydrogenase and microtiter plates. *Clin. Chim. Acta* 192:165–170.

Wendel, U., Özalp, I., Langenbeck, U. and Hummel, W. (1990b) Phenylketonuria in Turkey: Experience with an enzymatic colorimetric test for measurement of serum phenylalanine. *J. Inherited Metab. Disease* 13:295–297.

Wendel, U., Koppelkamm, M. and Hummel, W. (1991) Enzymatic phenylalanine estimation for the management of patients with phenylketonuria. *Clin. Chim. Acta* 201:95–98.

Yamazaki, Y., Hummel, W. and Kula, M.-R. (1987) Ein neues Verfahren zum direkten Nachweis mikrobieller Aminoacylaseaktivität auf Agarplatten. *Z. Naturforsch.* 42c:1082–1088.

Yamazaki, Y. and Kula, M.-R. (1987) Entwicklung neuer Plattentests zum Nachweis mikrobieller Hydrolysen von Estern und Oxidation von 2-Hydroxycarbonsäuren. *Z. Naturforsch.* 42c:1187–1192.

Frontiers in Biosensorics I
Fundamental Aspects
ed. by F. W. Scheller, F. Schubert and J. Fedrowitz
© 1997 Birkhäuser Verlag Basel/Switzerland

Phenol-oxidizing enzymes: mechanisms and applications in biosensors

M. G. Peter[1] and U. Wollenberger[2]

[1]Institut für Organische Chemie und Strukturanalytik and Interdisziplinäres Forschungszentrum für Biopolymere;
[2]Institut für Biochemie und Molekulare Physiologie der Universität Potsdam,
D-14469 Potsdam, Germany

Summary. Phenolic compounds are widely distributed in nature. Enyzmes which catalyze their
oxidation are monophenol monooxygenases, such as tyrosinases and laccases, and peroxidases.
Their metabolic role includes the decomposition of natural complex aromatic polymers as well
as polymerization of the oxidation products and the degradation of xenobiotics. Their catalytic
properties and broad availability gained impact on the development of biosenors for both environmentally important pollutants and clinically relevant metabolites.

Mechanisms for the phenol-oxidizine enzymes tyrosinases, laccases, and peroxidases are
reviewed and some examples for their use in the construction of phenol selective biosenors are
given.

Introduction

Enzymes catalyzing the oxidation of phenolic compounds are widespread in nature. They are of fundamental importance in all organisms
that produce polyphenolic compounds such as melanins (Prota, 1995) or
lignins (Dean and Erikson, 1994). Furthermore, phenolic oxidation is
involved in wound healing, immunological defense (Götz and Boman,
1985), and sclerotization of the cuticle in arthropods (Peter, 1993) as
well as in the biosynthesis of many secondary metabolites, such as alka-
loids, lignans, and tannins (Fig. 1). Many of the enzymatic reactions
have important clinical implications such as formation of melanin in the
substantia nigra of the brain and in the skin (Prota, 1992). The oxidative
polymerization of lignin precursors is of considerable technical interest
as is the enzymatic depolymerization of lignin in the bleaching of wood
pulp in the production of cellulose. In the past years, research on phenol
oxidizing enzymes has opened unexpected possibilities for technical
applications in the determination of environmentally and clinically
important phenolic compounds by means of biosenors. With the aim to
further the understanding of the molecular basis of their mode of action,
we shall briefly summarize current knowledge on the nature and mechanism of action of phenol oxidizing enzymes, i.e. tyrosinases, laccases,
and peroxidases (Tab. 1), and their application in the construction of
phenol-selective biosenors.

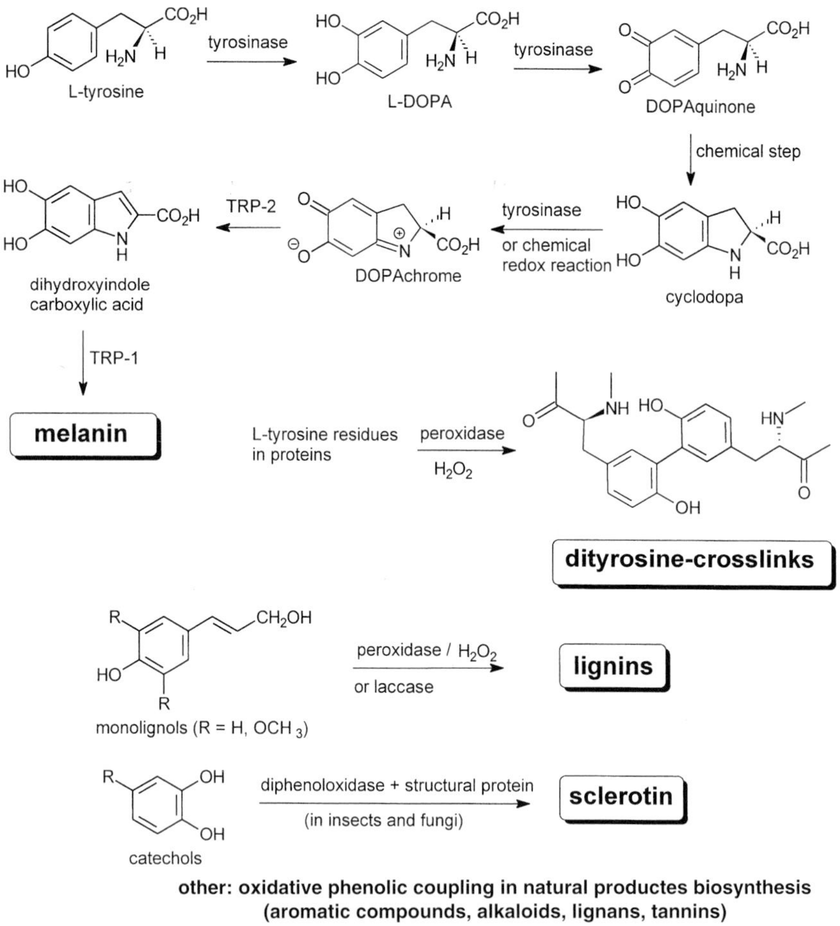

Figure 1. Examples of enzymatic phenol oxidation in natural products biosynthesis (TRP: tyrosinase-related protein).

Tyrosinase

Tyrosinase (EC 1.14.18.1/1.10.3.1) is a copper-containing monophenol monooxygenase that performs also a two-electron oxidation of catechols to o-quinones. The enzyme is widely distributed in nature where it has been found in bacteria, fungi, plants, and animals. It is distinctly different from tyrosine hydroxylase (EC 1.14.16.4) which requires tetrahydropteridine as a cofactor. Precise determination of molecular weights of tyrosinases are hampered by the fact that the enzymes are glycoproteins showing quite heterogenous glycosylation patterns. Tyrosinase from mouse melanocytes has a molecular weight of ca. 58,000 Da (Hearing and Jimenez, 1989; Yurkow and Laskin, 1989). DNA sequence analysis of the tyrosinase from the medaka fish *Oryzias latipes* indicates a protein of 540 amino acids, having

Table 1. Some comparative data on phenol-oxidizing enzymes.

	Tyrosinase (EC 1.14.18.1/1.10.3.1)	Laccase (EC 1.10.3.2)	Peroxidase (EC 1.11.1.7)
M_R subunits	44,000 (*Neurospora crassa*); single (*N. crassa*); tetramer (mushroom)	55–90,000 (fugi) single	44,000 (horseradish) single
3D(analogy, from x-ray)	hemocyanin	ascorbate oxidase	cyr. c. peroxidase
reaction[a]	$MH + H^+ + O_2 \rightarrow Q + H_2O$ $2\,DH_2 + 2\,O_2 \rightarrow 2\,Q + 4\,H_2O$	$2\,DH_2 + O_2 + \rightarrow 2\,D^\bullet + 2\,H_2O$	$2\,H_2O_2 \rightarrow 2\,H_2O + O_2$ $H_2O_2 + 2\,D \rightarrow 2\,H_2O + 2\,D^\bullet$
electron transfer	1 or $2 \times 2e^-$	$4 \times 1\,e^-$	$2 \times 1\,e^-$
metals	$Cu \xrightarrow{3\,\overset{o}{A}} Cu$ 1 Cu (II) pair for oxygen binding and electron transfer	$\begin{matrix} & \overset{o}{3.4\,A} \\ Cu & \!\!\!\!\!\text{—}\!\!\!\! & Cu \\ \overset{o}{3.9\,A}\!\!\searrow & & \swarrow\!\!\overset{o}{4.0\,A} \\ & Cu & \end{matrix}$ 1 Cu (II) pair for oxygen binding, 1 or 2 single Cu (II) for electron capture and transfer	 1 protohemin per subunit
substrates	monophenols catechols hydroquinone (poor)	monophenols hydroquinone methylhydroquinone catechols guajacol 2,6-dimethoxyphenol *p*-phenylendiamine syringaldazine iron and organ-iron complexes	ROOH (R = H, alkyl) monophenols diphenols aminophenols indophenols diamines leuco dyes *p*-hydroxyphenylacetic acid *p*-cresol *p*-anisidine metal complexes
inhibitors	autocatalytic inactivation carboxylates; mimosine; fusaric acid; 3-aminotyrosine; kojic acid; mercaptobenzothiazole; diethyldithiocarbamate; n-propylgallate; 1-methylimidazole-2-thiol; phenylthiourea; carbon monoxide; azide	naphthalenediol (inhibits syringaldazine oxidation); carbon monoxide; cinamic acids; cetyl triammonium bromide; azide; halogenide ions	H_2O_2; CN^-; HS^- CO is not an inhibitor

[a] D: diphenol; M: monophenol; Q: quinone.

five potential glycosylation sites and two copper-binding sites (Inagaki et al., 1994). Insects contain phenoloxidases in hemocytes, serum, integument, and fat body. A diphenoloxidase has been demonstrated in the integument of *Manduca sexta* that shows properties of a laccase. The tyrosinase from larval tissues of *Ceratitis capitata* and the integumental laccase show similar molecular weights of ca. 93,000 Da but differ in glycosylation (Charalambidis et al., 1994). Binuclear copper-binding sites are also present in the inactive prophenoloxidase that has been purified and cloned from a crayfish blood cell c-DNA library (Aspan et al., 1995). The deduced amino acid sequence codes for a polypeptide with a mass of 80,732 Da. Seqence comparisons show that the copper-binding sites are similar to the corresponding sites in arthropod hemocyanins and also, although the sequence similarities are less extensive, similar to tyrosinases from vertebrates and microorganisms. The occurrence of phenol-oxidizing enzymes in plants has been reviewed (Mayer and Harel, 1979; Mayer 1987).

Besides tyrosinase, there are additional enzymes which are important in the biosynthesis of melanins and which all belong to a larger family of binuclear copper proteins, containing two highly conserved copper-binding domains. Like tyrosinase, they are structurally related to the arthropodan and molluscan oxygen-binding hemocyanins. In particular, these are tyrosinase-related protein 2 (TRP-2) which functions as DOPA(3,4-dihydroxyphenylalanine)chrome tautomerase (EC 5.3.2.3), and TRP-1 the specific function of which is the oxidation of 5,6-dihydroxyindole-2-carboxylic acid (DHICA) to 5,6-indolequinone-2-carboxylate (Kobayashi et al., 1994; Morrison et al., 1994).

Tyrosinase mechanism

The mechanism of monophenol hydroxylation and diphenol oxidation by tyrosinase has been deduced mostly from structural studies on the tyrosinase of *Neurospora crassa* and comparison with hemocyanin as well as kinetic analysis of melanogenic enzymes from vertebrates (Wilcox et al., 1985; Sánchez-Ferrer et al., 1995) (Fig. 2). The results have led to the design and synthesis of metallo-organic model compounds which, in turn, yield further insight into the enzymatic mechanisms (Morooka et al., 1995; Solomon et al., 1992).

The enzyme from *Neurospora* consists of a single polypeptide chain of 32,000 Da which contains two copper ions. Each copper is coordinated by two strongly bound equatorial nitrogen atoms from imidazole nitrogen of histidine residues and one axial ligand, either weakly bound imidazole nitrogen of a third histidine or water ($\mu - \eta^2 : \eta^2$ geometry). In the resting stage, the enzyme is composed of 2–30% of the "blue" oxy form (E_{oxy}), the remainder being the colorless E_{met} form. Monophenol oxidation is initiated

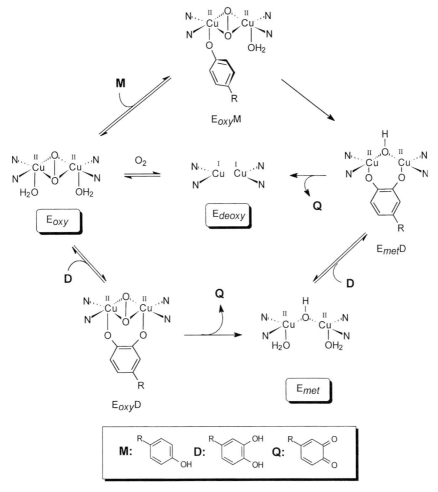

Figure 2. Simplified scheme for the reactions catalyzed by tyrosinase (adapted from Wilcox et al., 1985; Sánchez-Ferrer et al., 1995).

by binding of the substrate (normally L-tyrosine) to the E_{oxy} form to give via an $E_{oxy}M$ intermediate the $E_{met}D$ form. This decays into an o-quinone (normally DOPA quinone) and the E_{deoxy} form of the enzyme. Binding of oxygen completes the catalytic cycle of monophenol oxidation.

Diphenol oxidation is initiated by binding of a catechol to E_{oxy}. Decay of E_{oxy} D to a quinone and E_{met} is followed by binding of a second molecule of diphenol to give $E_{met}D$ which then undergoes the same transformations towards E_{deoxy} and E_{oxy} as in the case of monophenol oxidation.

When the course of the tyrosinase reaction is followed up by the accumulation of DOPAchrome, a characteristic lag phase is observed. Initially, a fraction of the enzyme is occupied by the monophenol. Decay of $E_{met}D$

towards E_{deoxy} yields DOPAquinone which at pH < 6 undergoes a slow cyclization to leukodopachrome ($\equiv$ cyloDOPA). However, when $E_{met}D$ decays into E_{met} and diphenol, not only a substrate for diphenolase (Fig. 2, lower cycle) E_{oxy} is generated but also a reduction equivalent for a chemical transformation of DOPAquinone to DOPA until a steady state concentration of the latter is reached. The lag phase is not observed during tyrosinase oxidation of non cyclizable diphenols (for details, see Sánchez-Ferrer et al., 1995).

Substrate specificity of tyrosinase

The best substrates for monophenol oxidation by mammalian enzyme are L-tyrosine and L-DOPA (Prota, 1992). Tyrosinase from cephalopods shows a ca. twofold higher affinity for D-tyrosine and D-DOPA than for the natural L-stereoisomers (Prota et al., 1981). Plant tyrosinases are involved in natural product synthesis as demonstrated recently by Todorova et al. (1994). Mushroom tyrosinase also oxidizes phenol and variety of other monophenols, including tyrosine residues in peptides (Yasunobu et al., 1959; Cory and Frieden, 1967; Ito et al., 1984). The enzyme from the moth, *Manduca sexta*, oxidizes phenolic substrates with catalytic efficiencies V_{max}/K_m in the order N-β-alanyldopamine > N-acetyldopamine > dopamine > DOPA (Aso et al., 1990). Resorcinols and phenols bearing electron withdrawing substituents or bulky substituents such as *tert*-butyl groups in the *o*- or *p*-position are not oxidized by tyrosinases. Electron-donating substituents enhance the reaction rate (reviewed by Prota, 1992). A compartive selection of data is presented in Table 2.

Table 2. Relative rates of oxidation of phenolic compounds by various enzymes

Compound	Tyrosinase[a]	Tyrosinase[b]	Tyrosinase[c]	Laccase[a]	Laccase[d]
N-acetyldopamine	100	–	–	100	–
N-β-alanyldopamine	150	100	100	63	100
DOPA	15	8	71	7	36
methylhydroquinone	2	0.4	0.4	241	197
4-methylcatechol	56	–	–	311	–
dopamine	83	–	–	27	–
tyrosine	0.3	< 0.01	8	0	< 0.01

[a] enzyme from *Calliphora vicina* (Barrett and Andersen, 1981).
[b] tyrosinase from *Manduca sexta* (Aso et al., 1984; Morgan et al., 1990).
[c] tyrosinase from mushrooms (Morgan et al., 1990).
[d] laccase from *Pyricularia oryzae* (Morgan et al., 1990).

Tyrosinase inhibitors

It has long been known that tyrosinase undergoes a slow autocatalytic inactivation during the catalytic cycle. It was first believed that this inhibition results from a covalent binding of a quinonoid reaction product to the enzyme (Wood and Ingraham, 1965). However, Lerch (1978; 1983) demonstrated that a histidine residue is destroyed oxidatively during catalysis. A number of competitive inhibitors are known which bind to the active site of the enzyme. The most efficient of those show either a nitrogen or an oxygen function in conjunction with another oxygen in a distance of about 2.7–2.8 Å which corresponds to a distance of about 3 Å of the copper atoms in the active site of the enzyme (Fig. 3). Some noteworthy examples are salicylhydroxamic acid (Allan and Walker, 1988), tropolone (Kahn and Andrawis, 1985), 3-amino-L-tyrosine (Maddaluno and Faull, 1988), and kojic acid (Cabanes et al., 1994) (Fig. 3). It is not always clear whether a particular compound inhibits monophenol hydroxylation, diphenol oxidation or both, though the latter case was reported with tropolone (Kahn and Andrawsi, 1985). A distinction between tyrosinases and laccases has been suggested based on the observation that the former are inhibited by cinnamic acids while the latter are inactivated by polyvinylpyrrolidone, some detergents such as cetyl triammonium bromide, and medium chain fatty acids (Walker and McCallion, 1980). Also certain thiols which most likely form complexes with the copper atoms are strong inhibitors of tyrosinase.

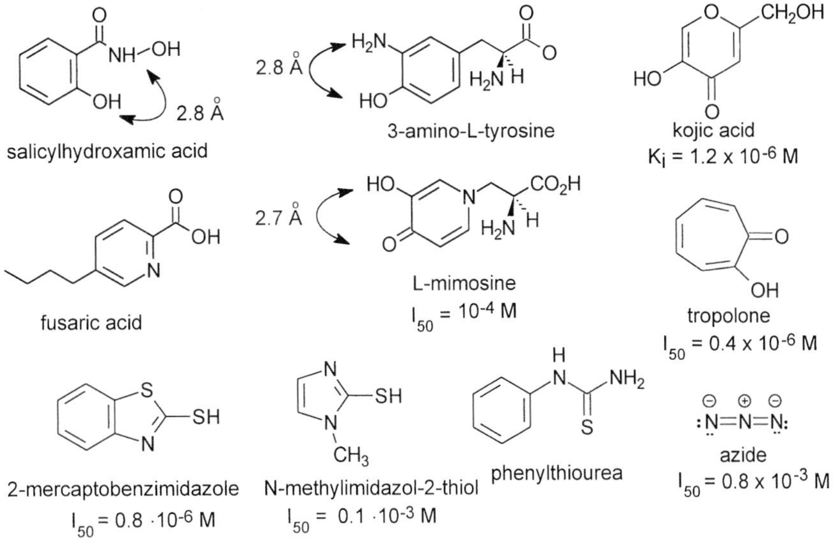

Figure 3. Some inhibitors of tyrosinase.

Laccase

Laccases (EC 1.10.3.2) are copper-containing 'blue' oxidases, which couple four one-electron oxidation processes of p-diphenols and catechols to
the four-electron reduction of dioxygen to water. In contrast to the tyrosinase reaction, also catechols, organic, and inorganic metall complexes are
oxidized, whereas tyrosine is not affected. The structure, properties, catalytic mechanism, and application of laccases are described in a number of
recent articles (Messerschmidt and Huber, 1990; Solomon, 1992; Thurston, 1994; Yaropolov et al., 1994). Like tyrosinases, laccases are glycoproteins; the carbohydrate content is about 10–45% of the total molecular
weight, which is 55–90,000 Da for fungal laccaes and 110–140,000 Da
for enzymes from plants. Only a few fungal enzymes have been cloned.

Laccases are widely distributed, though not ubiquitously, in plants
(Mayer and Harel, 1979; Mayer, 1987), fungi (Pelaez et al., 1995; Raghukumar et al., 1994), and insects (Andersen, 1985; Andersen et al., 1996),
but are apparently absent from higher organisms.

In plants, laccases are involved probably in the polymerization of
coumaryl, coniferyl, and sinapyl alcohol (monolignols) towards lignins
(Fig. 1). As Dean and Eriksson (1994) have pointed out, a distinction of
laccases and peroxidases on the basis of literature data is not always possible. Fungi produce extracellular and intracellular laccases. Expecially in
white rot basidiomycetes, laccases (Kawai et al., 1988) together with peroxidases (*vide infra*) effect the depolymerization of lignin and are also
involved in pigment formation and hardening of the spores (Dean and
Eriksson, 1994).

Tautomeric rearrangements of alkylquinones to quinone methides in
conjunction with phenol oxidation have been described also to be accelerated by laccases in insects (Andersen, 1989a, 1989b; Thomas et al., 1989).
This reaction which plays a fundamental role in formation of the exoskeleton
of insects (Peter, 1993; Andersen et al., 1996) and which is reminiscent to
the DOPAchrome tautomerase rearrangement (*vide supra*), transforms
N-acetyldopamine into the p-quinone methide, involving a highly stereoselective benzylic deprotonation (Peter and Merz, 1995).

Laccase mechanism

Less is known on the mechanism of laccase as compared with tyrosinase
catalysis. Most laccases are single chain "blue" copper proteins which, in
contrast to tyrosinase, contain three or four copper ions. There is some
structural and mechanistic analogy to ascorbate oxidase (EC 1.10.3.3)
which contains three copper ions (Messerschmidt and Huber, 1990). One
of the laccase copper ions is located in a so-called Type-1 copper-binding
site. It is EPR positive and shows an intensive absorption at ca. 600 nm,

thus being responsible for the blue color of the enzyme. The Type-2 copper-binding site contains also a single copper ion, while Type-3 copper consists of a tightly coupled Cu (II) ion pair that is coupled also with Type-2 copper in a trinuclear arrangement. A reaction mechanism for laccase has been proposed based on studies of the reduction of laccase from *Rhus vernicifera* by hydroquinone and ascorbic acid (Andreasson and Reinhammar, 1979). Electrons transferred from the substrate to Type-1 copper are sequestered by Type 3 copper and reduce oxygen to water via a tightly bound peroxide intermediate while Type-2 copper facilitates the breakage of the oxygen-oxygen bond in the latter. However, the detailed mechanism of both electron transfer steps from the substrate to the Type 1 copper site and the formation of oxygen intermediates is still quite unclear. A Report of a PQQ prosthetic group in *Phlebia radiata* laccase (Karhunen et al., 1990) should be confirmed as discussed by Dean and Erisson (1994).

Substrate specificity of laccase

Laccase oxidizes a wide variety of phenols and aromatic amines, such as guaiacol, 4-methylcatechol, α-naphthol, *p*-anisidine, *p*-phenylenediamine, 2,7-diaminofluorene, and syringaldazine. The latter compound is widely used for laccase assays. Furthermore, laccase oxidizes a number of inorganic and organic metal ion complexes among them ferrocyanide (Cenas and Kulys, 1988), ferrocenes (Ghindilis et al., 1995), and cytochrome c (Jin et al., 1996). At low pH catalytic rate constants in the order of 4×10^5 M^{-1} s^{-1} $- 1 \times 10^6$ $M^{-1}s^{-1}$ have been determined for various diphenols and ferrocyanide (Cenas and Kulys, 1988) with the laccase from *Polyporus anisoporus*. For the oxidation of ferrocytochrome c, rate constants of 9×10^3 $M^{-1}s^{-1}$ (Jin et al., 1996) and 125 $M^{-1}s^{-1}$ (Sakurai, 1992) for laccase from *Coriolus hirsutus* and *Rhus vernicifera* have been reported. The reaction rate parallels the redox potentials of the copper Type 1 of the respective laccase of 785 mV (at pH 5.4) and 394 mV (at pH 7.5) (Reinhammar, 1972).

Inhibitors of laccase

Long-known inhibitors are azide and halogenide ions which, in general, are unspecific inhibitors of the blue copper oxidases. Only a few selective inhibitors of laccases are known (Dean and Eriksson, 1994) The most specific on seems to be *N*-hydroxyglycine (Murao et al., 1992). Kojic acid and salicylhydroxamine acid (Fig. 3) not only inhibit tyrosinase but also laccase. Clearly, there is a need for specific laccase inhibitors and a most promising field of research seems to be natural products chemistry.

Peroxidase

Peroxidases (EC 1.11.1.7) effect, with participation of hydrogen peroxide as the ultimate electron acceptor, one-electron oxidations of a wide range of substrates. The majority of peroxidases are glycoproteins of 20–70 kD molecular weight, which contain ferric protoporphyrin IX as a prosthetic group. They are ubiquitously found in nature and named after their sources. Some of the peroxidases are well investigated, e.g. cytocrome c peroxidase from baker's yeast is the first peroxidase of which a high-resolution crystal structure has been obtained (Poulos et al., 1980).

Mechanism of peroxidase

The mechanism of peroxidase catalyzed reactions has been studied intensively (see reviews by Anni and Yonetani, 1992; Everse et al., 1991; Poulos, 1993; Ryan et al., 1994). The kinetics of catalysis reveal a ping-pong mechanism. In the first step, the peroxide binds to a free coordination site of iron (III) and is reduced to water in a rapid two-electron process, whereby compound I ($Fe(IV)=O$ $P^\bullet$) is formed as the stable primary intermediate according to equation (1):

$$POD(Fe\ III) + ROOH \xrightarrow{k_1} Compound\ I + ROH \tag{1}$$

P is a porphyrin or a protein residue and $P^\bullet$ is the corresponding radical formed as the oxidation product. In most peroxidases, one electron is abstracted from the iron atom and the other electron from heme macrocycle forming iron (IV) and a porphyrin cation radical. In addition, one oxygen atom derived from the peroxide is transferred to the heme iron. In the next step, equation (2), a substrate AH_2 reduces the porphyrin cation radical of compound I while the iron remains in its oxidized state ($Fe(IV)=O$). This stable intermediate is usually denoted as compound II. Compound II in turn reacts with another reducing substrate to regenerate the native state of the peroxidase, as shown in equation (3).

$$Compound\ I + AH_2 \xrightarrow{k_2} Compound\ II + AH^\bullet \tag{2}$$

$$Compound\ II + AH_2 \xrightarrow{k_3} POD(Fe\ III) + AH^\bullet \tag{3}$$

Formation of compound I is fast with an apparent second order rate constant in the order of 10^6–10^7 $M^{-1}s^{-1}$ (Anni and Yonetani, 1992). Furthermore, the oxidation of reducing substrates by compound I is 10–100 times faster than by compound II. Therefore, in the presence of excess of peroxide and AH_2-limitation, the overall reaction is controlled by k_3. The substrate radical AH$^•$ undergoes a nonenzymatic disproportionation as well as coupling of phenoxy radicals according to equations (4) and (5):

$$AH^• + AH^• \longrightarrow AH_2 + A \tag{4}$$

$$AH^• + AH^• \longrightarrow HA-AH \tag{5}$$

Studies of the structure and reaction mechanism of other peroxidase as well as mutant enzymes showed that the structure of the catalytic site and the reaction mechanism of cytochrome-c peroxidase is to a limited extend, applicable to other types of heme peroxidases, such as horseradish peroxidase (HRP) (Anni and Yonetani, 1992).

Substrate specificity

Depending on the nature of the peroxidase, a number of compounds reduce the higher oxidation state intermediates of the enzyme back to the native form. With the most commonly used HRP (horseradish peroxidase), virtually any organic and inorganic reducing agent may react in this way. Among the factors evaluated for the reaction of HRP with the reducing substrate, hydrophobic interactions and the redox potential are most important. These substrates bind at approximately the same site in the vicinity of the heme, in an orientation perpendicular to the heme plane, and interact mainly with the exposed part of the heme (Ator and Ortiz de Montellano, 1987). Among the best electron donors are hydroquinone and *o*-phenylene diamine with reaction rates of 3×10^6 M^{-1}s^{-1} and 5×10^7 M^{-1}s^{-1} (Barman, 1992). Tropolone which inhibits tyrosinase, is oxidized by HRP (Kahn, 1985) while cytochrome c peroxidase acts on cytochrome c. Also, melanogenic reactions are observed upon exposure of appropriate precursors to peroxidase (Gross and Sizer, 1959).

In the context of this article, some examples for oxidation of phenolic compounds shall be mentioned in particular. Both, mono- and diphenols are oxidized and synthetis lignins may be prepared with monolignols as substrates. In conjunction with laccases, peroxidases are involved in lignin degradation by white rot fungi. Indeed, screening of a large number of fungi revealed the presence of lignin peroxidases, manganese peroxidases and laccases (Katagiri et al., 1995; Pelaez et al., 1995; Raghukumar et al., 1994). These enzymes not only effect depolymerization by benzylic cleavage in the polymer but also ring cleavage of aromatic units (Miki et al., 1988; Umezawa and Higuchi, 1985, 1989). In addition, soybean peroxidase has been shown to effect the oxidation and (eventually) beta-ether cleavage of a lignin model dimer (McEldoon et al., 1995).

HRP is active in a number of organic solvents. The properties of the enzyme can be improved further by chemical modifications of the protein.

Inhibitors of peroxidase

Only few inhibitors of HRP are known. Sulfide and cyanide reversibly inibt HRP with I_{50} values of 50 and 676 µM. Furthermore, HRP is inhibited in presence of bivalent metal ions in the order $Mn^{2+} > Co^{2+} > Pb^{2+} > Fe^{2+} > Cu^{2+}X > Cd^{2+} > Ni^{2+}$ (Zollner, 1993). Inactivation of peroxidases during the oxidation of phenols in most likely caused by formation of polyphenols which bind nonspecifically of the enzyme. High concentrations of hydrogen proxide may result in complex formation of compound I and hydrogen peroxide, leading to a suicidal inactivation (Anni and Yonetani, 1992).

Applications of phenol-oxidizing enzymes in biosensors

The first report of an application of polyphenol oxidases in biosensors was published by Macholan and Schanel (1977) who measured oxygen consumption in determinations of phenolic substrates with enzymes from potato and mushroom. Since then a number of related papers have been published. Enzymes with a broad substrate spectrum, i. e. laccase, tyrosinase, polyphenol oxidase, and peroxidase, as well as specific enzymes, e. g. phenol hydroxylase and catechol-oxidase, may be used for the construction of enzyme electrodes (Scheller and Schubert, 1992). Alternatively, whole cells (Riedel et al., 1993) and plant tissue from various fruits (Wang and Lin, 1988) and vegetables (Uchiyama et al., 1988) have been combined with various electrodes.

Tyrosinase and laccase-based biosensors have been used for determinations of phenols, catechols, and catecholamines (Tab. 3). Related sensors have been also constructed for the analysis of tyrosinase inhibitors (Smit and Rechnitz, 1993; Stancik et al., 1995). Elimination of substances interfering with the electrochemical reaction in the determination of glucose in biological samples has been achieved (Maidan and Heller, 1992; Wang et al., 1993b; Wollenberger et al., 1986). More recently, considerable attention has been paid to systems operating in organic solvents (Hall et al., 1988; see also Stöcklein, this book) and in the gas phase (Dennison et al., 1995; Kaisheva et al., 1996). Miniaturization of biosensors for various phenols of clinical importance, for example neurotransmitters, is a formidable task (Pantano and Kuhr, 1995).

Phenol-oxidizing enzymes have been used in conjunction with oxygen electrodes, mediator-modified electrodes, pH-sensors, and unmodified electrode materials. Two basic detection principles can be outlined: (i) indication of oxygen consumption (Macholan and Schanel, 1977; Campanella et al., 1993; Pfeiffer et al., 1990) or (ii) measurement of the electroactive quinonoid reaction products. The quinones can be directly reduced at a cathodically poised electrode (Skladal, 1991; Ortega et al., 1994; Wang et al., 1994; Wasa et al., 1984, Ruzgas et al., 1995) or, alternatively, deter-

Table 3. Bioelectrocatalytic recycling electrodes for phenols, such as aminophenol, catechol, dopamine, adrenaline, noradrenaline, hydroquinone, and p-cresol.

Enzyme	Electrode material	Detection limit (nM)	Comment	Reference
mushroom	GC	5		Skaldal, 1991
tyrosinase	GC	100	AQ, measurements in organic solvents	Wang et al., 1993a
			accumulation	Wang and Chen, 1995a
	CP	10–40	bulk modification with PEI promotor	Gorton, 1995 Ortega et al., 1993
	CP	40–100	graphite epoxy	Önnerfjord et al., 1995
	CI	100	thick-film-electrode	Wang and Chen 1995b
	CP		graphite-epoxy,	Wang et al., 1994
fungal laccase	RVC	70–200	glutardialdehyde	Wasa et al., 1984
	GC	600	AQ, measurements in organic solvents	Wang et al., 1993a
	SG	< 1000	adsorption, + tyrosinase	Yaropolov et al., 1995
	CP	2	bulk modification	Wollenberger and Neumann, unpublished
HRP	SG	500–4000	adsorption	Ruzgas et al., 1995
	CP		bulk modification with lactitol	

RVC-reticulted vitreous carbon, GC-glassy carbon, SG-spectroscopic graphite, CP-carbon paste, CI-carbon ink.

mined via redox mediators (Bonakdar et al., 1989; Kulys and Schmid, 1990, Kotte et al., 1995), including a conductive polymer (Coche-Guerente et al., 1995). In general, the linear range of the sensors combining phenol oxidases and Clark-type oxygen electrodes is between µM and mM substrate concentrations.

A serious problem encountered in the detection of phenols by means of biosensors results from the inactivation of the enzymes, in particular tyrosinase and laccase, by reactive intermediates such as hydroxyl and phenoxy radicals, quinones and quinone methides, as well as from the adsorption of polymeric polyphenols to the protein and surface of the biosensors. This can be prevented, at least in part, by adding reducing agents such as hydrazine hydrochloride (Macholan, 1990) or ascorbate (Hasebe et al., 1995) which recycle the substrate chemically, thus increasing the sensitivity by 1–2 orders of magnitude. Other approaches use electrochemical methods (Rivas and Solis, 1994). Mushroom tyrosinase immobilized in a polyurethane hydrogel was applied in a methylphena-

Table 4. Enzymatic signal amplification in biosensors using phenol oxidases.

Analyte	Enzyme couple	Transducer	Amplification factor	Reference
adrenaline; p-aminophenol	laccase/ (PQQ)glucoseDH	oxygen electrode	10000 5000	Ghindilis et al., 1995
ferrocene derivatives		antimony pH electrode	1000	Eremenko et al., 1995b
phenol derivatives; adrenaline;	tyrosinase/ (PQQ)glucose dehydrogenase	oxygen electrode	100–1000	Eremenko et al., 1995a
tyrosine containing peptides			150–450	Makower et al., 1996
adrenaline; p-aminophenol	laccase/ oligosaccharide dehydrogenase	oxygen electrode	3000	Bier et al., 1996
Benzoquinone/ hydroquinone	laccase/fructose dehydrogenase	oxygen electrode	700	Wollenberger, unpublished results
Benzoquinone/ hydroquinone	laccase/cyt. b_2	oxygen electrode	500	Scheller et al., 1987

zonium-zeolite-modified enzyme sensor (Kotte et al., 1995). The most elegant approach consists of the coupling with reducing enzyme systems, such as (PQQ)glucose dehydrogenase, which shuttle the analyte in cyclic series of reduction/oxidation reactions accompanied by oxygen consumption (Tab. 4).

Another type of sensor is based on the competition of HRP and catalase for hydrogen peroxide. HRP uses hydrogen peroxide in the presence of phenols with the result of diminished oxygen formation by catalase. In this way phenol, bilirubin, and aminopyrine have been detected (Renneberg et al., 1982). The hydrogen peroxide required was either injected into the measuring cell or generated in the enzyme layer itself by co-immobilizing glucose oxidase with HRP.

An ultrasensitive biosensor was created when laccase from *Coriolus hirsutus* and PQQ(pyroloquinoline quinone)-dependent glucose dehydrogenase from *Acinetobacter calcoaceticus* were co-entrapped in a 10-μm layer of polyvinyl alcohol in front of a Clark-type oxygen electrode (Ghindilis et al., 1995). Owing to the broad spectrum of substrates for both enzymes, the sensor responds to various catecholamines and phenol derivatives. The highest sensitivity was observed for *p*-aminophenol and adrenaline, where the lower limit of detection (S/N 3:1) is 70 pM and 1 nM, respectively. The extraordinary efficiency of the bi-enzymatic amplification sensors is based on the excess of enzyme molecules compared with the concentration of the analyte molecule within the reaction layer (Wollenberger et al.,

1993). The current density of the membrane-covered sensor is almost three orders of magnitude higher than the bare electrode. The electrode can also be used for the development of electrochemical immunoassays.

The bioelectrocatalytic sensing principle is not restricted to phenolic substrates. Smit and Rechnitz (1993) demonstrated the application of a tyrosinase-based sensor for cyanide detection. The inhibition of tyrosinase results in a diminished rate of ferrocyanide oxidation and therefore a decrease of the sensor signal which depends, however, on the nature of the redox mediator. Recently, assays of inhibitors of the catecholase activity of tyrosinase, i.e. cholorophenols, atrazine, and carbamates, have been described (Besombes et al., 1995). Analysis of thiourea-derived inhibitors can also be performed in non-aqueous solutions (Stancik et al., 1995). Very recently, bioelectrocatalysis of chlorophenols and p-cresol has been accomplished with HRP (Ruzgas et al., 1995). This enzyme opens up an avenue to the determination of environmentally important chlorinated phenol derivatives not detected so far by tyrosinase and laccase.

Conclusions

Knowledge of enzyme mechanisms is one of the prerequisites for the rational development of biotechnical applications, *inter alia* biosensors which are gaining increasing importance in environmental and clinical analysis. In particular, substrate selectivity, enzyme inhibition, and the pathways and recycling of electrons in redox reactions offer possibilities for numerous variations in the final approaches. In the case of phenoloxidases, not only the mechanisms of enzymatic catalysis but also the chemistry of reactive intermediates and products cause often underestimated problems, due to autocatalytic inhibition and fouling of electrodes. Thus, the combination of the enzyme with chemical mediators and/or additional biocatalysts is of particular importance since it results in remarkable sensitivity enhancements.

Other efforts in future research will focus on new techniques of enzyme immobilization, including the development of new materials for ultrathin coating of the electrode surface. This is of particular importance in the miniaturization of biosensors required for detection of minute quantities of analytes in e.g. a single cell of a living organism. In connection with environmental analysis, disposable sensors are most useful in routine field applications. Further research is also needed in the manipulation of enzymatric substrate selectivity be solvent composition or by protein engineering.

Undoubtedly there is a great demand for innovative ideas. We hope that this review will stimulate research and development of phenol-selective biosensors and thus contribute to the improvement of environmetal monitoring and clinical analyis.

Acknowledgements

Work cited from our laboratories was supported by the Deutsche Forschungsgemeinschaft, the German Bundesministerium für Forschung und Technology, the Commission of the European Communities, and by the Fonds der Chemischen Industrie.

References

Allan, A.C. and Walker, J.R.I. (1988) The selective inhibition of catechol oxidases by salicyl-hydroxamine acid. *Phytocemistry* 27:3075–3076.

Andersen, S.O. (1985) Sclerotization and tanning of the cuticle. *In*: G.P. Kerkut and L.I. Gilbert (eds): *Comparative Insect Physiology, Biochemistry, and Pharmacology,* Vol. 3. Pergamon Press, New York, pp 59–74.

Andersen, S.O. (1989a) Enzymatic activities in locust cuticle involved in sclerotization. *Insect Biochem.* 19:59–67.

Andersen, S.O. (1989b) Enzymatic activities involved in incorporation of *N*-acetyldopamine into insect cuticle during sclerotization. *Insect Biochem.* 19:375–382.

Andersen, S.O., Peter, M.G. and Roepstorff, P. (1996) Cuticular sclerotization in insects. *Comp. Biochem. Physiol.* 113B:689–705.

Andreasson, L.E. and Reinhammar, B. (1979) The mechanism of electron transfer in laccase catalyzed reactions. *Biochim. Biophys. Acta* 558:145–156.

Anni, H. and Yonetani, T. (1992) Mechanism of action of peroxidases. *In*: H. Siegel and A. Siegel (eds): *Metal Ions in Biological Systems.* Marcel Dekker, New York, pp 219–241.

Aso, Y., Kramer, K.J., Hopkins, T.L. and Whetzel, S.Z. (1984) Properties of tyrosinase and DOPA quinone imine conversion factor from pharate pupal cuticle of *Manduca sexta* L. *Insect Biochem.* 14:463–472.

Aspan, A., Huang, T.S., Cerenius, L. and Soderhall, K. (1995) c-DNA cloning of prophenoloxidase from the fresh-water crayfish *Pacifastacus leniusculus* and its activation. *Proc. Natl. Acad. Sci. USA* 92:939–943.

Ator, M.A. and Ortiz de Montellano, P.R. (1987) Protein control of prosthetic heme reactivity. Reaction of substrates with the heme edge of HRP. *J. Biol. Chem.* 262:1542–1551.

Barman, T.E. (1992) (ed) *Enzyme Handbook*, Vol. 1. Springer Verlag, Berlin, pp 234–235.

Barrett, F.M. and Andersen, S.O. (1981) Phenoloxidases in larval cuticle of the blowfly, *Calliphora vicina. Insect Biochem.* 11:17–23.

Besombes, J.L., Cosnier, S., Labbe, P. and Reverdy, G. (1995) A biosensor as warning device for the detection of cyanide, chlorophenols, atrazine and carbamate pesticides. *Anal. Chim. Acta* 311:255–263.

Bier, F.F., Ehrentreich-Förster, E., Bauer, C. and Scheller, F. (1996) High-sensitive competitive immunodetection of 2,4-dichlorophenoxyacetic acid using enzymatic amplification with electrochemical detection. *Fres. J. Anal. Chem.* 354:861–865.

Bonakdar, M., Vilechez, J.L. and Mottola, H.A. (1989) Bioamperometric sensor for phenol based on carbon paste electrodes. *J. Electroanal. Chem.* 354:861–865.

Cabanes, J., Chazarra, S. and García-Carmona, F. (1994) Kojic acid, a cosmetic skin whitening agent, is a slow-binding inhbitor of catecholase activity of tyrosinase. *J. Pharm. Pharmacol.* 46:982–985.

Campanella, L., Beone, T., Sammartino, M.P. and Tomassetti, M. (1993) Determination of phenol in wastes and water using an enzyme sensor. *Analyst* 118:979–986.

Cenas, N.K. and Kulys, J.J. (1988) *Fermentativnuyi perenos electrona* (russ., *Enzymatic electron transfer*) Mokslas, Vilnius.

Charalambidis, N.D., Bournazos, S.N., Zervas, C.G., Katsoris, P.G. and Marmaras, V.J. (1994) Glycosylation and adhesiveness differentiate larval *Ceratitis capitata* tyrosinases. *Arch. Insect Biochem. Physiol.* 27:235–248.

Coche-Guerente, L., Cosnier, S. and Innocent, C. (1995) Poly(amphiphilic pyrrole)-PPO electrodes for organic-phase enzymatic assay. *Anal. Lett.* 28:1005–1016.

Cory, J.G. and Frieden, E. (1967) Differential reactivites of tyrosine residues of proteins to tyrosinase. *Biochemistry* 6:121–126.

Dean, J.F.D. and Eriksson, K.-E.L. (1994) Laccase and the deposition of lignin. *Holzforschung* 48 (Suppl.):21–33.

Dennison, M.J., Hall, J.M. and Turner, A.P.F. (1995) Gas-phase microbiosensor for monitoring phenol vapor at ppb levels. *Anal. Chem.* 67:3922–3927.

Eremenko, A.F., Makower, A., Wen, J., Rüger, P. and Scheller, F.W. (1995a) Biosensor based on an enzyme modified electrode for highly-sensitive measurement of polyphenols. *Biosens. Bioelectron.* 10:717–722.

Eremenko, A.F., Makower, A. and Scheller, F.W. (1995b) Measurement of nanomolar diphenols by substrate recycling coupled to a pH-sensitive electrode. *Fresenius Z. Anal. Chem.* 351:729–731.

Everse, J., Everse, K.E. and Grisham, M.B. (eds) (1991) *Peroxidases in Chemistry and Biology,* Vol. 1 and 2. CRC Press, Boca Raton.

Ghindilis, A.L., Makower, A. and Scheller, F.W. (1995) Nanomolar determination of the ferrocene derivatives using a recycling enzyme electrode – development of a redox label immunoassay. *Analytical Lett.* 28:1–11.

Gorton, L. (1995) Carbon paste electrodes chemically modified with enzymes, tissues, and cells. A review. *Electroanal.* 7:23–45.

Götz, P. and Boman, H.G. (1985) Insect immunity. *In:* G.A. Kerkut and L.I. Gilbert (eds): *Comprehensive Insect Physiology, Biochemistry and Pharmacology,* Vol. 3. Pergamon Press, Oxfod, pp 453–485.

Gross, A.J. and Sizer, I.W. (1959) The oxidation of tyramine, tyrosine, and related compounds by peroxidase. *J. Biol. Chem.* 234:1611–1614.

Hall, G.F., Best, D.A. and Turner, A.P.F. (1988) Amperometric enzyme electrode for the determination of phenols in chloroform. *Enzyme Microb. Technol.* 10:543–546.

Hasebe, Y., Hirano, T. and Uchiyama, S. (1995) Determination of catecholamines and uric acid in biological fluids without pretreatment, using chemically amplified biosensors. *Sensor. Actuator. B* 24–25:94–97.

Hearing, V.J. and Jimenez, M. (1989) Analysis of mammalian pigmentation at the molecular level. *Pigment Cell Res.* 2:75–85.

Inagaki, H., Bessho, Y., Koga, A. and Hori, H. (1994) Expression of the tyrosinase-encoding gene in a colorless melanophore mutant of the medaka fish, *Oryzias latipes. Gene* 150: 319–324.

Ito, S., Kato, T., Shinpo, K. and Fujita, K. (1984) Oxidation of tyrosine residues in proteins by tyrosinase – formation of protein-bonded 3,4-dihydroxyphenylalanine and 5-S-cysteinyl-3,4-dihydroxyphenylalanine. *Biochem. J.* 222:407–411, in press.

Jin, W., Wollenberger, U., Bier, F.F., Makower, A. and Scheller, F. (1996) Electron transfer between cytochrome c and laccase. *Bioelectrochem. Bioenerg.* 39:221–225.

Kahn, V. (1985) Tropolone – a compound that can aid in differentiating between tyrosinase and peroxidase. *Phytochemistry* 24:915–920.

Kahn, V. and Andrawis, A. (1985) Inhibition of mushroom tyrosinase by tropolone. *Phytochemistry* 24:905–908.

Kaisheva, A., Iliev, I., Kazareva, R., Christov, S., Petkova, J., Wollenberger, U. and Scheller, F. (1996) Enzyme/gas diffusion electrodes for determination of phenol. *Sensor. Actuator.* in press.

Karhunen, E., Niku-Paavola, M.L., Viikari, L., Haltia, T., van der Meer, R.A. and Duine, J.A. (1990) A novel combination of prosthetic groups in a fungal laccase; PQQ and two copper atoms. *FEBS Lett.* 267:6–8.

Katagiri, N., Tsutsumi, Y. and Nishida, T. (1995) Correlation of brightening with cumulative enzyme-activity related to lignin biodegradation during biobleaching of kraft pulp by white-rot fungi in the solid-state fermentation system. *Appl. Environm. Microbiol.* 61: 617–622.

Kawai, S., Umezawa, T., Shimada, M. and Higuchi, T. (1988) Aromatic ring cleavage of 4,6-di-*tert*-butylguaiacol, a phenolic lignin model compound, by laccase of *Coriolus versicolor. FEBS Lett.* 236:309–311.

Kobayashi, T., Urabe, K., Winder, A. Jimenez-Cervantes, C., Imokawa, G., Brewington, T., Solano, F., García-Borron, J.C. and Hearing, V.J. (1994) Tyrosinase-related protein-1 (Trp1) functions as a DHICA oxidase in melanin biosynthesis. *EMBO J.* 13: 5818–5825.

Kotte, H., Grundig, B., Vorlop, K.D., Strehlitz, B. and Stottmeister, U. (1995) Methylphenazonium-modified enzyme sensor-based on polymer thick-films for subnanomolar detection of phenols. *Anal. Chem.* 67:65–70.

Kulys, J. and Schmid, R. (1990) A sensitive enzyme electrode for phenol monitoring. *Anal. Lett.* 23:589–597.

Lerch, K. (1978) Amino acid sequence of tyrosinase from *Neurospora crassa*. *Proc. Natl. Acad. Sci. USA* 75:3635–3639.

Lerch, K. (1983) *Neurospora* tyrosinase: Structural, spectroscopic and catalytic properties. *Mol. Cell. Biochem.* 52:125–138.

Macholan, L. (1990) Phenol-sensitive enzyme electrode with substrate recycling for quantification of certain inhibitory aromatic acids and thio compounds. *Coll. Czech. Chem. Commun.* 55:2152–2159.

Macholan, L. and Schanel, L. (1977) Enzyme electrode with immobilized polyphenol oxidase for determination of phenolic substrates. *Coll. Czech. Chem. Commun.* 42: 3667–3675.

Maddaluno, J.F. and Faull, K.F. (1988) Inhibition of mushroom tyrosinase by 3-amino-L-tyrosine: Molecular probing of the active site of the enzyme. *Experientia* 44:885–887.

Maidan, R. and Heller, A. (1992) Elimination of electrooxidizable interferant-produced currents in amperometric biosensors. *Anal. Chem.* 64:2889–2896.

Makower, A., Eremenko, A.V., Streffer, K., Wollenberger, U. and Scheller, F.W. (1996) Tyrosinase-glucose dehydrogenase substrate recycling biosensor. Highly sensitive measurement of phenolic compounds. *J. Chem. Tech. Biotech.* 65:39–44.

Mayer, A.M. (1987) Polyphenoloxidase in plants – recent progress. *Phytochemistry* 26:11–20.

Mayer, A.M. and Harel, E. (1979) Polyphenoloxidases in plants. *Phytochemistry* 18:193–215.

McEldoon, J.P., Pokora, A.R. and Dordick, J.S. (1995) Lignin peroxidase-type activity of soybean peroxidase. *Enzyme Microb. Technol.* 17:359–365.

Messerschmidt, A. and Huber, L. (1990) The blue oxidases, ascorbate oxidase, laccase and ceruloplasmin. Modelling and structural relationship. *Eur. J. Biochem.* 187:341–352.

Miki, K., Kondo, R., Renganathan, V., Mayfield, M.B. and Gold, M.H. (1988) Mechanism of aromatic ring cleavage of a β-biphenylyl ether dimer catalyzed by lignin peroxidase of *Phanerochaete chrysospori*. *Biochemistry* 27:4787–4794.

Morgan, T.D., Thomas, B.R., Yonekura, M., Czapla, T.H., Kramer, K.J. and Hopkins, T.L. (1990) Soluble tyrosinase from pharate pupal integument of the tobacco hornworm, *Manduca sexta* (L.). *Insect Biochem.* 20:251–260.

Morooka, Y., Fujisawa, K. and Kitajima, N. (1995) Transition-metal peroxo complexes relevant to metalloproteins. *Pure Appl. Chem.* 67:241–248.

Morrison, R., Mason, K. and Frost-Mason, S. (1994) A cladistic analysis of the evolutionary relationships of the members of the tyrosinase gene family using sequence dat. *Pigm. Cell Res.* 7:388–393.

Murao, S., Hinode, Y., Matsumura, E., Numata, A., Kawai, K. Ohishi, H., Jin, H., Oyama, H. and Shin, T. (1992) A novel laccase inhibitor, *N*-hydroxyglycine, produced by *Penicillium citrinum* YH-31. *Biosci. Biotech. Biochem.* 56:987–988.

Önnerfjord, P., Emneus, J., Marko-Varga, G. and Gorton, L. (1995) Tyrosinase graphite-epoxy based composite electrodes for detection of phenols. *Biosens. Bioelectron.* 10:607–619.

Ortega, F., Dominguez, E., Burestedt, E., Emneus, J., Gorton, L. and Marko-Varga, G. (1994) Phenol oxidase-based biosensors as selective detection units in column liquid chromatography for the determination of phenolic compounds. *J. Chromatogr.* 675:65–78.

Ortega, F., Dominguez, E., Jönsson-Pettersson, G. and Gorton, L. (1993) Amperometric biosensor for the determination of phenolic compounds using a tyrosinase graphite electrode in a flow injection system. *J. Biotech.* 31:289–300.

Pantano, P. and Kuhr, W.G. (1995) Enzyme-modified microelectrodes for in vivo neurochemical measurements. *Electroanal.* 7:405–416.

Pelaez, F., Martinez, M.J. and Martinez, A.T. (1995) Screening of 68 species of basidiomycetes for enzymes involved in lignin degradation. *Mycol. Res.* 99:37–42.

Peter, M.G. (1993) Die molekulare Architektur des Exoskeletts von Insekten. *Chem. uns. Zeit* 27:189–197.

Peter, M.G. and Merz, A. (1995) Stereoselective benzylic deprotonation in the enzymatic rearrangement of *N*-acetyldopamine derived *o*-quinone to the *p*-quinone methide. *Tetrahedron Asymmetry* 6:839–842.

Pfeiffer, D., Wollenberger, U., Makower, A., Scheller, F., Risinger, L. and Johansson, G. (1990) Amperometric amino acid electrodes. *Electroanal.* 2:517–523.

Poulos, T.L. (1993) Peroxidases. *Current Opinion Biotech.* 4:484–489.

Poulos, T.L., Freer, S.T., Alden, R.A., Edwards, S.L., Skoglund, U., Takio, K., Eriksson, B., Yuong, N.-H., Yonetani, T. and Kraut, J. (1980) The crystal structure of cytochrome c peroxidase. *J. Biol. Chem.* 255:575–580.

Prota, G. (1992) *Melanins and Melanogenesis.* Academic Press, London.

Prota, G. (1995) The chemistry of melanins and melanogenesis. *Progr. Chem. Org. Nat. Prod.* 64:93–148.

Prota, G., Ortonne, J.P., Voulot, C., Khatchadourian, C., Nardi, G. and Palumbo, A. (1981) Occurrence and properties of tyrosinase in the ejected ink of Cephalopods. *Comp. Biochem. Physiol.* 68B:415–419.

Raghukumar, C., Raghukumar, S., Chinnaraj, A., Chandramohan, D., Dsouza, T.M. and Reddy, C.A. (1994) Laccase and other lignocellulose modifying enzymes of marine fungi isolated from the coast of India. *Bot. Marina* 37:515–523.

Reinhammar, B.R.M. (1972) Oxidation-reduction potenials of the electron acceptors in laccases and stellacyanin. *Biochim, Biophys. Acta* 275:245–259.

Renneberg, R., Pfeiffer, D., Scheller, F. and Jänchen, M. (1982) Enzyme sequence and competition electrodes based on immobilized glucose oxidase, peroxidase and catalase. *Anal. Chim. Acta* 134:359–364.

Riedel, K., Hensel, J., Rothe, S., Neumann, B. and Scheller, F. (1993) Microbial sensors for determination of aromatics and their chloro derivatives. *Appl. Microbiol. Biotechnol.* 502–506.

Rivas, G.A. and Solis, V.M. (1994). Electrochemical quantification of phenol using mushroom tyrosinase – determination of the kinetic parameters of the enzyme. *Electroanalysis* 6, 1136–1140.

Ruzgas, T. Emneus, J., Gorton, L. and Marko-Varga, G. (1995) The development of a peroxidase biosensor for monitoring phenol and related aromatic compound. *Anal. Chim. Acta* 311:245–253.

Ryan, O., Smyth, M.R. and O'Fagain, C. (1994) Horseradish peroxidase: the analyst's friend. *Essay in Biochemistry* 28:129–146.

Sakurai, T. (1992) Kinetics of electron transfer between cytochrome-c and laccase. *Biochemistry* 31:9844–9847.

Sánchez-Ferrer, Á., Rodríguez-López, J.N., García-Cánovas, F. and García-Carmona, F. (1995) Tyrosinase: a comprehensive review of its mechanism. *Biochim. Biophys. Acta.* 1247:1–11.

Scheller, F. and Schubert, F. (1992) *Biosensors,* Elsevier, Amsterdam.

Scheller, F., Wollenberger, U., Schubert, F., Pfeiffer, D. and Bogdanovskaya, V.A. (1987) Amplification and switching by enzymes in biosensors. *GBF Monographs* 10:39–49.

Skladal, P. (1991) Mushroom tyrosinase-modified carbon paste electrode as an amperometric biosensor for phenols. *Coll. Czech. Chem. Commun.* 56:1427–1433.

Smit, M.H. and Rechnitz, G.A. (1993) Toxin detection using a tyrosinase-coupled oxygen electrode. *Anal. Chem.* 65:380–385.

Solomon, E.I., Baldwin, M.J. and Lowery, M.D. (1992) Electronic structures of active sites in copper proteins. *Chem. Rev.* 92:521–542.

Stancik, L., Macholan, L. and Scheller, F. (1995) Biosensing of tyrosinase inhibitors in nonaqueous solvents. *Electroanal.* 7:649–651.

Thomas, B.R., Yonekura, M., Morgan, T.D., Czapla, T.H., Hopkins, T.L. and Kramer, K.J. (1989) A trypsin-solubilized laccase from pharate pupal integument of the tobacco hornworm, *Manduca sexta. Insect Biochem.* 19:611–622.

Thurston, C.F. (1994) The structure and function of fungal laccases. *Microbiology* 140:19–26.

Todorova, M., Werner, C. and Hesse, M. (1994) Enzymatic phenol oxidation and polymerization of the spermine alkaloid aphelandrine. *Phytochemistry* 37:1251–1256.

Uchiyama, S., Tamata, M., Tofuku, Y. and Suzuki, S. (1988) A catechol electrode based on spinach leaves. *Anal. Chim. Acta* 208:287–290.

Umezawa, T. and Higuchi, T. (1985) A novel C_α-C_β cleavage of a β-O-4 lignin model dimer with rearrangement of the β-arygl group by *Phaneroachaete Chrysosporium. FEBS Lett.* 192:147–150.

Umezawa, T. and Higuchi, T. (1989) Cleavages of aromatic ring and β-O-4 bond of synthetic lignin (DHP) by lignin peroxidase. *FEBS Lett.* 242:325–329.

Walker, J.R.L. and McCallion, R.F. (1980) The selective inhibition of ortho- and para-diphenol oxidases. *Phytochemistry* 19:373–377.

Wang, J., and Chen, Q. (1995a) Highly sensitive biosensing of phenolic compounds using bioaccumulation/chronoamperometry at a tyrosinase electrode. *Electroanal.* 7:746–749.

Wang, J., and Chen, Q. (1995 b) Microfabricated phenol biosensors based on screen printing of tyrosinase containing carbon ink. *Anal. Lett.* 28:1131–1142.

Wang, J., and Lin, M.S. (1988) Mixed plant tissue-carbon paste bioelectrode. *Anal. Chem.* 60:1545–1548.

Wang, J., Lin, Y., Eremenko, A.V., Ghindilis, A.L. and Kurochkin, I.N. (1993 a) A laccase electrode for organic-phase enzymatic assays. *Anal. Lett.* 26:197–207.

Wang, J., Naser, N. and Wollenberger, U. (1993 b) Use of tyrosinase for enzymatic elimination of acetaminophen interference in amperometric sensing. *Anal. Chim. Acta* 285:19–24.

Wang, J., Fang, L., and Lopez, D. (1994) Amperometric biosensor for phenols based on a tyrosinase-graphite-epoxy biocomposite. *Analyst* 119:455–458.

Wasa, T., Akimoto, K., Yao, T. and Murao, S. (1984) Development of laccase membrane electrode by using carbon electrode impregnated with epoxy resin and is response characteristics. *Nippon Kagaku Koishi* 9:1398–1403.

Wilcox, D.E., Porras, A.G., Hwang, Y.T., Lerch, K., Winkler, M.E. and Solomon, E.I. (1985) Substrate analogue binding to the coupled binuclear copper binding site in tyrosinase. *J. Am. Chem. Soc.* 107:4015–4027.

Wollenberger, U., Scheller, F., Pfeiffer, D., Bogdanovskaya, V.A., Tarasevich, M.R. and Hanke, G. (1986) Laccase/glucose oxidase electrode for determination of glucose. *Anal. Chim. Acta* 187:39–45.

Wollenberger, U., Schubert, F., Pfeiffer, D. and Scheller, F. (1993) Enhancing biosensor performance using multienzyme systems. *Trends in Biotechnology* 11:255–262.

Wood, B.J.B. and Ingraham, L.L. (1965) Labeleld tyrosinase from labelled substrate. *Nature* 205:291–292.

Yaropolov, A.I., Kharybin, A.N., Emneus, J., Marko-Varga, G. and Gorton, L. (1995) Flow-injection analysis of phenols at a graphite electrode modified with co-immobilized laccase and tyrosinase. *Anal. Chim. Acta* 308:137–144.

Yaropolov, A.I., Skorobogatko, O.V., Vartanov, S.S. and Varfolomeyev, S.D. (1994) Laccase – properteis, catalytic mechanism, and applicability. *Appl. Biochem. Biotechnol.* 49:257–280.

Yasunobu, K.T., Peterson, E.W. and Mason, H.S. (1959) The oxidation of tyrosine-containing peptides by tyrosinase. *J. Biol. Chem.* 234:3291–3295.

Yurkow, E.J. and Laskin, J.D. (1989) Purification of tyrosinase to homogeneity based on its resistance to sodium dodecyl sulfate proteinase-K digestion. *Arch. Biochem. Biophys.* 275:122–129.

Zollner, H. (1993) *Handbook of Enzyme Inhibitors*, Part A. VCH, Weinheim, pp 367–368.

Frontiers in Biosensorics I
Fundamental Aspects
ed. by F. W. Scheller, F. Schubert and J. Fedrowitz
© 1997 Birkhäuser Verlag Basel/Switzerland

Enzymes and antibodies in organic media: analytical applications

W. F. M. Stöcklein and F. W. Scheller

Institute of Biochemistry and Molecular Physiology, University of Potsdam, c/o Max-Delbrück-Center of Molecular Medicine, D-13122 Berlin, Germany

Summary. This chapter outlines the influence of organic solvents on antibodies, enzymes, and their reactions, the different kinds of enzyme-solvent systems and the various advantages of organic solvents compared with water, with respect to analytical purposes. Examples for electrochemical, optical and thermometric assays in organic solvents are given. The potential of organic solvents for the modification of immunoassays is exemplified, opening up new applications.

Introduction

Water is not suitable as a solvent for most processes in industrial chemistry. Many organic compounds of commercial interest are poorly soluble or unstable in water. In contrast to chemical processes, biocatalyses have been carried out for a long time solely in aqueous media, as enzymes obviously act only in those. However, at a closer look, it turns out that many enzymes are membrane bound, partially as enzyme complexes, in a rather hydrophobic micro-environment. This means that the water concentration is significantly reduced in the vicinity of these enzymes. Therefore it is not absurd to include more hydrophobic solvents, when working with enzymes, antibodies and other biofunctional units. The idea to use organic solvents in connection with enzymes can be found in literature dating back more than 80 years (Bourquelot and Bridel, 1911). First descriptions of the influence of organic solvents on enzymes were published in the sixties, especially concerning chromatographic separation methods (Singer, 1962). However it was not until the eighties, that one discovered that organic media have a large potential for biotechnological applications (Zaks and Klibanov, 1985). Especially hydrolytic enzymes were successfully applied for the production of esters in high enantiomeric excess (Dordick, 1989). Analytical applications include oxidoreductases like peroxidase and tyrosinase, which are used as biocomponents in organic phase enzyme electrodes (Saini et al., 1991).

This review gives a short overview of the present state of knowledge on the behaviour of enzymes and antibodies in organic solvents and highlights the application of organic solvents in biochemical analysis.

Classification of organic solvents and solvent-systems

Various investigations have dealt with correlations between physico-chemical solvent parameters and the activity, specificity and stereo-selectivity of enzymes or catalytic antibodies (abzymes). A correlation was found for many biocatalysts between their activity and the logarithm of the partition coefficient P of the solvent between octanol and water, $log\ P$ (Laane et al., 1987). Therefore $log\ P$ will be used herein for classification of organic solvents. Simplified, hydrophilic solvents have a low, hydrophobic solvents a high $log\ P$-value. When the water content was taken into account, or the solubility of the solvent in water, a better correlation was found for enzyme activity and $log\ P$ in the range 0.5 to 1.5. The water content is given by the water activity a_w, which is measure of the concentration of free water in the system. So-called anhydrous solvents are those with a water content of not more than about 0.01 %. Other correlation studies of enzyme activity or antibody-hapten binding include the hydrogen-acidity α, hydrogen basicity β, polarity/polarizability π^*, dielectric constant ε, polarity index of *Snyder*, molecular mass or length of the solvent molecule.

Enzymes and other proteins are hardly soluble in organic solvents, in contrast to many substrates, e. g. lipids. They can be exposed to organic solvents in several ways:

a) Lyophilized or precipitated protein can be suspended in "water-free" solvents by rigorous stirring or sonication.

b) By immobilization (Taylor, 1991). Suitable matrices are membranes, spherical beads, gels or polymer films. The most frequently used immobilization techniques are physical adsorption, ionic binding, covalent binding, entrapment into a porous matrix, e. g. polyacrylamide or Ca-alginate, chemical crosslinking and metal-chelate-binding.

c) In two-phase-systems, like water-in-oil emulsions, which separate into two layers, if not stirred. Recently, a modified two-phase system was described, consisting of a mixture of two special polymers in chloroform. Emulsions are formed in the mixed state, which favor enzyme reactions with water-insoluble substrates. Two layers appear in the resting state, containing each predominantly one polymer (Otamiri et al., 1994). Ideally, enzyme and product can be separated into the two phases.

d) In water-in-oil microemulsions, which are formed after addition of detergents. They are optical transparent colloidal dispersions with a water content of less than about 20% by volume. The term "reversed micelles" is used for water/detergent ratios (w_o or R) of less than about 15 (Luisi et al., 1988). The optical transparency allows the photometric detection of enzyme reactions in reversed micelles. The rapid exchange of the micelle contents after collision makes reversed micelles applicable for enzyme reactions, with separate preparation of enzyme- and substrate-containing micelles.

Enzyme crystals have been subjected to x-ray structure analysis, in order to learn about solvent effects on the molecular level (Yennawar et al., 1994).

Analogous nonaqueous systems are supercritical fluids and gas phases. The hydrophobicity of supercritical fluids can be easily manipulated by pressure changes. Recently, biosensors have been described for the enzymatic detection of phenol (Karayannis et al., 1994) and the immunochemical detection of parathion and cocaine (Suleiman and Guilbault, 1992) in the gas-phase.

Enzymes in organic solvents

Among natural proteins enzymes have been studied most intensively with respect to their behavior in organic solvents (Gupta, 1992). Therefore some fundamental properties of enzymes and organic solvents have to be described prior to the applied aspects.

Solubility: Enzymes are not entirely insoluble in nonaqueous solvents. For example, lyophilized lysozyme was shown to be soluble (> 10 mg/ml) in all solvents that are protic, hydrophilic and polar, and in binary mixtures thereof (Chin et al., 1994). The solubility of enzymes in less polar solvents can be improved by chemical modification with fatty acids or polyethyleneglycol, by complex formation with amphiphilic polymers or detergent addition. Enzyme solubility is also pH dependent.

Activity: Organic solvents affect both the Michaelis constant K_m and the catalytic constant $k_{cat.}$ The measured K_m depends on substrate solubility, partition coefficient (in two-phase systems) and solvation of substrate and enzyme. Therefore, the catalytic efficiency k_{cat}/K_m is determined preferentially.

The amount of essential water needed for enzyme activity is several hundred, but for some enzymes even less than 50, water molecules per enzyme molecule (Klibanov, 1989). This water shell is attacked by solvents, depending on their hydrophilicity, the hydrophilic solvents being more detrimental to the enzymes (Tab. 1). Typically, the decrease of enzyme activity as a function of the concentration of water-miscible solvents follows a sigmoidal course. The denaturation capacity DC can be calculated from the threshold concentration of the solvent causing 50% inhibition of enzyme activity (Belova et al., 1991). However, an enhancement of activity was shown for several enzymes at a specific concentration range of water-miscible cosolvents (Batra and Gupta, 1994).

Enzymes have a "memory" for the buffer (pH, substrates, inhibitors, lyoprotectors or salts) from which they were lyophilized (Klibanov, 1989). This phenomenon of enzyme conditioning or molecular imprinting is explained by the rigidity of the enzyme structure in nonaqueous solvents. Recently, Klibanov et al. (1995) found that dehydration (e.g. lyophilization) results in profound reversible changes in the protein conformation, especi-

Table 1. Classification of solvents

$log\,P$	Solubility in water	Removal of essential water	Example ($log\,P$)[a]	k_{cat}/K_m[b] $(M^{-1}min^{-1})$	%H_2O bound/ enzyme[c]
<2	>0.4	strong	acetone (-0.23)	0.022	1.2
$2 < log\,P < 4$	$0.04–0.4$	medium	toluene (2.5)	4.4	2.3
>4	<0.04	weak	octane (4.5)	63	2.5

[a] $log\,P$ after Laane et al. (1987); [b] catalytic efficiency of α-chymotrypsin in "water-free" organic solvent, after Klibanov (1989); [c] the amount of water bound to the enzyme (w/w).

ally a drop in the α-helix content and an increase in the β-sheet content. The catalytic properties, including selectivity, in nonaqueous solvents depend on the mode of enzyme dehydration.

Stability: The increase of enzyme thermostability in nonaqueous solvents can be enormous, compared to aqueous media. Plausible reasons are the increase of the number of intramolecular hydrogen bonds and the rigidity of the protein molecule (Klibanov, 1989). In mixtures of water and water-miscible solvents the (chemical) stability of enzymes decreases with increasing solvent content and may have a distinct maximum at low water contents, as for nonaqueous solvents.

There are several ways to increase the stability of enzymes in organic solvents: chemical modification, addition of stabilizing agents, e.g. polyhydroxy-compounds, immobilization or protein engineering. Thermostable enzymes are also possible candidates for application in organic solvents, as a correlation was found between thermostability and solvent stability in two-phase systems. Recently, solvent-tolerant bacteria have been shown to be another source of solvent stable enzymes.

Specificity: The appropriate solvent for a given enzyme reaction can be selected following general rules, regarding the $log\,P$ values of the reactants and the medium (Laane et al., 1987). The specificity of any enzyme reaction should be predictable, based on the solvent-water partition coefficients of the substrates. Wescott and Klibanov (1994) found an inversion of substrate specificity, with hydrophobic substrates being preferred in hydrophilic media and vice versa, reflecting the impact of hydrophobic effects in enzyme-substrate binding. Enantiomeric solvents like (R)- and (S)-carvone can also affect the enantioselectivity of enzymatic ester formation differently.

Bioanalytical reactions in organic solvents

Enyzmatic analysis

Most of the published applications of enzyme reactions in organic solvents deal with organic synthesis, especially ester synthesis with lipases or proteases, where the equilibrium of hydrolytic reactions is shifted towards the ester by replacement of water (Dordick, 1989). However, the analytical aspects have attained increasing attention since the first description of an organic solvent enzyme electrode (OPEE) by Hall et al. (1988).

The use of organic solvents for enzymatic analysis has several advantages (Saini et al., 1991)

- lipophilic substrates or inhibitors become better accessible,
- reactions can be performed which hardly occur in water (e.g. reversal of hydrolytic reactions),
- stability may be improved, due to the antimicrobial conditions and the general increase of chemical and thermal stability of enzymes in non-polar solvents,
- specificity can be influenced (improved in terms of a specific task) by choice of the appropriate solvent.

Of particular importance for organic phase enzyme electrodes are:

- enzymes and mediators can be immobilized by simple adsorption or en-trapment,
- side reactions can be suppressed, e.g. phenol-polymerization on electrode surface,
- electrode response is not affected by hydrophilic interfering substances,
- application is extended, e.g. to direct biosensing in chemical reactors,
- new possibilities for the design and fabrication of electrodes are opened up,
- the potential window for electron transfer reactions is extended.

Examples for biosensor applications in organic solvents are listed in Table 2.

Enzyme electrodes
Typical organic phase enzyme electrodes (OPEEs) are based on ampero-metric (Clark-type oxygen electrodes and other 2- or 3-electrode systems) or potentiometric sensors (Fig. 1).

*Amperometric sensors.*They consist of a working-, counter- and/or reference electrode. Modified Clark oxygen electrodes are used to measure the activity of molecular oxygen consuming enzymes, e.g. tyrosinase. In-activation of enzymes, especially by polymerization reactions of phenol, can be avoided by substitution of organic solvents for water. Another advantage of organic solvents compared with water is the higher solubility of oxygen. Enzymes can be immobilized by adsorption to glass fiber filter membranes or other materials. It may be useful to place an oxygen-permeable PTFE membrane between the enzyme layer and the electrode surface.

Other 2- or 3-electrode systems were used in combination with horsera-dish peroxidase, tyrosinase, laccase and other enzymes. The electrons are

Table 2. Analytical application of enzymes in organic solvents

Enzyme	Analyte	Sample	Solvent	Detection	Ref.
Acetylcholin-esterase	paraoxon (pesticide)	water apples	various[1]	electrode	Mionetto (1993)
Chlorophyllase	chlorophyll	–	water/ acetone	photometer	Khamessan et al. (1993)
Cholesterol oxidase	cholesterol	butter, margarine	chloroform/ hexane	electrode:	Hall and Turner (1991)
				fluorimeter:	Valencia-González and Díaz-Garcia (1994)
				photometer:	Kazandjian et al. (1986)
α-Chymotrypsin (stereospecific), esterase	aminoacid ester L/D	buffer (process control)	DMF/H_2O – 50%	thermistor	Hundeck et al. (1993)
Glucose oxidase	glucose	–	acetonitrile/ H_2O (9 + 1)	electrode	Iwuoha and Smyth (1994)
HRP[2]	lauroyl-peroxide (peroxide-value)	plant oils	chloroform	electrode/ FIA[3]	Mannino et al. (1994)
HRP[2]	H_2O_2	–	e.g. chloro-form/H_2O	electrode	Schubert et al. (1992) Dong and Guo (1994)
HRP[2]	hydroper-oxides	–	ethanol, acetonitrile	micro-electrode	Wang et al. (1991)
Laccase	catechols	–	alcohols/ H_2O	electrode/ FIA[3]	Wang et al. (1993b)
β-Lactamase (and others)	penicillin	–	ethanol (toluene...)	thermistor	Flygare and Danielsson (1988)
Tyrosinase	phenols	olive oil	chloroform	electrode/ FIA[3]	Wang et al. (1992)
			n-hexane	electrode	Campanella et al. (1992)
Tyrosinase	inhibitor (thio-urea derivatives)	–	n-hexane	electrode	Stančík et al. (1995)
Tyrosinase	water (traces)	solvents	various	electrode/ FIA[3]	Wang et al. (1993)

[1] from ethyleneglycol to hexadecane ($log\ P$ = – 1.93 to + 8.8); [2] HRP: horseradish peroxidase; [3] FIA: flow injection analysis.

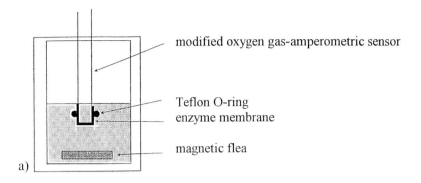

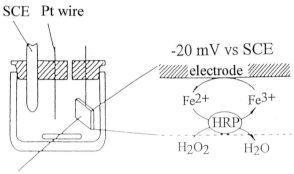

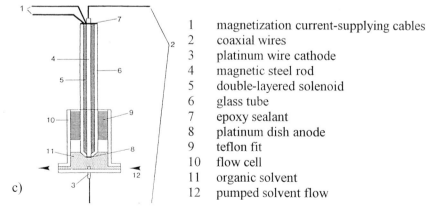

1   magnetization current-supplying cables
2   coaxial wires
3   platinum wire cathode
4   magnetic steel rod
5   double-layered solenoid
6   glass tube
7   epoxy sealant
8   platinum dish anode
9   teflon fit
10 flow cell
11 organic solvent
12 pumped solvent flow

Figure 1. Organic phase enzyme electrodes (OPEE). (a) modified oxygen electrode with immobilized tyrosinase for the detection of phenols, after Campanella et al. (1992), (b) electrochemical cell for the detection of hydrogen peroxide, after Schubert et al. (1991). The enzyme horseradish peroxidase (HRP) and the water-soluble mediator ferricyanide are adsorbed to the graphite foil. (c) potentiometric electrode for monitoring esterification catalyzed by chymotrypsin in organic solvents, with flow through cell for flow injection analysis, after Miyabayashi et al. (1989).

transferred from the enzyme to the electrode indirectly, via mediators, or directly. For example, hydrogen peroxide was detected with peroxidase and the mediator ferricyanide (Schubert et al., 1991). Phenol and phenolic analytes can be detected by electrochemical reduction of the quinone products of the tyrosinase reaction, both in aqueous and in organic media (Wang et al., 1993 a; Schubert et al., 1992). Inhibitors of tyrosinase, including several herbicides, which are poorly soluble in water, were also detected with tyrosinase electrodes, and the selectivity was found to be inversed when water was replaced with an organic solvent (Stančík et al., 1995).

Most OPEEs operate in solvents saturated or premixed with water. Therefore, a special immobilization technique was developed for analysis in water-free solvents, where the necessary water is provided by a cryohydrogel (Dong and Guo, 1994).

The low conductivity of organic solvents can be overcome by the addition of special salts like tetraalkylammonium salts (e. g. TBAB, TEATS). As an example, flow injection analysis was combined with electrochemical detection for the detection of enzyme inhibitors in acetonitrile containing 2% buffer and 0.1 M TEATS (Wang et al., 1993 a). These conducting salts are soluble in the more polar solvents, but insoluble in nonpolar solvents. For such solvents, closely spaced microelectrodes in combination with thin ionically conducting films were developed (Saini and Turner, 1995).

Direct heterogeneous electron transfer between microperoxidase-11 and naked glassy carbon electrodes in an organic solvent (DMSO with $1-3.6\%$ H_2O and conducting salt) was demonstrated by Mabrouk (1995).

Potentiometric electrodes
They are used to detect the pH or potential change caused by an enzyme reaction. As an example, the ester formation catalyzed by chymotrypsin in toluene was monitored in this way (Miyabayashi et al., 1989). Here, the enzyme was immobilized on magnetic particles, which could be fixed to the tip of the electrode by switching on a magnetic field, thus enabling rapid and easy exchange of "consumed" enzyme.

The lower limit of detection of OPEEs is about 1 µM for substrates (Hall et al., 1988) and 10 nM for inhibitors (Stančík et al., 1995), with preconcentration even 10 pM (Mionetto, 1993). Further (potential) analytical applications include food (fats and oils), environment (pesticides, phenol, oil contaminants), petrochemical industry and military (toxic gases).

Recently we could show that synthetic biomimetic catalysts can replace or complement enzymes in OPEEs. Alkanes were oxidized in an organic solvent with substituted metalloporphyrins mimicking the widespread cytochrome P450 enzymes, developed in the working group of Mansuy (Mansuy et al., 1982). The consumption of the oxidant of the biomimetic reaction, an organic peroxide, was detected enzymatically. It should be possible also to detect the product of the biomimetic reaction by enzymes or direct electrochemical reactions. The bioanalytical potential of metallo-porphyrins will be investigated in more detail in a proposed project with European partners.

Spectrophotometry
Cholesterol was detected with a cholesterol oxidase reactor/room tempera-ture phosphorescence (RTP) sensor system, operating in a continuous organic flowing stream. The oxygen consumption during the enzymatic reaction was followed via the changes in the RTP of an oxygen-sensitive metal chelate. The RTP was measured with a fluorescence spectrometer. A second enzyme, horseradish peroxidase, was included in order to remove hydrogen peroxide (Valencia-González and Díaz-Garcia, 1994).

Other photometric assays in organic solvents have been published, but as they do not contribute to biosensors they are not discussed here. Assays which were performed in the presence of low concentrations of water-miscible solvents will be discussed in the immunoassay section.

Thermometry
Enzyme thermistors have been developed for on-line detection of analytes with flow injection analysis. It could be shown that the sensitivity of a thermometric analysis was increased in organic solvents, due to their lower heat capacity, compared with water (Flygare and Danielsson, 1988). How-ever, negative effects of organic solvents on thermometric analysis should also be considered, e.g. low enthalpy changes, reversible enzyme inhibition or equilibrium shifts.

Photoacoustic spectrometry
An even more positive effect is expected for the substitution of organic solvents for water in photoacoustic spectrometry, due to not only the lower heat capacity, but also the higher thermal expansion coefficient of non-aqueous solvents. This was demonstrated for the enzymatic detection of glucose and urea (Flygare and Danielsson, 1988).

Antibody-antigen interaction

Influence of organic solvents
Russell et al. (1989) described for the first time the binding of a hapten, aminobiphenyl, to immobilized antibodies in various polar solvents. In the meantime only few publications have appeared on antibody-antigen/hapten binding in organic solvents. Generally, a change of specificity (or cross-reactivity) and decrease of affinity was observed for various haptens in both polar water-miscible and nonpolar water-immiscible solvents. The inhi-bition of antibody-hapten binding was investigated in more detail for the hapten testosterone (Giraudi and Baggiano, 1993). With increasing solvent

concentration the binding is increasingly inhibited, first reversibly, at higher solvent concentrations irreversible precipitation occurs. An increased binding was observed for some solvents at low mole fractions. The observed effects depend largely on the differences in solvation of the hapten (and probably also antibody) with water or organic solvent molecules. The effects of organic solvents also depend on how the individual noncovalent binding forces contribute to the antibody-antigen/hapten binding. From the practical point of view, the slow phase transfer of the analyte between the solvent phase and immobilized antibody causes problems, especially when porous carriers are used.

Manipulations of the antibody affinities by organic solvents find application in immunoassays and in the chromatographic purification of antibodies or antigens.

Solvent modified immunoassays

There are several remarkable applications of organic solvents in immunoassays, although no breakthrough has been obtained for routine analysis.

A homogeneous immunoassay for thyroxin was developed, which is based on size changes of reverse micelles (Kabanov et al., 1989). Antibody and hapten-enzyme-conjugate are prepared in reverse micelles separately. Binding of antibody to conjugate leads to an increase in catalyst size. Consequently, the enzyme activity decreases, as the activity has an optimum when the size of the reverse micelle equals the size of the catalyst. Therefore, activity is proportional to the concentration of free hapten, which competes with conjugate for the antibody binding site.

Another application of organic solvents was described for enzyme-linked immunosorbent assays (ELISA) on microtiter plates (Freudenberg et al., 1989). The coating efficiency could be improved and the unspecific binding of antibodies was reduced, when hydrophobic antigens dissolved in chloroform/methanol instead of aqueous antigen solutions were used for coating.

There is a large potential of organic solvents to improve immunoassays:

(i) *Hydrophobic analytes:* Immunoassays can be used as screening method to select positive aqueous samples for further analysis by instrumental analysis, e.g. HPLC or GC/MS. However, many analytes, such as toxic soil or food contaminants, are hydrophobic and poorly soluble in water. They have to be extracted, usually with organic solvents, prior to analysis. Therefore, the direct immunochemical detection of analytes in nonaqueous solvents would extend the screening potential of immunoassays to such samples. An ELISA protocol for detection of parathion in hexane has been published recently (Francis and Craston, 1994). The enzyme label used for this ELISA rendered the assay rather insensitive. Therefore it seems to be essential to consider the rules menitoned above for selection of solvents according to the *log P* values of substrates and products, or haptens and conjugates. A more hydrophobic, low molecular weight label (e.g. fluorescence dye or redox marker) should be superior to enzymes for this kind of assay, and no solvent-buffer exchange should be necessary during the assay. Electrochemical detection of low molecular weight labels is also the basis for the development of new homogeneous immunoassays in aqueous and organic media.

(ii) *Elution modifiers*: Analytes like biotechnologically produced hormones or pesticides in aqueous samples can be purified or enriched for analysis by immunochromatographic methods. The antigen or hapten is bound to immobilized antibodies at neutral pH and eluted with buffers of extreme pH or high concentrations of chaotropic ions. These extreme conditions are more or less detrimental for the antibodies and antigens. The (partial) replacement of such pH and salt extremes by the addition of organic solvents may preserve antibody and analyte and, moreover, may allow direct detection of labels in flow injection immunoanalysis. For example, conjugates bound to immobilized antibodies – their amount depending on free analyte concentration – are usually detected in a substrate addition step and then eluted in a second step. If the elution buffer is compatible with the enzymatic reaction, e.g. thanks to the use of organic solvents, one step can be saved and the assay becomes faster and simpler. With laccase as a model tracer, which is active in acidic elution buffers, this principle could be demonstrated for the detection of a coumarin derivative (Stöcklein and Scheller, 1995). The addition of 10% ethanol allowed the pH to be raised by 0.3 pH units.

(iii) *Sensitivity:* The sensitivity of competitive or displacement immunoassays can possibly be increased by the addition of organic solvents, which influence the antibody affinities for the hapten and the hapten-enzyme conjugate differently. However, the apparent increase in sensitivity may be counteracted by side effects, e.g. release of antibodies from coated microtiter plates causing decreased overall signals.

(iv) *Specificity:* The specificity of antibodies can be modified by organic solvents in order to render the antibodies more specific for a selected hapten. Moreover, analytes can be determined individually in mixtures of crossreacting substances, if the assay is performed in the presence of different solvent/water mixtures causing different crossreactivity patterns.

Catalytic antibodies

A new branch of biotechnology is the development and application of catalytic antibodies (abzymes). More than 50 reactions have been described, which are speeded up in the presence of abzymes (Benkovic, 1992). Tailor-made abzymes catalyze reactions for which no equivalent enzymatic or chemical catalysts are known.

In analogy to hydrolytic enzymes, abzyme reactions were investigated also in organic solvents. For example, a tetrahydrofuran derivative was pepared on the gram scale using abzymes in a biphasic system of hexane and 5% buffer (Shevlin et al., 1994). The ring closure reaction was regio- and stereoselective. This example demonstrates also some benefits of a biphasic system: product inhibition was avoided, as well as substrate limitation (insolubility in water), and the product recovery was easier compared with a monophasic aqueous system.

Catalytic antibodies are not used as analytical tools at present. However, we have started a project to develop catalytic antibodies against urea derivatives (pesticides). We use suitably designed transition state analogues of the target molecules for the immunization of mice in order to obtain antibodies with the desired catalytic activity. One of the intended cleavage products (e. g. phenols) can be detected with well-developed bioanalytical methods. The urea derivatives used in this project are poorly water soluble, and the analysis of these analytes in soil and other solid samples affords their extraction with organic solvents. Therefore, we will also investigate the reactions of the obtained catalytic antibodies in organic solvents and screen the antibodies (catalytic and non-catalytic) for a selected specificity in various aqueous-organic media.

Conclusion

Biochemical analyses cover a wide range of applications. The respective assay system can be optimized by engineering of the biological component on the one hand, and adjustment of the assay conditions on the other hand. Whereas it is common to determine the optimum pH, buffer (including supplements), ionic strength and temperature, the medium itself has been only a stepchild of investigations. Now we begin to become acquainted with the idea of putting enzymes or antibodies into organic solvents, and the first experiences of surprise will give way to systematic work and specific usage of organic solvents not only in organic synthesis, but also in biochemical analysis.

References

Batra, R. and Gupta, M.N. (1994) Enhancement of enzyme activity in aqueous-organic solvent mixtures. *Biotechnol. Lett.* 16:1059–1064.
Belova, A.B., Mozhaev, V.V., Levashov, A.V., Shergeeva, M.V., Martinek, K. and Khmelnitskii, Y.L. (1991) Relationship between the physicochemical characteristics of organic solvents and their denaturing capacity towards proteins. *Biochemistry (Moscow)* 56:1357–1373.
Benkovic, S.J. (1992) Catalytic antibodies. *Ann. Rev. Biochem.* 61:29–54.
Bourquelot, E. and Bridel, M. (1911) *J. Pharm. Chim.* 4:385–390 (cited in Klibanov, 1989).
Campanella, L., Sammartino, M.P. and Tomassetti, M. (1992) The effect of organic solvent properties on the response of a tyrosinase enzyme sensor. *Sensors Actuators B* 7:383–388.
Chin, J.T., Wheeler, S.L. and Klibanov, A.M. (1994) *Biotechnol. Bioengin.* 44:140–145.

Dong, S. and Guo, Y. (1994) Organic phase enzyme electrode operated in water-free solvents. *Anal. Chem.* 66:3895–3899.

Dordick, J.S. (1989) Enzymatic catalysis in monophasic organic solvents. *Enzyme Microb. Technol.* 11:194–211.

Flygare, L. and Danielsson, B. (1988) Advantages of organic solvents in thermometric and optoacoustic enzymatic analysis. *Annals New York Acad. Sci.* 542:485–496.

Francis, J.M. and Craston, D.H. (1994) Immunoassay for parathion without its prior removal from solution in hexane. *Analyst* 119:1801–1805.

Freudenberg, M.A., Fomsgaard, A., Mitov, I. and Galanos, C. (1989) ELISA for antibodies to lipid A, lipopolysaccharides and other hydrophobic antigens. *Infection* 17:56–62.

Giraudi, G. and Baggiano, C. (1993) Solvent effect on testosterone-antitestosterone interaction. *Biochim. Biophys. Acta* 1157:211–216.

Gupta, M.N. (1992) Enzyme function in organic solvents. *Eur. J. Biochem.* 203:25–32.

Hall, G.F. and Turner, A.P.F. (1991) An organic phase enzyme electrode for cholesterol. *Anal. Lett.* 24:1375–1388.

Hall, G.F., Best, D.J. and Turner, A.P.F. (1988) The determination of p-cresol in chloroform with an enzyme electrode used in the organic phase. *Anal. Chim. Acta* 213:113–119.

Hundeck, H.G., Weiß, M., Scheper, T. and Schubert, F. (1993) Calorimetric biosensor for the detection and determination of enantioimeric excesses in aqueous and organic phases. *Biosens. Bioelectron.* 8:205–208.

Iwuoha, E.I. and Smyth, M.R. (1994) Organic-phase application of an amperometric glucose sensor. *Analyst* 119:265–267.

Kabanov, A.V., Khrutskaya, M.M., Eremin, S.A. and Klyachko, N.L. (1989) A new way in homogeneous immunoassay: reversed micellar systems as a medium for analysis. *Anal. Biochem.* 181:145–148.

Karayannis, M., O'Sullivan, C., Turner, A.P.F., Hall, J., Dennison, M., Saini, S., Hobbs, B. and Aston, W.J. (1994) The development of microbiosensors for monitoring hazardous gases in the environment. *In*: P. Bennetto and J. Büsing (eds): EC Report *Biosensors for environmental monitoring.* European Commission DG 12, Brussels, pp 65–71.

Kazandjian, R.Z., Dordick, J.S. and Klibanov, A.M. (1986) Enzymatic analysis in organic solvents. *Biotechnol. Bioengin.* 28:417–421.

Khamessan, A., Kermasha, S., Khalyfa, A. and Marsot, P. (1993) Biocatalysis of chlorophyllase from the alga *Phaeodactylum tricornutum* in a water-miscible organic-solvent system. *Biotechnol. Appl. Biochem.* 18:285–298.

Klibanov, A.M. (1989) Enzymatic catalyis in anhydrous organic solvents. *Trends Biochem. Sci. (TIBS)* 14:141–144.

Klibanov, A.M., Desai, U.R. and Griebenow, K. (1995) Enzyme structure under nonaqueous conditions. *In*: Abstract book of *Enzyme Engineering XIII,* San Diego.

Laane, C., Boeren, S., Vos, K. and Veeger, C. (1987) Rules for optimization of biocatalysis in organic solvents. *Biotechnol. Bioeng.* 30:81–87.

Luisi, P.L., Giomini, M., Pileni, M.P. and Robinson, B.H. (1988) Reverse micelles as hosts for proteins and small molecules. *Biochim. Biophys. Acta* 947:209–246.

Mabrouk, P.A. (1995) First direct interfacial electron transfer between a biomolecule and a solid electrode in non-aqueous media: direct electrochemistry of microperoxidase-11 at glassy carbon in dimethyl sulfoxide solution. *Anal. Chim. Acta* 307:245–251.

Mannino, S., Cosio, M.S. and Wang, J. (1994) Determination of peroxide value in vegetable oils by an organic-phase enzyme electrode. *Anal. Lett.* 27:299–308.

Mansuy, D., Bartoli, J.-F. and Momenteau, M. (1982) Alkane hydroxylation catalyzed by metalloporphyrins: evidence for differnt active oxygen species with alkylhydroperoxides and iodosobenzene as oxidants. *Tetrahedron Lett.* 23:2781–2784.

Mionetto, N. (1993) *Biocapteur amperometrique pour la detection d'insecticides a action anticholinesterasique.* These de doctorat de l'Universite de Perpignan.

Miyabayashi, A., Reslow, M., Adlercreutz, P. and Mattiasson, B. (1989) A potentiometric enzyme electrode for monitoring in organic solvents. *Anal. Chim. Acta* 219:27–36.

Otamiri, M., Adlercreutz, P. and Mattiasson, B. (1994) Polymer-polymer organic solvent two-phase system – a new type of reaction medium for bioorganic synthesis. *Biotechnol. Bioeng.* 43:987–994.

Russell, A.J., Trudel, L.J., Skipper, P.L., Groopman, J.D., Tannenbaum, S.R. and Klibanov, A.M. (1989) Antibody-antigen binding in organic solvents. *Biochem. Biophys. Res. Comm.* 158:80–85.

Saini, S. and Turner, A.P.F. (1995) Multi-phase bioelectrochemical sensors. *Trends Anal. Chem.* 14:304–310.

Saini, S., Hall, G.F., Downs, M.E.A. and Turner, A.P.F. (1991) Organic phase enzyme electrodes. *Anal. Chim. Acta* 249:1–15.

Schubert, F., Saini, S. and Turner, A.P.F. (1991) Mediated amperometric enzyme electrode incorporating peroxidase for the determination of hydrogen peroxide in organic solvents. *Anal. Chim. Acta* 245:133–138.

Schubert, F., Saini, S., Turner, A.P.F. and Scheller, F. (1992) Organic phase enzyme electrodes for the determination of hydrogen peroxide and phenol. *Sensors Actuators B* 7:408–411.

Shevlin, C.G., Hilton, S. and Janda, K.D. (1994) Automation of antibody catalysis: a practical methodology for the use of catalytic antibodies in organic synthesis. *Bioorg. Med. Chem. Lett.* 4:297–302.

Singer, S.J. (1962) The properties of proteins in nonaqueous solvents. *Adv. Protein Chem.* 17:1–68.

Stančík, L., Macholán, L. and Scheller, F. (1995) Biosensing of tyrosinase inhibitors in nonaqueous solvents. *Electroanalysis* 7(7):649–651.

Stöcklein, W.M.F. and Scheller, F.W. (1995) Effects of organic solvents on semicontinuous immunochemical detection of coumarin derivatives. *Sensors Actuators B* 24–25:80–84.

Suleiman, A.A. and Guilbault, G.G. (1992) Piezoelectric immunosensors and their applications. *In:* F. Scheller and R.D. Schmid (eds): GBF Monographs Vol. 17. *Biosensors: Fundamentals, Technologies and Applications.* Verlag Chemie, Weinheim, pp 491–500.

Taylor, R.F. (ed.) (1991) *Protein Immobilization: Fundamentals and Applications.* Marcel Dekker, New York.

Valencia-González, M.J. and Díaz-Garcia, M.E. (1994) Enzymatic reactor/roomtemperature phosphorescence sensor system for cholesterol determination in organic solvents. *Anal. Chem.* 66:2726–2731.

Wang, J. and Reviejo, A.J. (1993) Organic-phase enzyme electrode for the determination of trace water in nonaqueous media. *Anal. Chem.* 65:845–847.

Wang, J., Wu, L.-H. and Angnes, L. (1991) Organic-phase enzymatic assays with ultramicroelectrodes. *Anal. Chem.* 63:2993–2994.

Wang, J., Reviejo, A.J. and Mannino, S. (1992) Organic-phase enzyme electrode for the determination of phenols in olive oils. *Anal. Lett.* 25:1399–1409.

Wang, J., Dempsey, E., Eremenko, A. and Smyth, M.R. (1993a) Organic-phase biosensing of enzyme inhibitors. *Anal. Chim. Acta* 279:203–208.

Wang, J., Lin, Y., Eremenko, A.V., Ghindilis, A.L. and Kurochkin, I.N. (1993b) A laccase electrode for organic-phase enzymatic assays. *Anal. Lett.* 26:197–207.

Wescott, C.R. and Klibanov, A.M. (1994) The solvent dependence of enzyme specificity. *Biochim. Biophys. Acta* 1206:1–9.

Yennawar, N.H., Yennawar, H.P. and Farber, G.K. (1994) X-ray crystal structure of γ-chymotrypsin in hexane. *Biochemistry* 33:7326–7336.

Zaks, A. and Klibanov, A.M. (1985) Enzyme-catalyzed processes in organic solvents. *Proc. Natl. Acad. Sci. USA* 82:3192–3196.

Frontiers in Biosensorics I
Fundamental Aspects
ed. by F. W. Scheller, F. Schubert and J. Fedrowitz
© 1997 Birkhäuser Verlag Basel/Switzerland

Nucleic acid based sensors

F. F. Bier[1] and J. P. Fürste[2]

[1]Institute for Biochemistry and Molecular Physiology, University of Potsdam,
 c/o Max-Delbrück-Center of Molecular Medicine, D-13122 Berlin, Germany;
[2]Institute of Biochemistry, Free University of Berlin, D-14195 Berlin, Germany

Summary. Nucleic acids may be analyte or molecular recognition elements in biosensors. Both aspects merge in the genosensor approach, where detection of special sequences is facilitated by hybridization of a target nucleic acid to a complementary immobilized template. All three roles of nucleic acids in biosensors are discussed and the state of sensor development reviewed. With the invention of evolutionary synthesis strategies applied to nucleic acids new types of biomolecular receptors are accessible. The impact of aptamers and ribozymes on biosensor development is discussed.

Introduction

Compared to the rapidly increasing number of publications on biosensors regarding the biomolecular recognition process involved, nucleic acids are still a largely unexplored theme. This concerns sensors to detect nucleic acids as well as sensors using nucleic acids as molecular recognition elements. The late introduction of one of the best characterized biomolecules into the field of biosensors is likely due to problems of isolation and proper handling of nucleic acids. In addition, the inherent complexity of even small genomes (e. g. viruses or plasmids) is so enormous that it seems to be inappropriate to try to apply a simple sensor. It seems unlikely to envisage a sensor able to address the sequencing of a whole genome. There are, however, many situations in which the complex information as a whole is not needed but only the presence or amount of nucleic acid has to be determined. Examples are the screening for impurities in biopharmaceuticals (Briggs and Panfili, 1991) or the search for the unwanted release of genetically engineered microorganisms in the environment (Smith et al., 1995; Jansson, 1995). One of the most often asked questions is to assay a specific sequence within a sample, e. g. a virus. Other questions concern the comparison of an unknown sample sequence with a known one, e. g. looking for mutations. The technique to search for a complementary sequence by providing a well-defined template to form a double-strand now called hybrid, is well known as hybridization and has proven to be a very specific tool in molecular biology. For this application the typical sensor response could be sufficient to yield a quantitative signal. Since analysis of DNA is now becoming a routine method in areas as different as medical health care and plant breading (Alford and Caskey, 1994), it is necessary to develop short and simple detection schemes.

Hybridization is linked to the first role of nucleic acid, that is the storage of information. The second role of nucleic acid in nature is the processing of information, that is the capacity to specifically bind to ligands. Ligand binding of nucleic acids covers a very broad range of different interactions as counterion binding and highly (sequence) specific protein-nucleic acid interactions e.g. promoter and repressor recognition and binding. Thus we can imagine two ways of using nucleic acids in sensors: First, sensors capable of determining nucleic acids, either detecting the presence of nucleic acids, distinguishing between DNA and RNA, or searching for a special sequence. Second, sensors using nucleic acids as molecular recognition elements, which bind either drugs or proteins. The binding of nucleic acids by nucleic acids, called annealing or hybridization, falls into both categories and is at the moment the most advanced field of investigation.

The present article will start to review simple biosensors and will then proceed to the more complex ones. The second chapter deals with the determination of nucleic acid mainly with the view on quantitation. In the third chapter the broadest field of present activity, hybridization-based sensors, is scanned; we will guide the reader by presenting the approaches reported up to now by following the various transduction schemes. The fourth chapter describes nucleic acid-ligand interactions. This field of scientific activities is less directed to biosensor development but is a field of biosensor application; here we focus on the analytical potential of these investigations and show some promising new developments and how they might influence the biosensor world. In the concluding section, we will summarize the perspectives of future developments.

Sensors for quantification of nucleic acids

Nucleic acids are usually quantified by spectrophotometric monitoring at 260 nm. Spectrophotometry with bench top standard photometers is limited to about 0.01 O.D., which corresponds to 0.3 µg/ml DNA. This method, however, can only be a rough estimate when the extinction coefficient of the specific nucleic acid is not precisely known. DNA and RNA cannot be distinguished and nucleotide impurities interfere with the measurement. At present, isotope labeling or chemiluminescence-supported enzyme labeling are applied to quantify nucleic acids. These methods need skilled personnel and appropriate facilities. The rapid development of molecular biology raises the need for a more precise measurement of small amounts in small volumes. Reports on electrochemical methods for the analysis of nucleic acids date back to the late 50s (reviewed in Palecek, 1996). The application as an analytical tool to quantify the total amount of nucleic acids, however, was proposed very recently. Wang et al. (1995) reported for tRNA a detection limit of 10 pg (400 amol) using potentiometric stripping analysis on carbon paste electrodes.

The biosensor approach to nucleic acid-quantitation employs unspecific DNA-binding proteins. Kung et al. (1990) have adapted the light-addressable protentiometric sensor (LAPS) to DNA quantitation. The LAPS, developed by the Molecular Devices Corp., Menlo Park, CA, was constructed originally for immunodetection and thus is an example of how to transform immunodetection schemes into nucleic acid determination (Guesdon, 1992). Briefly, the method runs as follows: A pH shift is determined potentiometrically, which is caused by bound urease. For DNA quantitation a single-stranded DNA binding protein (SSB from *E. coli*), which is biotin-labeled, and an anti-DNA antibody urease conjugate are added to a denatured DNA sample to form a sandwich type assay. At the sensor surface the biotin-labeled complexes are caught via streptavidin and the amount of urease thus fixed is proportional to the DNA present in the sample. For a more detailed description of the method the reader is referred to a recent review by Owicki et al. (1994). The signal gained from this procedure depends strongly on fragment length and is limited to fragments longer than 90 nt (nt = nucleotides). The detection limit for a 872-nt sample was 2 pg. A similar approach has been applied by the same authors for a more sequence specific analysis (see below, Olson et al. 1991). The binding of DNA to DNA binding proteins has also been observed using surface plasmon resonance (SPR) (see below); especially the SSB-protein binding to single-stranded DNA has been investigated by Fisher et al. (1994). In this study as well as in all other publications related to protein-nucleic acid interactions, quantitation of DNA was not the aim of the authors. Thus it only can be stated that SPR as well as other affinity transducers might be adapted to quantify DNA in a similar way.

Sensors for hybridization – gene sensors

Searching for a special sequence is routinely done using Watson-Crick base pairing, called hybridization. One of the single strands, DNA or RNA, is labeled and catches the complementary sequence, the target, out of a sample, which again may be DNA or RNA. This type of hybridization assay ("gene-probe-assay") is one of the most specific and easy to perform assays in the molecular biologist's laboratory (Nickerson, 1993); high sensitivity is achieved using the polymerase chain reaction (PCR) combined with isotope labeling or chemiluminescence with enzyme labels. However, getting the information is still time consuming and laborious. To combine this powerful technique with the sensor approach has been recognized to be very promising since the development of affinity sensors started about 15 years ago; however, it was not until very recently that reports were given on successful hybridization on sensor surfaces.

The hybridization procedure

In principle all non-radioactive label techniques developed for hybridization detection on any kind of solid support, including filter membranes, may serve as basis for a biosensor development. Mainly immunolabeling has been investigated intensively as reviewed by Guesdon (1992). All immunolabel based assays can build the basis for a nucleic acid sensor. An example for this is the light-addressable potentiometric sensor mentioned above, which also has been adapted to nucleic acid hybridization by Olson et al. (1991).

The typical hybridization protocol includes 5 steps (Sambroock, 1989): 1) bind sample on support (e.g. nitrocellulose, filter membrane); 2) denature and remove not bound single-stranded DNA; 3) bind labeled probe by base pairing, this step brings all significance into the experiment and is the essential hybridization; 4) wash stringently – to reduce probability of mismatched double strands; 5) develop label – this may be a multi-step procedure, if necessary, attach messenger, e.g. antibodies, add substrate for enzyme label etc. In general, the probe is a specifically known sequence, which characterizes a gene, a species or whatever is of interest. The washing procedure has to be optimized in every special case for each probe to give highest significance.

Gene probe sensors may shorten this procedure, first, by helping to find the optimum conditions and much of the work published today falls into that category; and second, by avoiding some of the steps described above. Which step may be avoided depends on the sensor technology applied, e.g. surface characterizing techniques like evanescent wave sensors usually reduce or even allow to omit completely washing steps; label free techniques omit all developing steps linked with the use of labels.

However, the hybridization protocol has to be changed for a sensor, because it is no longer feasible to immobilize the sample; it is the probe which has to be linked to the solid support and signal generation has to be gained from the binding event of the sample. Figure 1 shows various approaches to facilitate signal generation by forming the double strand. Figure 1 (a) visualizes the obviously most simple case of direct determination of the target by immobilization of the template. Figure 1 (b) demonstrates the use of a labeled probe; in this case the sample could be determined in a competitive assay format. Figure 1 (c) shows an extension of this idea, in which two probes, one immobilized on the sensor surface, the second labeled, work like "primers" in the PCR: A signal will occur only if the complementary sequence for both probes is present in the sample. Original PCR products can be determined in a similar way if both primers, which are now at the different strands of the duplex, are labeled. Additionally, Figure 1 (d) shows the use of double strand specific ligands; the figure shows the example of intercalators.

Before we discuss these variants in more detail with published examples, we have to focus more on attachment of the template to the sensor surface.

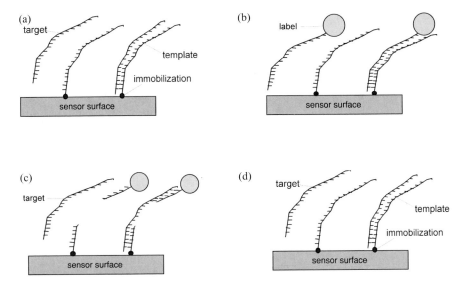

Figure 1. Modes of detection in gene-sensors. (a) Hybridization on the surface detected without any label, (b) use of labeled targets, (c) two short templates, labeled and immobilized, are bridged by the target, (d) use of double-strand specific ligands, e. g. intercalators.

Immobilization of the template sequence

As mentioned above the template on the sensor surface has the same role as the probe in the usual hybridization protocol. It is therefore a completely arbitrary and known sequence, which can be manipulated, just as the labeling of the probe in usual hybridization experiments.

Bridges: the avidin-biotin system

The avidin-biotin system is a widely used labeling procedure and both biotinylation e. g. at the 5′-end, as well as avidin (or streptavidin) immobilization on any kind of transducer surface are following common protocols. For instance, 5′-biotinylation of the template is achieved by incorporating a biotinylated phosphoramidite during the synthesis of the oligomer. Therefore, this convenient bridge is used in several studies especially in connection with optical sensors (Nilsson et al., 1995a, 1995b; Watts et al., 1995; Abel, 1995; Persson et al., 1995; Bier and Scheller, 1996, Kleinjung et al., 1996 (submitted), and all work reported on protein-nucleic acid interaction cited below). The regenerability of the template, however, was found to be limited: after several regeneration cycles the signal of hybridization is decreased significantly (Abel, 1995; Bier and Kleinjung, unpublished results); in contrast, Watts et al. (1995) observed 10-regeneration cycles without loss of hybridization capability using 10 mM NaOH for regeneration. Other strategies, like photobiotinylation, may cause problems in the hybridization reaction if the template is an oligomer of several ten nt; however, when using longer templates (several thousand nt) it might be useful.

Other bridges established as labeling systems like digoxigenin-anti-digoxigenin antibody may be adapted as well, however, no report has been given up to now; an advantage of using an immunolabel for immobilization might be a higher regenerability of the sensor surface, but for the price of loosing the template.

Covalent coupling

Covalent coupling of nucleic acid has been reported by Graham et al. (1992), using glutaraldehyde and terminally amino-functionalized oligomers (3'- and 5') on silanized glass surfaces. Millan and Mikkelsen (1993) immobilized an oligomer, a 20-mer oligo(deoxythymidylic acid), on an oxidized glassy carbon electrode, using succinimidyl enhanced carbodiimide chemistry to couple guanidine residues specifically. For this approach the template has to be elongated enzymatically to add guanidine nucleotides (up to 100). In an extension of this work Millan et al. (1994) immobilized oligomers on carbon paste electrodes, the carbon paste had to be modified by stearic acid to facilitate guanidine coupling.

Most elegant is the direct synthesis of the template on the sensor surface. This has been demonstrated by Piunno et al. (1995) and Abel (1995) using solid phase synthesis as introduced by Merrifield (1986). Abel also demonstrates high regenerability of several 100 times for template oligonucleotides synthesized directly on the activated fiber surface.

Adsorption

Adsorption without any chemical modification of the nucleic acid would be the most simple way of immobilizing DNA single strands. However, under physiological conditions adsorption of nucleic acid, neither double-stranded nor single-stranded, does occur (in contrast to many proteins), and special conditions have to be chosen to facilitate adsorption of nucleic acid (Wang et al., 1995). Su et al. (1994) used Palladium-coated quartz crystals to adsorb plasmid DNA of 4000 nt length, which differs much from the oligomers used in all other experiments described before. The attached DNA was found to be rather stable and hybridizes the complement even after adsorption significantly.

Thiol chemistry

Hashimoto et al. (1994b) used the well-investigated self-assembly effect of thiol derivatives on gold surfaces to bind DNA to gold electrodes. They used a modified DNA-containing mercaptohexyl group at the 5'-phosphate end of the probe (20-mer) and achieved a stable coupling of the single-stranded DNA, however, data on regeneration as indicator for the stability of the binding are not provided.

Gene probe sensors

The problem of getting a signal from a mere binding event has been investigated extensively for immunochemical-based bioaffinity sensors. Some of
these approaches have been successfully applied to nucleic acid determination (see below). However, the special situation of base-pairing and duplex
formation during hybridization gives the opportunity for a very special kind
of binding assay; e.g. the binding of ligands like intercalators, which are
specific for double-stranded DNA, allows very sensitive recognition of
dsDNA even in the presence of RNA or single-stranded oligonucleotides.

Evanescent field based sensors – label free

There are several reasons why optical transducers are preferred in investigating gene probe sensors. Especially label-free techniques have the
advantage of ease of performance, facilitated by commercially available
instrumentation, and a clear understanding of the surface interactions
involved. These label-free techniques are surface plasmon resonance
(SPR), the resonant mirror and the grating coupler (other configurations
like interferometers discussed for transduction in the literature have not
been applied to nucleic acids yet).

All these label-free measurements are based on the same physical
phenomenon: the influence of the refractive index of a thin layer on the sensor surface, generated by the bound species, on the electromagnetic (light)
wave reflected under a certain angle, which is facilitated through the evanescent wave.

Surface plasmon resonance (SPR) was adapted to bioaffinity sensing by
Liedberg et al. (1983) and Flanagan and Pantell (1984). First attempts in
nucleic acids determination were reported by Pollard-Knight et al. (1990).
Today SPR is well introduced for bioaffinity measurements by the com-
mercialization of this technique in the BIAcore instruments by Pharmacia,
Uppsala, Sweden. The technique is described in many reviews, e.g. by
Jönsson and Malmqvist (1992).

Nilsson et al. (1995a) used SPR and presented data on the hybridization
of several oligomers to one template (22-mer to 24-mer). The work reported covers a rather broad scope of "gene assembly" experiments, as
routinely performed by molecular biologists, this includes hybridization
at various sites of a 69-nt template. The authors claim to have observed
differences in hybridization kinetics depending on the position of the
template with respect to the biotin anchor. However, it has to be pointed
out that there are differences in the sequences compared. Work by
Persson et al. (1995) showed, that SPR can be used to observe point mutations within 8-mer hybridization. The binding signal decreases dramati-
cally if a mismatch occurs and the amount of decrease can be correlated
to the position of the mismatch within the 8-mer. Nilsson et al. (1995b)
demonstrated that a single base mismatch within a 17-nt sequence can be

Table 1. Comparison of gene-sensor performances

Transducer	Target size (nt)	Detection limit/ minimum amount	Regeneration	Comment	Ref.
Surfaces plasmon resonance	22–24	–/20 pmol	50 mM NaOH	gene assembly experiments	Nilsson et al., 1995
Surfaces plasmon resonance	8	31 nM/1.5 pmol	dissoziation	mismatch position, affinity constants	Persson et al., 1995
Resonant mirror	40	9.2 nM/0.92 pmol	10 mM NaOH	PCR primer	Watts et al., 1995
Grating coupler	250	5 nM/1.4 pmol	50 mM NaOH	poly (dA)	Bier and Kleinjung, unpubl.
Grating coupler	22	50 nM/150 pmol	50 mM NaOH	contains EcoRI recognition site	Bier and Scheller, 1996
Fiber optic fluorescence + fluor. label	20	1 nM/–	85 °C	labeled probe	Graham et al., 1992
Fiber optic fluorescence + intercalator	20	13 nM	85 °C	oligo(dT)/ EtBromide	Piunno et al., 1995
Fiber optic fluorescence + fluor. label	16	75 fM	50% urea	labeled probe 950 repeats	Abel, 1995
Fiber optic fluorescence + intercalator	16	2.1 pM	50% urea	Ru-complex	Abel, 1995
Fiber optic fluorescence + intercalator	22	30fM/3.2 mol	50 mM NaOH	bis-inter-calator	Kleinjung et al., 1996
Piezoacoustic	4000	2 ng	–	adsorptive template hybridizing at 65 °C	Su et al., 1995
Voltammetric glassy carbon electrode	4000	2.5 ng	90 °C	poly (dA)	Millan and Mikkelsen, 1993
Voltammetric carbon paste electrode	16	–	90 °C	cystic fibrosis marker	Millan et al., 1994
Voltammetric gold electrode	20	40 aM/400 zmol	none		Hashimoto et al., 1994

distinguished by kinetic analysis. For this study the authors have chosen a sequence from an oncogenic suppressor p53 exon.

Watts et al. (1995) demonstrated hybridization on another evanescent field transducer, the resonant mirror (Cush et al., 1993; Buckle et al., 1993), which is now commercialized by Fisons, UK (IAsys). They investigated more deeply the immobilization procedure concluding that avidin-biotin coupling is the most effective of the compared methods. Watts investigated hybridization of a 40-mer target to a 19-mer template, which is a PCR primer of *Legionella pneumophila*. They found that hybridization is strongly reduced if the immobilized probe sequence is neighbored by non-complementary sequences.

A third evanescent field transducer, the grating coupler (Nellen et al., 1988, Tiefenthaler and Lukosz 1989; the method is reviewed by Tiefenthaler, 1992), is available by Artificial Sensing Instruments (ASI, Zurich, Switzerland). Bier and Scheller (1996) used grating couplers in an instrumentation developed by Brandenburg and Gombert (1993) (IpM Fraunhofer Gesellschaft, Freiburg, Germany) and applied them to nucleic acid determination. They showed the rapid hybridization of 8-, 15- and 22-mers to a 24-mer template, which is completed within 1 min.

As mentioned above, all these direct measuements, sometimes called biomolecular interaction analysis (BIA), are based on the same physical principle and thus the sensitivities of the methods are comparable. The detection limit found in all reported work is about 1 ng (5–10 nM). Thus for high sensitive determination of DNA the label-free approach needs improvement.

In principle the BIA-approach allows the evaluation of kinetic data: however, data available today are not consistent. The binding process of hybridization is a multi-site multi-step process including base-pairing and building of a helical structure; therefore the usual kinetic approach in terms of association and dissociation rate constants might not be sufficient to describe nucleic acid binding correctly.

Evanescent field based sensors using fluorochromes
Fluorochromes are usually introduced for highest sensitivity combined with high specificity. This is also the case for the sensor approach in nucleic acid determination. All fiber optical approaches reported use the evanescent field to distinguish between bound and unbound nucleic acid. Additionally, fluorochromes are necessary to detect a signal; they can be introduced in several ways.

The fluorochrome can be a label connected to the target sequence (Fig. 1 (b)) as reported by Graham et al. (1992) as well as by Abel (1995). Graham found a detection limit of 1.2 nM using a fluoresceine labeled probe-target, i.e. a tracer (a real target cannot be labeled) of 16 nt hybridizing to a 16-mer template, which was covalently bound to the fiber surface. Abel, also using a 16-mer template and a fluoresceine labeled tracer, found a much lower detection limit of 75 fM. However, he demonstrated the competitive binding assay for the real target and found a detection limit of 1.1 nM for the target.

To our knowledge there has been no attempt up to now to use weaker binding tracers, with shorter sequences compared to the target or with arbitrarily introduced mismatches, which could enhance the sensitivity of the competitive assay.

An extension of the use of tracers is the use of two probes both binding to the same target at different sites (Fig. 1(c)). One probe is the template immobilized on the sensor surface to catch the target, the second probe is labeled, in this example with a fluorochrome, but others would also be possible. Only if the target contains complementary sequences to both probes to bridge them, and thus bringing the label in close proximity to the sensing surface, can a signal occur. This approach might be more specific than using only one probe. No report was found in the literature for the realization of this type of gene sensor using fluorophores.

The special situation encountered with nucleic acids is the knowledge we have about very specific ligands, mainly to dsDNA. Many of them were found or designed as anti-cancer drugs. But fluorochromes are available for a broad range of excitation and emission wave lengths developed of fluorescence microscopy or fluorescence-activated cell sorting (FACS). Piunno et al. (1995) demonstrated the feasibility of this approach using ethidium bromide, a well-known intercalator for DNA staining; intercalators are planar aromats slipping between base pairs bound by stacking forces (Waring, 1968). The merit of Piunno et al. is to have published this approach first, however, in terms of sensitivity much has to be done. They reported a detection limit of 13 nM for the very unspecific dA_{20}/dT_{20} system. Abel (1995) also showed the use of intercalators to trace dsDNA on a fiber surface. He used a ruthenium complex (ruthenium phenantroline dipyridophenazine) and found a detection limit of 2 pM. This detection limit is higher than what he found for the labeled tracer; however, in this case a direct monitoring of the target is possible and the detection limit has to be compared with the competitive assay, which is less sensitive.

The recently developed intercalator PicoGreen (Molecular Probes, Eugene, OR) has the advantage of being very sensitive. It has the same excitation and emission wavelengths as fluorescein, one of the most often used fluorochromes. Very sensitive DNA quantitation was thus achieved with a remarkably simple experimental setup by Kleinjung et al. (1996) (submitted). These authors found a detection limit of 30 fM for a 22-mer as well as for 250 nt poly (dT). Thus the absolute amount in a 100-μl sample volume was only 2×10^6 copies of the target sequence.

Electrochemical detection
Similar to the optical transduction schemes various approaches have been described to link the hybridization procedure to an electrochemical signal. Again the detection schemes visualized in Figure 1 can serve as a guide.

As already mentioned in the 2nd paragraph (nucleic acid quantitation) the LAPS has also been adapted to sequence specific DNA determination (Olson et al., 1991). Olson used a two-probe approach as visualized in Figure 1(c); the template is linked to the sensor surface, a membrane, via biotin-streptavidin coupling and the labeled probe sequence, the tracer, is fluorescein-labeled. An anti-fluorescein antibody urease conjugate was employed for signal generation, i.e. the pH shift, which is determined poten-tiometrically (as described above). The detection limit for a target of 114 nt length was found at 30 amol (1 pg or 2×10^7 copies), corresponding to pM range. The target lenght influences the signal, thus longer targets were detected with less sensitivity than shorter ones, which has been attributed to steric hindrance of binding reactions during hybridization as well as for the capture and label enzyme conjugate steps. The whole assay time was 75 min and thus significantly longer than all optical measurements.

DiCesare et al. (1993) used a two-probe approach (Fig. 1(b)) for the determination of PCR products. They developed a FIA-system based on magnetic beads combined with electrochemiluminescence, which now is available by Perkin Elmer, Wilton, CT. One of the probes was labeled with biotin, the other with a ruthenium chelate, which takes part in an electro-generated chemiluminescence reaction. The hybridization takes place in solution and the biotinylated end of the hybrid was captured via streptavidin onto the beads, which were held magnetically in front of a working elec-trode to generate the luminescence signal. This procedure allowed the determination of 10 amol of a PCR product including the labeled probes as primers. As in the LAPS experiment, the authors stated an inverse length dependency of the signal.

Electroactive DNA-ligands. Many well-known DNA-ligands, anti-cancer drugs or fluorochromes like daunomycin or acridine orange, are duplex specific. Some of them have been found to be electrochemically active and this activity is influenced by the binding process, at least the binding changes the diffusion rate of the ligand and thus peak currents of voltammetric mea-surements are enhanced compared to the free ligand. This is the basis of electrochemical hybridization assays (Fig. 1(d)).

Mikkelsen and coworkes investigated glassy carbon and carbon paste electrodes in combination with tris(2,2'-bipyridyl)cobalt(III) and similar complexes, which had been shown to associate reversibly with the minor groove of double-stranded helical DNA. While DNA alone is not electro-chemically active in the relevant potential range, the cobalt complex is reversibly electroactive at 110 mV vs. Ag/AgCl. Thus, the preconcentration of cobalt complexes at a DNA-modified electrode results in higher peak currents than expected for a diffusing redox couple (Millan and Mikkelsen, 1993). The same authors also adapted the DNA-modified carbon paste electrodes for the detection of a genetic damage, cystic fibrosis, which is characterized by a three-base deletion. The minimum quantity detectable in this work was investigated for a 4000-nt poly(dA) and found at 2.5 ng (Millan et al., 1994).

Also Hashimoto et al. (1993, 1994 a) used graphite electrodes combined with intercalators as hybridization indicators. Much higher sensitivity, however, was reported by the same authors (Hashimoto et al., 1994 b) using gold electrodes modified with a 20-mer oligonucleotide complementary to an oncogene v-myc region. The bisbenzimide Hoechst 33258 was used as electroactive species, it is well known as a duplex-specific fluorochrome, which preferably binds to double-stranded DNA in the minor groove. Hashimoto found a detection limit of 0.1 pg/ml; this corresponds to 4×10^4 copies in 1 ml sample volume.

Piezoacoustic sensors

Despite the fact that the notion of hybridization sensors was first discussed in combination with piezoacoustic transducers (Fawcett et al., 1988), the data provided at that time were hardly reliable. A more careful analysis by Su et al. (1994), which allowed the measurement of binding on the Pd-surfaces of a quartz crystal in solution, demonstrated the interfacial viscosity change by duplex formation. Denatured linearized plasmid DNA of 4400 nt (pPT-2 from *E. coli*) was adsorbed on the Pd surface and the response of hybridization to the same denatured sample was observed in the presence of a high amount (25 times excess) of non-interfering DNA. The resonance frequency shift of the quartz crystal was in good agreement with results determined by isotope-labeled samples. The detection limit found was 2 ng and the assay needs several hours to complete hybridization, which is mainly due to the long DNA fragment used in this study.

Nucleic acid ligand interaction

Nucleic acid ligand interactions cover a great variety of biochemical interactions. Small ligands, including multivalent ions, different types of intercalators and binders to the minor or major groove, interact with nucleic acids more or less specific to a conformation or a sequence. Protein-nucleic acid interactions also vary in strength and specificity according to the role of the proteins in the cell. A completely new class of nucleic acid ligand interactions has been discovered with the development of evolutionary design of oligonucleotides, which enables to produce new nucleic acid structures that bind to arbitrarily chosen compounds. Thus the "ligand" becomes an "analyte".

Despite the fact that from the analytical point of view small ligands, which strongly bind to DNA and therefore mostly affect DNA processing, are mutagenic or cell toxic and should be of interest, to our knowledge, no sensor for such ligands based on nucleic acids has been developed up to now. A recent report from Weetall and coworkers (Horvath et al., 1995) on intercalation of polyaromatic carcinogens using fluorescent polarization studies showed the analytical potential of this kind of ligand binding. The authors developed a displacement assay using acridine orange, a fluorescent intercalator with binding modes for single-stranded and double-strandes DNA, as a tracer and found detection limits down to 50 nM for benzo[a]pyrene and benzo[j]fluoranthene. This investigation might serve as a stimulus for biosensor development.

Protein-nucleic acid interaction analysis

The physical and chemical aspects of protein-nucleic acid interactions are widely investigated. The contribution of biosensor research and development to this theme is mainly to be a new and versatile tool for real time interaction analysis. The analytical potential of these interactions is difficult to foresee from today's point of view. Many applications will be derived from a better understanding of the interaction itself. We will give here a selection of examples, which in our opinion bears analytical relevance. It is not by chance that all examples use evanescent field transducers, mainly SPR. It allows the label-free observation of the binding events in real time. There are two types of investigations: 1) the binding studies focusing on kinetic data, especially association and dissociation rate constants, and 2) investigations on the activity of DNA-modifying enzymes. All studies used immobilized DNA or, in a few examples RNA, bound via biotin-streptavidin.

Kinetic analysis of protein nucleic acid binding

As already demonstrated in the nucleic acid quantitation section the binding of the single-strand binding protein from *E. coli* (SSB) can be used to quantify single-stranded DNA. Fisher et al. (1994a) turned around the system and used immobilized polydeoxythymidylic acid (poly(dT)$_{70}$) to characterize the binding behavior of the SSB protein. They found a steric cooperativity for the SSB protein binding.

All other investigations on nucleic acid binding proteins are more specific: Bondeson et al. (1993) performed a kinetic analysis of the lactose repressor protein to its operator-DNA. Fisher et al. (1994b) investigated the influence of mutations on the binding kinetics of recombinant oncoproteins (derived from the retrovirus E26) to various DNA-substrates. Parsons et al. (1995) demonstrated the effect of a co-repressor on the binding kinetics of the *E. coli* methionine repressor-operator complex. Wu et al. (1995) observed the binding of zinc finger proteins to several DNA sequences including HIV (human immunodeficiency virus) DNA. They randomized the amino acid composition of the binding site, the "fingers" using phage display; the authors discuss potential therapeutic applications of optimized proteins. Zinc finger proteins are uniquely capable of binding both, DNA and RNA.

As RNA-binders they are involved in the encapsulation of viral RNA. DeRocquigny et al. (1995) investigated the binding kinetics of the nucleocapsid protein NCp7 of HIV to genomic RNA using enzymatically elongated poly(A) HIV-RNA to capture the genome with a biotinylated poly(T). The aim of this investigation is to develop an assay for the screening of NCp7-RNA inhibitors.

The number of examples will increase in the near future, and the role of biosensor technology will be to support these investigations by rapid and reliable kinetic analyses. Since many of these investigations aim at the development of new drugs for targeting cancer cells or viruses, it is likely that diagnostic tools will be derived from this research.

DNA-modifying enzymes

Enyzmatic activity can be characterized by three steps: Binding of the substrate, catalytic reaction, and release of the product. Only the first and the last step can be observed directly using evanescent field transducers. At the end it is again the binding that is analyzed. Nilsson et al. (1995a) observed the action of T4 DNA ligase, T7 DNA polymerase and the Klenow fragment as well as the action of the restriction endonuclease XhoI. Bier et al. (unpublished) observed the action of EcoRI in more detail by switching the enzyme activity after binding by the addition of bivalent cations. Such observations give a direct view of the action of enzymes. The result of the catalytic process can be determined, and it can be extrapolated that this approach will help to rapidly find optimum conditions for any nucleic acid modifying enzyme to be investigated with respect to base composition of the substrate nucleic acid as well as to activators or inhibitors.

Evolutionary design of nucleic acid ligands

The future development of new biosensors is likely to benefit from recent advancements in the understanding of the basic nature of nucleic acids. This understanding has come about with the rejection of the idea that nucleic acids are mere strings of sequence, and their realization as molecules that fold into defined three-dimensional structures. In 1990, three publications described the invention and application of evolutionary design (Tuerk and Gold, 1990; Ellington and Szostak, 1990; Robertson and Joyce, 1990). The power of this method derives from the inherent nature of some nucleic acids to combine both, a genotype (nucleotide sequence) and a phenotype (ligand binding or catalytic activity), into one molecule. Taking advantage of this unique property, oligonucleotide ligands with high affinities for a striking variety of molecular targets have been identified through directed evolution. The procedure for identifying oligonucleotide ligands is called SELEX (Systematic Evolution of Ligands by EXponential enrichment; Tuerk and Gold, 1990). The specific oligonucleotides emerg-

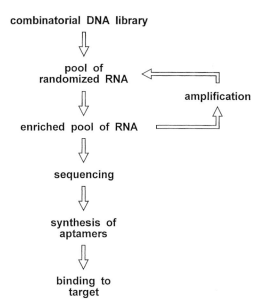

combinatorial DNA library

pool of
randomized RNA

amplification

enriched pool of RNA

sequencing

synthesis of
aptamers

binding to
target

Figure 2. Schematic diagram of the SELEX process for the *in vitro* selection and production of aptamers.

ing from this method have been called "aptamers" (Ellington and Szostak, 1990).

The *in vitro* selection of RNA aptamers follows in general the SELEX scheme shown in Figure 2. The SELEX procedue starts with chemical synthesis of a DNA library. Taking advantage of solid phase synthesis, stretches of random sequence are generated. Repeating the random cycles for n positions can produce a stochastic collection of up to 4^n variant sequences. The total size of the avarage library is in the order of 10^{12} to 10^{18} unique molecules. RNA is generated from this DNA library through *in vitro* transcription. The pool of variant RNA molecules is then subjected to selection for a physical property of interest, such as binding to a target structure. RNA molecules that have steric and electronic complementarity to the target structure can be partitioned from nonfunctional variants by their binding properties. The selected variants are amplified by reverse transcription, polymerase chain reaction (PCR) and transcription. This process of selection and amplification is then repeated until the sequence complexity of the pool is sufficiently reduced. The specific sequences, individual sequences of functional RNA molecules from the enriched pool, are finally obtained through molecular cloning and sequencing of their DNA copies.

There are several advantages of SELEX compared to other strategies employing combinatorial libraries (Janda, 1994; Gold, 1995). First, the dual nature of nucleic acids, having a defined sequence and shape allows to identify functional molecules from a large pool of variants. Through repeated amplification and selection, the binding variants are efficiently enriched, and the synthetic history of the process is simply determined by sequencing the DNA copies. This obsoletes procedures like tagging of the library that might interfere with the binding properties of variants (Janda, 1994). Second, unlike short peptides, small oligonucleotides form stable intramolecular structures (Olivera et al., 1990; Gold et al., 1995). The stability of short oligonucleotides is likely due to complex interactions of the standard nucleotides, such as Watson-Crick- and non-Watson-Crick-base pairs, bulges, hairpin loops, pseudoknots and G quartets. Functional oligonucleotides obtained by SELEX are often as small as 25 nucleotides and have binding affinities that rival those of large proteins. Finally, the diversity of sequences can be expanded throughout the process. Through error-prone amplification in the individual cycles, new variants from the enriched sequences can be obtained. Therefore, new RNA molecules with the desired properties are isolated through repeated cycles of variation, selection and replication, much like the process of natural selection.

The obvious limitation associated with the application of aptamers in biosensors is their instability in the presence of biological media. This problem has been addressed by using nucleotide analogs that render the ligands more stable against nuclease degradation (reviewed in Eaton and Pieken, 1995). The studies focused on substitutions that are compatible with the enzymatic steps of *in vitro* selection. Employing 2′-aminopyrimidine ribonucleoside triphosphates, modified RNA ligands were selected that confer similar affinities and specificities as unmodified RNA ligands (Jellinek et al., 1995).

RNA aptamers that bind to small molecules
RNA aptamers have been selected which are specific for a variety of small molecules (reviewed in Lorsch and Szostak, 1996). The target structures consist of organic dyes (Ellington and Szostak, 1990); nucleotides, such as ATP (Sassanfar and Szostak, 1993) and GDP/GMP (Connell and Yarus, 1994); and amino acids, such as arginine (Famulok, 1994), valine (Mayerfeld and Yarus, 1994) and tryptophan (Famulok and Szostak, 1992). The tryptophan-specific aptamer demonstrated stereo-specificity of the interaction. While D-tryptophan-agarose is recognized with a dissociation constant K_d of 18 µM, L-tryptophan agarose is not recognized within detection limits ($K_d > 12$ mM). Lauhon and Szostak (1995) have successfully isolated an aptamer that is specific for nicotinamide mononucleotide (NMN) and nicotinamide adeninen dinucleotide (NAD).

The aptamer discriminates between different redox states of the cofactor, having a K_d of 2.5 µM for NAD but recognizing NADH with a K_d of about 40 µM. Furthermore, SELEX has been applied to small molecules of pharmacological value, like vitamine B_{12} with a K_d of 90 nM, and the beta-2 agonist theophylline, with a K_d of 0.1 µM (Jenison et al., 1994). In their selection of theophylline-specific aptamers, Jenison et al. (1994) applied a new method. The specificity of the oligonucleotide ligand was improved by counterselection to avoid cross-reaction with caffeine. The selection was performed by partitioning the RNA library using immobilized theophylline. Prior to eluting with theophylline, cross-reacting RNAs were depleted by elution with caffeine. The resulting aptamers were able to discriminate between the two xanthine derivatives with 10-fold higher specificity than monoclonal antibodies against theophylline (Jenison et al., 1994; Poncelet et al., 1990).

RNA aptamers that bind to proteins
The first target of SELEX was bacteriophage T4 DNA polymerase (Tuerk and Gold, 1990). Since this initial application, a variety of protein-specific aptamers have been described (reviewed in Gold et al., 1995). The targets include proteins that naturally interact with RNA. For example, Schneider et al. (1992) raised aptamers to the bacteriophage R17 coat protein (K_d of 1 nM. Ringquist et al. (1995) obtained aptamers for ribosomal protein S1 (K_d of 4 nM) and Brown and Gold. (1995) successfully isolated aptamers for bacteriophage Qβ replicase (K_d of 5 nM). The method was applied to mammalian RNA-specific proteins. These studies primarily focused on projects with the potential to lead to the development of new therapeutics. They include the HIV-1 Rev-protein (Bartel et al., 1991, the HIV-1 tat protein (Tuerk and MacDougal-Waugh, 1993) and the reverse transcriptase of HIV (Tuerk et al., 1992). The binding constants determined for these protein specific-aptamers rivaled or exceeded those obtained for the natural RNA-protein interaction (K_d of $1-5$ nM).

Aptamers have also been raised to proteins that are not known to bind naturally to RNA. The binding constants obtained were in the low nanomolar range for thrombin (Kubik et al., 1994), the IgE class of antibodies and mouse monoclonal antibody raised against the human insulin receptor (Doudna et al., 1995). Surprisingly, Jellinek et al. (1993 and 1994) described binding constants for basic fibroblast growth factor (bFGF) and vascular endothelial growth factor (VEGF) in the picomolar range, thus far exceeding those obtained for known nucleic acid-binding proteins.

DNA aptamers
Specific ligands can also be derived from DNA libraries, since DNA molecules are also able to fold into the necessary complex structures (Ellington and Szostak, 1992; Bock et al., 1992). The binding constants obtained for DNA ligands were in the same range as those achieved for RNA ligands. The selection for DNA aptamers is simpler, since PCR amplification is the sole enzymatic step used for the selection of functional ligands. The main difficulty arises from producing single-stranded DNA during or after PCR amplification. This has been overcome by asymmetric PCR or removing a biotinylated strand by base denaturation and subsequent separation on an

avidin-agarose column. DNA aptamers might be more suitable for potential sensors than RNA molecules because of the greater stability of DNA.

Ligand-binding DNAs were selected for small dye molecules (Ellington and Szostak, 1992), ATP (Huizenga and Szostak, 1995) and proteins, like HIV-1 reverse transcriptase, elastase and human thrombin (Bock et al., 1992). The thrombin-specific ligands have anticlotting activity and are thus candidates for new lead compounds. The structure of this aptamer was solved by NMR, revealing a central binding motif composed of an intra-molecular G quartet. The interactions of thrombin and aptamer were also evaluated by applying surface plasmon resonance, either by attaching a biotinylated DNA aptamer to a sensor chip and measuring thrombin binding or by immobilizing thrombin on the sensor surface and monitoring the binding of aptamers.

Potential new ribozymes
Ribozymes are catalytic RNAs. The name derived from a combination of RIBOnucleic acids and enZYMES. Catalysis by biopolymers was originally considered to be performed only by proteins. Early proposals on the cata-lytic function of RNA date back to 1964; however, not until the finding of T. Cech and S. Altman did one suspect that catalytic RNAs still exist as functional molecules in present days biosphere (Kruger et al., 1982; Guerrier-Takada et al., 1983). Even though there are indications for a possible catalytic involvement of ribosomal RNA in peptide bond formation (Noller et al., 1992), the well-characterized activities of natural ribozymes are restricted to splicing, transesterification and cleavage of phosphodiester bonds. It has been shown that metal ions are essential for their activities. In particular, the activities of the group I intron from *Tetrahymena*, the Rnase P and several small ribozymes, like hammerhead and hairpin ribozyme, are completely dependent upon the presence of divalent metal ions. Yarus (1993) proposed in his conjecture that RNAs and proteins can fix metal ions in space with similar versatility and that RNAs should therefore be able to perform any reaction also carried out by protein metalloenzymes. Ribonucleic acids might thus be able to function in such diverse mecha-nisms as electrophilic catalysis, general acid/base catalysis or the catalysis of redox reactions.

Several investigators have successfully modified the activity of natural ribozymes, such as the group I intron from *Tetrahymena* (Robertson and Joyce, 1990), and the hairpin ribozyme (Berzal-Herranz, 1992). Further-more, the combinatorial library approach has been applied to select new ribozymes from random libraries. Pan and Uhlenbeck (1992) identified RNAs that catalyzed Pb^{2+}-dependent self-cleavage by a similar mechanism as described for ribonucleases. Bartel and Szostak (1993) selected an RNA catalyst capable of ligating a nucleoside triphosphate to the 3'-terminus of an oligonucleotide, thus mimicking the basic mechanism of a polymerase. By modifying an ATP-specific aptamer, Lorsch and Szostak (1994) isolated a new ribozyme with polynucleotide kinase activity. In addition, RNAs were identified that catalyze aminoacylation (Illangsekhare et al., 1994) or the cleavage of an amide bond (Dai et al., 1995). The catalytic potential of nucleic acids, however, is not restricted to RNA. Breaker and Joyce (1994) described a new deoxyribozyme, capable of cleaving the phosphodiester bond of an oligonucleotide. Most recently, Cuenoud and Szostak (1995) reported the isolation of a DNA metalloenzyme with DNA ligase activity.

Once properly designed for a special analytical problem, it is obvious that oligonucleotide catalysts could function as molecular recognition elements in metabolism-based biosensors. Since new ribozymes can be selected for arbitrarily chosen reactions, they might supplement the set of available catalysts for biosensors in the near future.

Perspectives

The DNA sequences of genes and genomes are the basis for further advances in biotechnology and biomedical applications. Sophisticated methods have already been developed to facilitate rapid and automatic sequencing (for review see for example Chen, 1994). Such methods are not likely to be substituted by sensor technology. However, the rapid development of miniaturization, for example by capillary electrophoresis, may overcome such limitations (Effenhauser et al., 1994; Fung and Yeung 1995; Woolley and Mathies, 1995). The sequencing by hybridization approach (Drmanac et al., 1993; Drmanac et al., 1994) requires an array of systematically varied oligunucleotides that are simultaneously hybridized under identical conditions. Starting with macroscopic arrays on filter membranes, efforts were made to facilitate the "on chip" synthesis with high spatial resolution (recent reviews have been given by Noble, 1995, and McIntyre, 1996). At present, spots of 100 μm diameter can be accomplished and the aim is to arrive at a spot size of 10 μm in the near future. More than 16,000 spots containing arbitrary sequences as templates have already been placed on a single silicon wafer. Up to now, the readout of arrays was performed by fluorescence microscopy supported by image analysis and thus requiring target labels. New transduction schemes will be necessary to miniaturize the readout of smaller arrays.

As this review emphasizes, the trend of development is still towards several more sensitive systems. The most sensitive detection schemes reported are in the aM range capable of detecting as few as 10^4 copies of a target sequence. Such low detection limits might enable the development of DNA-based sensors to applications that may overcome the use of amplification techniques like PCR. The natural limit is obviously the detection of a single molecule. Recently, Eigen and Rigler (1994) reported the monitoring of single molecule interactions by fluorescent correlation spectroscopy (FCS). The application of this method of DNA hybridization was demonstrated by Kinjo and Rigler (1995), who used a fluorescently labeled primer hybridizing to single-stranded M13 phage DNA. The method is applicable to small volumes (less than 10 µl), works homogeneously and does not require any immobilization. It is, however, restricted to fluorescently labeled probes, which have to be significantly smaller than the target molecule.

The scope of sensor applications is likely to be broadened by the introduction of oligonucleotide ligands as molecular recognition elements. Aptamers might complement or even exceed the analytical potential of monoclonal antibodies. There is no obvious reason to exclude certain molecules from the spectrum of potential targets and the binding events are not necessarily restricted to physiological conditions. Since aptamers are often as small as 25 nucleotides, they can be prepared by solid phase chemical synthesis. This allows to incorporate modified nucleotides at defined positions to directly monitor a binding reaction or even improve the binding constant through covalent interactions. Jensen et al. (1996) observed upon photo-crosslinking a significant improvement in the binding characteristics of aptamers for the rev-protein that contained 5-iodo-uridine in place of uridine. In addition, potential cross-reactions can be significantly suppressed applying counterselection. The obvious problem connected to oligonucleotides, however, is their instability in biological media. The future application of oligonucleotide sensors will thus highly depend on modifications that render the molecules resistant to nuclease degradation.

References

Abel, A.P. (1995) *Faseroptischer Evaneszentfeld-Biosensor zum sequenzspezifischen Nachweis von Oligonukleotiden*, Dissertation, Universität Basel.

Alford, R.L. and Caskey, C.T. (1994) DNA analysis in forensic, disease and animal/pant identification. *Current Opinion in Biotechnology* 5:29–33.

Bartel, D.P. and Szostak, J.W. (1993) Isolation of new ribozymes from a large pool of random sequences. *Science* 262:1411–1418

Bartel, D.P., Zapp, M.L., Green, M.R. and Szostak, J.W. (1991) HIV-1 Rev regulation involves recognition of non-Watson-Crick pairs in viral RNA. *Cell* 67:529–536.

Berzal-Herranz, A., Joseph, S. and Burke, J.M. (1992) *In vitro* selection of active hairpin ribozymes by sequential RNA-catalyzed cleavage and ligation reactions. *Genes and Development* 6:129–134.

Bier, F.F. and Scheller, F.W. (1996) Label-free observation of DNA-hybridisation and endonuclease activity on a wave guide surface using a grating coupler. *Biosens. Bioelectron.* 11:669–679.

Bock, L.C., Griffin, L.C., Latham, J.A., Vermaas, E.H. and Toole, J.J. (1992) Selection of single-stranded DNA molecules that bind and inhibit human thrombin. *Nature* 35:564–566.

Bondeson, K., Frostell-Karlsson, A., Fägerstam, L. and Magnusson, G. (1993) Lactose repressor-operator DNA interactions: Kinetic analysis by a surface plasmon resonance biosensor. *Anal. Biochem.* 214:245–251.

Brandenburg, A. and Gombert, A. (1993) Grating couplers as chemical sensors: A new optical configuration. *Sensor. Actuator. B* 17:35–40.

Breaker, R.R. and Joyce, G.F. (1994) A DNA enzyme that cleaves RNA. *Chem. Biol.* 1:223–229.

Briggs, J. and Panfili, P.R. (1991) Quantitation of DNA and protein impurities in biopharma-ceuticals. *Anal. Chem.* 63:850–859.

Buckle, P.E., Davies, R.J., Kinning, T., Yeung, D., Edwards, P.R., Pollard-Knight, D. and Lowe, C.R. (1993) The resonant mirror: a novel optical sensor for direct sensing of biomolecular interactions. Part II: Applications. *Biosens. Bioelectron.* 8:355–363.

Chen, E.Y. (1994) The efficiency of automated DNA sequencing. *In*: M.D. Adams, C. Fields and J.C. Venter (eds): *Automated DNA sequencing and analysis.* Acad. Press, London, pp 11–16.

Connell, G. and Yarus, M. (1994) RNAs with dual specificity and dual RNAs with similar specificity. *Science* 264:1137–1141.

Cuenoud, B. and Szostak, J.W. (1995) A DNA metalloenzyme with DNA ligase activity. *Nature* 375:611–614.

Cush, R., Cronin, J.M., Stewart, W.J., Maule, C.H., Molloy, J. and Goddard, N.J. (1993) The resonant mirror: a novel optical biosensor for direct sensing of biomolecular interactions, Part I: Principles of operation and associated instrumentation. *Biosens. Bioelectron.* 8:347–353.

Dai, X., De Mesmacker, A. and Joyce, G.F. (1995) Cleavage of an amide bond by a ribozyme. *Science* 267:237–240.

de Rocquigny, H., Delaunay, T., Dong, C.Z., Grosclaude, J., Fournie-Zaluskiand, M.C. and Roques, B.P. (1995) Interaction of the HIV-1 nucleoprotein NCp7 and derived peptide with (1–415) genomic RNA monitored by real time BIA. *Book of Abstracts of the 5th European BIAsymposium, Stockholm, September, 27th–29th, 1995.*

DiCesare, J., Grossman, B., Katz, E., Picozza, E., Ragua, R. and Woudenberg, T. (1993) A high-sensitivity electrochemiluminescence-based detection system for automated PCR product quantitation. *Bio Techniques* 15:152–157.

Doudna, J.A., Cech, T.R. and Sullenger, B.A. (1995) Selection of an RNA molecule that mimics a major autoantigenic epitope of human insulin receptor. *Proc. Natl. Acad. Sci. USA* 92:2355–2359.

Drmanac, R., Drmanac, S., Strezoska, Z., Paunescu, T., Labat, I, Zeremski, M., Snoddy, J., Funkhouser, W.K., Koop, B., Hood, L. and Crkvenjakov, R. (1993) DNA sequence determi-nation by hybridization: A strategy for efficient large-scale sequencing. *Sience* 260: 1649–1652.

Drmanac, R., Drmanac, S., Jarvis, J. and Labat, I. (1994) Sequencing by hybridisation. *In*: M.D. Adams, C. Fields and J.C. Venter (eds): *Automated DNA sequencing and analysis.* Acad. Press, London, pp 3–10.

Eaton, B.E. and Pieken, W.A. (1995) Ribonucleosides and RNA. *Ann. Rev. Biochem.* 64:837–863.

Effenhauser, C.D., Paulus, A., Manz, A. and Widmer, H.M. (1994) High speed separation of antisense oligonucleotides on a micromachined capillary electrophoresis device. *Anal. Chem.* 66:2949–2953.

Eigen, M. and Rigler, R. (1994) Sorting single molecules: Application to diagnostics and evolutionary biotechnology. *Proc. Natl. Acad. Sci. USA* 91:5740–5757.

Ellington, A.D. and Szostak, J.W. (1990) *In vitro* selection of RNA molecules that bind specific ligands. *Nature* 346:818–822.

Ellington, A.D. and Szostak, J.W. (1992) Selection *in vitro* of single-stranded DNA molecules that fold into specific ligand-binding structures. *Nature* 355:850–852.

Famulok, M. (1994) Molecular recognition of amino acids by RNA-aptamers: an L-citrulline binding motif and its evolution into a L-arginine binder. *J. Am. Chem. Soc.* 116:1698–1706.

Famulok, M. and Szostak, J.W. (1992) Stereospecific recognition of tryptophan agarose by *in vitro* selected RNA. *J. Am. Chem. Soc.* 114:3990–3991.

Fawcett, N.C., Evans, J.A., Chien, L.-C. and Flowers, N. (1988) Nucleic acid hybridization detected by piezoelectric resonance. *Anal. Lett.* 21:1099–1114.

Fisher, R.J., Fishav, M., Casas-Finet, J., Bladen, S. and McNitt, K.L. (1994a) Real-time BIA-core measurements of *Escherichia coli* single-stranded DNA binding (SSB) protein to poly-deoxythymidylic acid reveal single-state kinetics with steric cooperativity. *Methods: A Com-panion to Methods in Enyzmology* 6:121–133.

Fisher, R.J., Fishav, M., Casas-Finet, J., Erickson, J.W., Kondoh, A., Bladen, S.V., Fisher, C., Watson, D.K. and Papas, T. (1994b) Real time DNA binding measurements of the ETS1 recombinant oncoproteins reveal significant kinetic differences between the p42 and p52 iso-forms. *Protein Sci.* 3:257–266.

Flanagan, M.T. and Pantell, R.H. (1984) Surface plasmon reasonance and immunosensors. *Electron. Lett.* 20:968–970.

Fung, E.N. and Yeung, E.S. (1995) High-speed DNA sequencing by using mixed poly(ethylene oxide) solutions in uncoated capillary columns. *Anal. Chem.* 67:1913–1919.

Gold, L. (1995) Oligonucleotides as research, diagnostic, and therapeutic agents. *J. Biol. Chem.* 270:13581–13584.

Gold, L., Polisky, B., Uhlenbeck, O.C. and Yarus, M. (1995) Diversity of oligonucleotide functions. *Ann. Rev. Biochem.* 64:763–797.

Graham, C.R., Leslie, D. and Squirrell, D.J. (1992) Gene probe assays on fibre-optic evanescent wave biosensor. *Biosens. Bioelectron.* 7:487–493.

Guerrier-Takada, C., Gardiner, K., Marsh, T., Pacre, N. and Altman, S. (1983) The RNA moiety of Ribonuclease P is the catalytic subunit of the enzyme. *Cell* 35:849–857.

Guesdon, J.-L. (1992) Immunoenzymatic techniques applied to the specific detection of nucleic acids. *J. Immunol. Meth.* 150:33–49.

Hashimoto, K., Miwa, K. Goto, M. and Ishimori, Y. (1993) DNA sensor: A novel electrochemical gene detection method using carbon electrode immobilized DNA probes. *Supramolec. Chem.* 2:265–270.

Hashimoto, K., Ito, K. and Ishimori, Y. (1994a) Novel DNA sensor for electrochemical gene detection. *Anal. Chim. Acta.* 286:219–224.

Hashimoto, K., Ito, K. and Ishimori, Y. (1994b) Sequence-specific gene detection with a gold electrode modified with DNA probes and an electrochemically active dye. *Anal. Chem.* 66:3830–3833.

Horvath, J.J., Gueguetchkeri, M., Gupta, A., Penumatchu, D. and Weetall, H.H. (1995) A new method for the detection and measurement of polyaromatic carcinogens and related compounds by DNA intercalation. *In*: K.R. Rogers, A. Mulchandani and W. Zhou (eds): *Biosensor and Chemical Sensor Technology*, cap. 5, ACS Symp. Ser., Vol. 613, Washington DC, pp 44–60.

Huizenga, D.E. and Szostak, J.W. (1995) A DNA aptamer that binds adenosine and ATP. *Biochemistry* 34:656–665.

Illangsekhar, M., Sanchez, G., Nickels, R. and Yarus, M. (1994) Aminoacyl-RNA synthesis catalyzed by an RNA. *Science* 267:643–647.

Janda, K.D. (1994) Tagged versus untagged libraries: methods for the generation and screening of combinatorial chemical libraries. *Proc. Natl. Acad. Sci. USA* 91:10799–10785.

Jansson, J.K. (1995) Tracking genetically engineered microorganisms in nature. *Current Opinion in Biotechnology* 6:275–283.

Jellinek, D., Lynott, C.K., Rifkin, D.B. and Janjic, N. (1993) High-affinity RNA ligands to basic fibroblast growth factor inhibit receptor binding. *Proc. Natl. Acad. Sci. USA* 90:11227–11231.

Jellinek, D., Green, L.S., Bell, C. and Janjic, N. (1994) Inhibition of receptor binding by high-affinity RNA ligands to vascular endothelial growth factor. *Biochemistry* 33:10450–10456.

Jellinek, D., Green, L.S., Bell, C., Lynott, K., Gill, N., Vargeese, C., Kirschenheuter, G., McGee, D.P.C., Abdesinghe, P., Pieken, W.A., Shapiro, R., Rifkin, D.B., Moscatelli, D. and Janjic, N. (1995) Potent 2′-amino-2′-deoxypyrimidine RNA inhibitors of basic fibroblast growth factor. *Biochemistry* 34:11363–11372.

Jenison, R.D., Gill, S.C., Pardi, A. and Polisky, B. (1994) High-resolution molecular discrimination by RNA. *Science* 263:1425–1429.

Jensen, K.B., Atkinson, B.L., Willis, M.C., Koch, T.H. and Gold, L. (1995) Using *in-vitro* selection to direct the covalent attachment of human-immunodeficiency-virus type-1 rev protein to high-affinity RNA ligands. *Proc. Natl. Acad. Sci. USA* 92:12220–12224.

Jönsson, U. and Malmqvist, M. (1992) Real time biospecific interaction analysis: the integration of surface plasmon resonance detection, general biospecific interface chemistry and microfluidics into one analytical system. *In*: A.P.F. Turner (ed.): *Advances in Biosensors Vol.* 2, JAI Press, London, pp 291–336.

Kinjo, M. and Rigler, R. (1995) Ultrasensitive hybridization analysis using fluorescence correlation spectroscopy. *Nucl. Acids. Res.* 10:1795–1799.

Kruger, K., Grabowski, P.J., Zaug, A.J., Sands, J., Gottschling, D.F. and Cech, T.R. (1982) Self splicing RNA: autoexcision and autocatalyzation of the ribosomal RNA intervening sequence of *Tetrahymena*. *Cell* 31:147–157.

Kubik, M.F., Stephens, A.W., Schneider, D.A., Marlar, R. and Tasset, D. (1994) High-affinity RNA ligands to human alpha-thrombin. *Nucleic Acids Res.* 22:2619–2626.

Kung, V.T., Panfili, P.R., Sheldon, E.L., King, R.S., Nagainis, P.A., Gomez jr., B., Ross, D.A., Briggs, J. and Zuk, R.F. (1990) Picogram quantitation of total DNA using DNA-binding proteins in a silicon sensor-based system. *Anal. Biochem.* 187:220–227.

Lauhon, C.T. and Szostak, J.W. (1995) RNA aptamers that bind flavin and nicotinamide cofactors. *J. Am. Chem. Soc.* 117:1246–1257.

Liedberg, B., Nylander, B. and Lundström, I. (1983) Surface plasmon resonance for gas detection and biosensing. *Sensor. Actuator. B* 4:299–304.

Lorsch, J. and Szostak, J.W. (1994) *In vitro* evolution of new ribozymes with polynucleotide kinase activity. *Nature* 371:31–36.

Lorsch, J. and Szostak, J.W. (1996) *In vitro* selection of nucleic acid sequences that bind small molecules. In: R. Cortese (ed): *Combinatorial libraries. Synthesis, screening and application potential.* Walter de Gruyter, Berlin/New York, pp 69–86.

Mayerfeld, I. and Yarus, M. (1994) An RNA pocket for an aliphatic hydrophobe. *Nat. Struct. Biol.* 1:287–292.

McIntyre, P.M. (1996) Microfabrication technology for DNA sequencing. *Trends Biotechnol.* 14:69–73.

Merrifield, B. (1986) Solid Phase Synthesis. *Science* 232:341–347.

Millan, K.M. and Mikkelsen, S.R. (1993) Sequence-selective biosensor for DNA based on electroactive hybridization indicators. *Anal. Chem.* 6:2317–2323.

Millan, K.M., Saraullo, A. and Mikkelsen, S.R. (1994) Voltammetric DNA biosensor for cyclic fibrosis based on a modified carbon paste electrode. *Anal. Chem.* 66:2943–2948.

Nellen, P.M., Tiefenthaler, K. and Lukosz, W. (1988) Integrated optical input grating couplers as biochemical sensors. *Sensor. Acatuator. B* 15:285–295.

Nickerson, D.A. (1993) Gene probe assays and their detection. *Current Opinion in Biotechnology* 4:48–51.

Nilsson, P., Persson, B., Uhlén, M. and Nygren, P.-A. (1995a) Real-time monitoring of DNA manipulations using biosensor technology. *Anal. Biochem.* 224:400–408.

Nilsson, P., Nygren, P.-A., Persson, B., Larsson, A. and Uhlén, M. (1995b) Mutation detection by hybridization kinetics analysis, using biosensor technology. *Book of Abstracts of the 5th European BIAsymposium, Stockholm, September, 27th–29th, 1995.*

Noble, D. (1995) DNA sequencing on a chip. *Anal. Chem.* 67:201A–209A.

Noller, H.F., Hoffarth, V. and Zimniak, L. (1992) Unusual resistance of peptidyl transferase to protein extraction procedures. *Science* 256:1416–1419.

Olivera, B.M., River, J., Clark, C., Ramilo, C.A., Corpuz, G.P., Abogadie, F.C., Mena, E.E., Woodward, S.R., Hillyard, D.R. and Cruz, L.J. (1990) Diversity of conus neuropeptides. *Science* 249:257–263.

Olson, J.D., Panfili, P.R., Zuk, R.F. and Sheldon, E.L. (1991) Quantitation of DNA hybridization in a silicon-based system: application to PCR. *Mol. Cell Probes* 5:351–358.

Owicki, J.C., Bousse, L.J., Hafeman, D.G., Kirk, G.L., Olson, J.D., Wada, H.G. and Parce, J.W. (1994) The light-addressable potentiometric sensor: principles and biological applications. *Annu. Rev. Biophys. Biomol. Struct.* 23:87–113.

Palecek, E. (1996) From Polarography of DNA to microanalysis with nucleic acid-modified electrodes. *Electroanalysis* 8:7–14.

Pan, T. and Uhlenbeck, O.C. (1992) *In vitro* selection of RNAs that undergo autocatalyc cleavage with Pb^{2+}. *Biochemistry* 31:3887–3895.

Parsons, I.D., Persson, B., Mekhalfia, A., Blackburn, G.M. and Stockley, P.G. (1995) Probing the molecular mechanism of action of co-repressor in the *E. coli* methionine repressor-operator complex using surface plasmon resonance (SPR). *Nucl. Acids Res.* 23:211–216.

Persson, B., Nilsson, P., Nygren, P.-A., Larsson, A., Uhlén, M. and Malmqvist, M. (1995) Affinity measurement of DNA hybridization using 8-mer probes. *Book of Abstracts of the 5th European BIAsymposium, Stockholm, September, 27th–29th, 1995.*

Piunno, P.A.E., Krull, U.J., Hudson, R.H.E., Damha, M.J. and Cohen, H. (1995) Fiber-optic DNA sensor for fluorimetric nucleic acid determination. *Anal. Chem.* 67:2635–2643.

Pollard-Knight, D., Hawkins, E, Yeung, D, Pashby, D.P., Simpson, M., McDougall, A., Buckle, P. and Charles, S.A. (1990) Immunoassays and nucleic acid detection with a biosensor based on surface plasmon resonance, Annales dedetection with a biosensor based on surface plasmon resonsance. *Annales de Biologie Clinique* 48:642–646.

Poncelet, S.M., Limet, J.N., Noel, J.P., Kayaert M.C., Galanti, L. and Collet-Cassart, D. (1990) Immunoassay of theophylline by latex particel counting. *Journal of Immunoassay* 11:77–88.

Ringquist, S., Jones, T., Snyder, E.E., Gibson, T., Boni, I. and Gold, L. (1995) High affinity ligands to *Escherichia coli* ribosomes and ribosomal protein S1: Comparison of natural and unnatural binding sites. *Biochemistry* 34:3640–3648.

Robertson, D.L. and Joyce, G.F. (1990) Selection *in vitro* of an RNA enzyme that specifically cleaves single-stranded DNA. *Nature* 344:467–468.

Sambroock, J., Fritsch, E.F. and Maniatis (1989) *Molecular Cloning, – A laboratory manual*, 2nd ed. Cold Spring Harbor Press, New York.

Sassanfar, M. and Szostak, J.W. (1993) An RNA motif that binds ATP. *Nature* 364:550–553.

Schneider, D., Tuerk, C. and Gold, L. (1992) Selection of high affinity RNA ligands to the bacteriophage-R17 coat protein. *J. Mol. Biol.* 228:862–869.

Smith, C.L., Kricka, L. and Krull, U.J. (1995) The development of single molecule environmental sensors. *Genetic Analysis: Biomolecular Engineering* 12:33–37.

Su, H., Kallury, K.M.R., Thompson, M. and Roach, A. (1994) Interfacial nucleic acid hybridization studied by random primer 32P labeling and liquid-phase acoustic network analysis. *Anal. Chem.* 66:769–777.

Tiefenthaler, K. (1992) Integrated optical couplers as chemical waveguide sensors. *In*: A.P.F. Turner (ed): *Advances in Biosensors*, Vol. 2, JAI Press, London, pp 261–289.

Tiefenthaler, K. and Lukosz, W. (1989) Sensitivity of grating couplers as integrated-optical chemical sensors. *J. Opt. Soc. Am. B* 6:209–220.

Tuerk, C. and Gold, L. (1990) Systematic evolution of ligands by exponential enrichment: RNA ligands to bacteriophage T4 DNA polymerase. *Science* 249:505–510.

Tuerk, C. and MacDougal, S. (1993) *In vitro* evolution of functional nucleic acids: high-affinity RNA ligands of HIV-1 proteins. *Gene* 137:33–39.

Tuerk, C., MacDougal, S. and Gold, L. (1992) RNA pseudoknots that inhibit human immunodeficiency virus type 1 reverse transcriptase. *Proc. Natl. Acad Sci. USA* 89:6988–6992.

Wang, J., Cai, X., Wang, J., Jonsson, C. and Palecek, E. (1995) Trace measurements of RNA by potentiometric stripping analysis at carbon paste electrodes. *Anal. Chem.* 67:4065–4070.

Waring, M. (1968) Drugs which affect the structure and function of DNA. *Nature* 219:1320–1323.

Watts, H.J., Yeung, D. and Parkes, H. (1995) Real-time detection and quantification of DNA hybridization by an optical biosensor. *Anal. Chem.* 67:4283–4289.

Woolley, A.T. and Mathies, R.A. (1995) Ultra-high-speed DNA sequencing using capillary electrophoresis chips. *Anal. Chem.* 67:3676–3680.

Wu, H., Yang, W.-P. and Barbas III, C.F. (1995) Building zinc fingers by selection: Toward a therapeutic application. *Proc. Natl. Acad. Sci. USA* 92:344–348.

Yarus, M. (1993) How many catalytic RNAs? Ions and the Cheshire cat conjecture. *FASEB* 7:31–39.

Frontiers in Biosensorics I
Fundamental Aspects
ed. by F. W. Scheller, F. Schubert and J. Fedrowitz
© 1997 Birkhäuser Verlag Basel/Switzerland

Receptor based chemical sensing

M. Sugawara[1], H. Sato[2], T. Ozawa[1] and Y. Umezawa[1]

[1]Department of Chemistry, School of Science, University of Tokyo, Tokyo 113, Japan;
[2]Department of Chemistry, Faculty of Science, Hokkaido University, Sapporo 060, Japan

Summary. This mini-review describes our recent approach of mimicking transmembrane and intracellular signalings displayed by various receptors in biomembranes for the development of new sensing membranes. Several important modes of receptor signaling have been utilized exploiting bio and synthetic receptors; (i) Ca^{2+} signaling by calmodulin, (ii) active transport of target and relevant compounds displayed by Na^+/D-glucose cotransporter and Na^+,K^+-ATPase, (iii) membrane permeability changes induced by glutamate receptor ion channel proteins and (iv) membrane potential changes induced by synthetic receptors. The newly designed sensing systems are demonstrated and discussed in terms of their novel mode of signal transduction, sensitivity and selectivity.

Introduction

Molecular recognition by receptors, followed by signal transduction, leading to transmembrane signaling, has been found to provide useful new principles for chemical sensors (Odashima et al., 1991; Sugawara et al., 1993). Some examples of our previous work using bioreceptors include L-glutamate sensors based on a glutamate receptor ion channel protein (Uto et al., 1990; Minami et al., 1991a; Minami et al., 1991b), D-glucose sensors based on a Na^+/D-glucose cotransporter (Sugao et al., 1993), and ATP sensors based on Na^+, K^+-ATPase (Adachi et al., 1993). Biomimetic ion channel and active transport sensors have also been proposed, utilizing various natural and artificial receptors; valinomycin (Sugawara et al., 1987; Sugawara et al., 1992; Sugawara et al., 1994), macrocyclic polyamines (Nagase et al., 1990), lipophilic derivatives of β-cyclodextrin (Odashima et al., 1993; Odashima et al., 1994), calix-arenes (Yagi et al., 1996), tri-n-octylmethylammonium chloride (Uto et al., 1986), dicylohexyl-18-crown-6 (Sugawara et al., 1988; Sugawara et al., 1991) and lasalocid A (Sugawara et al., 1996a). We recently succeeded in developing new sensing systems based on receptor signaling: a Ca^{2+} sensing system based on a calmodulin-mediated calcium signaling (Ozawa et al., submitted) and potentiometric bilayer lipid membrane (BLM) sensing systems based on a variety of synthetic receptors (Sato et al., in preparation). In this mini-review, several examples of such sensing systems together with our earlier work are demonstrated and discussed in view of novel modes of signal transduction, sensitivity and selectivity.

A Ca^{2+} sensor based on calmodulin

Among many Ca^{2+}-sensor proteins, calmodulin (CaM: 148 amino acid residues 16.7 kDa) is the ubiquitous Ca^{2+}-binding protein which serves as a multifunctional Ca^{2+}-sensor in a variety of cellular processes (Alberts et al., 1994). After the selective recognition by CaM of Ca^{2+} ion, in many cases, they interact with target enzymes and proteins, while the Ca^{2+}-free forms do not bind the targets. Conformational transitions of these proteins induced by Ca^{2+} binding enable the binding with the target and lead to the activation and regulation of the target proteins. The molecular mechanisms of such a fine tuning of protein conformation, which is responsible for a Ca^{2+}-dependent on/off switch for cellular processes, may have useful implications for the development of ion-selective sensors. We have thus developed a sensing system for Ca^{2+} by using CaM as a primary receptor and a synthetic peptide M13 as the target protein (Ozawa et al., submitted). The peptide M13 consists of 26 residues of amino acids comprising a calmodulin binding domain of myosin light chain kinase (MLCK).

In Figure 1 the principle of the present sensing system is shown. The target peptide M13 is immobilized in the dextran matrix attached to the surface of a gold film. The immobilization of M13 is achieved with amine coupling by reaction of the activated carboxylic groups of the dextran matrix with primary amines of the target peptide. Sample solutions containing CaM and Ca^{2+} are allowed to flow over the surface of the target protein-immobilized dextran matrix. The CaM-Ca^{2+}-target protein interaction and resulting formation of a ternary complex is detected by means of surface plasmon resonance (SPR).

The Ca^{2+}-induced change in SPR signal was defined as a difference of the two SPR signals: one is measured after injection of Ca^{2+} ions and the other is successively measured after injection of a running buffer containing EGTA. The dependence of the SPR signals on the concentration of free Ca^{2+} is shown in Figure 2. The SPR signal was found to increase selectively with the concentration of Ca^{2+}, due to the formation of the Ca^{2+}-CaM-M13 ternary complex in the dextran matrix. The Ca^{2+}-dependent SPR signals were obtained in the concentration range of Ca^{2+} from 3.2×10^{-8} M to 1.1×10^{-5} M. No change in the SPR signal was induced by Mg^{2+}, K^{+} and Li^{+} at concentrations as high as 1.0×10^{-1} M. Only Sr^{2+} at 5.1×10^{-4} M gave a response that indicates formation of a Sr^{2+}-CaM-M13 ternary complex. The detection limit for free Ca^{2+} of the present sensing system is 1.6×10^{-8} M (S/N = 3), which is comparable to a fluorescence-labeled calmodulin sensor (Blair et al., 1994), but lower than that with an optical sensor based on a Ca^{2+} ionophore (Morf et al., 1992).

The present sensing system utilized a Ca^{2+} signaling pathway by CaM and its target peptide M13 for detecting Ca^{2+}. The unique feature of this

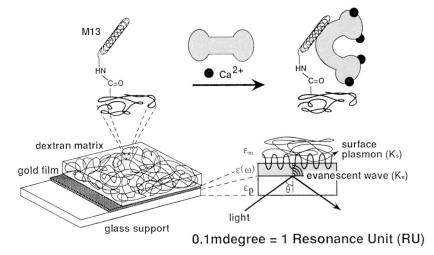

Figure 1. Principle of an optical Ca²⁺ ion sensor based on a calmodulin (CaM) mediated Ca²⁺ signaling pathway and surface plasmon resonance.

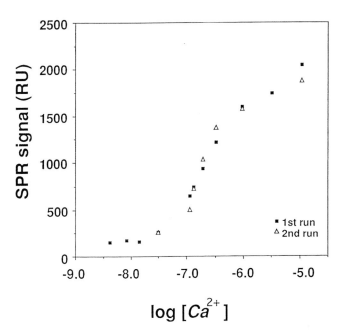

Figure 2. The SPR signal intensities as a function of concentration of free Ca²⁺ ions in 150 mM NaCl containing 0.5 mM EGTA and 10 mM HEPES buffer (pH 7.5).

approach is the use of a target protein to which a Ca^{2+}-CaM complex is bound reversibly. The ternary complex thus formed has a mass much larger than the immobilized M13 or Ca^{2+} ions in solution, enabling sensitive detection of small ions or molecules like Ca^{2+} by the SPR method. This approach can be applied to other receptor signaling pathways, such as hormonal activation of enzymes. The development of highly sensitive and selective sensing systems that will enable detecting bioactive small molecules by the SPR method is thus promising.

Active transport sensors based on Na$^+$/D-glucose cotransporter and Na$^+$, K$^+$-ATPase

The Na$^+$/D-glucose cotransporter protein is known to display active transport of D-glucose across the cell membrane by using an electrochemical Na$^+$ gradient as the driving force. The energy conversion proceeds in a coupled transport with a coupling stoichiometry of Na$^+$:D-glucose $= 1:1$ or $2:1$ (Alberts et al., 1994). Considering the efficiency and specificity of glucose transport displayed by Na$^+$/D-glucose cotransporter, a purified preparation of the cotransporter from the small intestine of anesthetized guinea pigs was immobilized into a planar bilayer lipid membrane (BLM) for constructing a highly sensitive BLM sensing system for D-glucose (Sugao et al., 1993). The principle of the D-glucose sensor based on a Na$^+$/D-glucose cotransporter is shown in Figure 3, where the electrochemical Na$^+$ gradient as the driving force for the active transport of D-glucose is provided by applying a given potential difference across the BLM facing to the solution containing Na$^+$ ions at a given concentration. The incorporation of purified Na$^+$/D-glucose cotransporter into the BLM, formed by the folding method (Fig. 4), was achieved by fusion of its proteoliposome. The concentration of D-glucose in sample solutions can be detected as the transmembrane Na$^+$ ion current generated by the symport of D-glucose and Na$^+$ through the BLM at a constant electrochemical Na$^+$ gradient. The Na$^+$ current coupled with the D-glucose transport was observed at 10^{-9} M concentration of D-glucose, followed by a steep rise in the response up to 5×10^{-8} M. At higher concentration of D-glucose, the current increased more gradually and it virtually leveled off above 10^{-6} M. Compared to the conventional glucose sensors based on glucose oxidase, which typically show a detection limit of 1 μM to several hundred μM, the detection limit of 10^{-9} M attained by the present sensing system can be regarded as extremely low. Among several related monosaccharides tested, i.e., L-glucose, D-fructose, D-mannose and D-galactose, only D-galactose (10^{-6} M), which is known to competitively bind to the Na$^+$/D-glucose cotransporter protein, induced a change in the response. The observed amperometric selectivity is consistent with the binding and transport selectivities displayed by this transporter protein in biomembranes (Umbach et al., 1990; Ikeda et al., 1989).

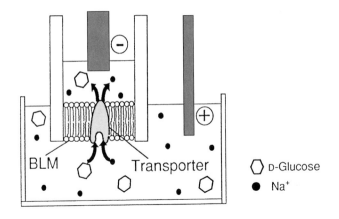

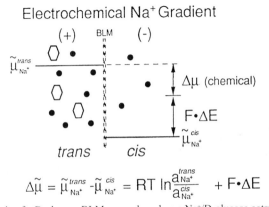

Figure 3. Principle of a D-glucose BLM sensor based on a Na$^+$/D-glucose cotransporter.

An adenosine-5′-triphosphate (5′-ATP) sensor, fabricated with a purified preparation of the Na$^+$, K$^+$-ATPase embedded into a BLM (Adachi et al., 1993), is another example of active transport sensors. This potentiometric sensor enabled to detect 5′-ATP over a range from 1.0×10^{-6} to 1.0×10^{-3} M with a detection limit (S/N $\geq$ 3) of 6.3×10^{-6} M. The selectivity for several nucleotides as well as acetyl phosphate was found to be in the order of 5′-ATP > 5′-UTP, 5′-GTP > acetyl phosphate, 5′-CTP $\gg$ 5′-ADP, which is parallel to that for the rate of hydrolysis by the ATPase in biomembranes (Beaugé and Berberián, 1984; Campos et al., 1988).

The active transport sensors described above utilized two different modes of active transport, i.e. secondary and primary active transport, displayed by cotransporter and pump proteins embedded in BLMs, respectively. These approaches can be applied to other cotransport, antiport and pump systems for the development of highly sensitive and selective sensing systems for substrates (analytes) such as sugars other than D-glucose and amino acids. Recently, a lactose sensor has been described by utilizing ISFETs modified with supported BLMs, containing a H⁺/lactose cotransporter (Ottenbacher et al., 1993).

Ion channel sensors based on glutamate receptor ion channel proteins

We have fabricated an ion channel sensor for L-glutamate by using glutamate receptor (GluR) ion channel proteins isolated from the synaptic membrane of rat brain (Uto et al., 1990; Minami et al., 1991 a; Minami et al., 1991 b). A purified preparation of the receptor ion channel proteins was incorporated into planar bilayer lipid membranes (BLMs) and the channel currents triggered by L-glutamic acid (L-Glu) were measured. Experimental set-ups for these measurements were virtually the same to those for the active transport sensor based on Na^+/D-glucose cotransporter (Fig. 4). In these systems, the concentration of L-Glu can be determined on the basis of the integrated transmembrane Na^+ ion current, which corresponds to the number of ions passed through the open channel (the period of the L-Glu binding). Hence, this current is a direct measure of the signal amplification by the channel protein. The relationship between the concentration of L-Glu and the response as an integrated channel current for the multi-channel type sensor (most likely contained ≥ 10 GluR ion channel proteins) is shown in Figure 5(a). A sharp concentration dependence was observed up to ca 1.5×10^{-7} M of L-Glu. The detection limit of this sensing system can be regarded as being lower than the lowest concentration tested, i. e., 3×10^{-8} M (30 nM). A high sensitivity to L-Glu

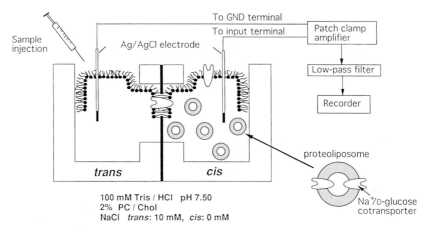

Figure 4. Experimetal set-ups used for formation of planar BLMs by the folding method, fusion of proteoliposomes and current measurements.

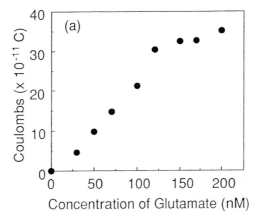

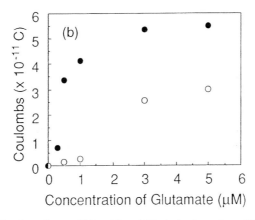

Figure 5. Concentration dependence of (a) multi- and (b) single-channel type BLM sensors based on glutamate receptor ion channel proteins. The single-channel type sensors contained only one (open circle) and two (closed circle) active GluRs.

with the detection limit of lower than 3×10^{-7} M was also observed for the single channel type sensors containing only one and two active GluR ion channel proteins (Fig. 5(b)). The present GluR ion channel sensor exhibited a high selectivity for the L-isomer of Glu as compared with the D-isomer, reflecting a combined effect of the relative strength of L- and D-Glu in binding and the relative potency to induce channel current.

The high sensitivities of the present sensing systems are evidently based on the intense signal amplification function of the GluR ion channel protein; the amplification factor was estimated to be ca. 2.1×10^5 on the basis of the average number of ions which pass through the channel during a single channel-open period (average 7.6 ms) (Uto et al., 1990). Considering the importance of GluR at postsynaptic membrane in mammalian brain, we are trying to evaluate chemical selectivity of this protein for various agonists and antagonists with GluR embedded in BLMs (Sugawara et al., 1996b).

Potentiometric BLM sensors based on synthetic receptors

Planar bilayer lipid membranes (BLMs) containing synthetic receptors are another important class of sensing membrane assemblies. We have described valinomycin-incorporated BLMs that generate membrane potentials responding to K^+ ions in a Nernstian manner, providing a basis for quantitative bilayer measurements (Minami et al., 1991 c). Recently, we succeeded in extending this approach to the cases for other synthetic receptors (ionophores) capable of selectively recognizing alkali and alkaline earth metal ions (Sato et al., in preparation). The formation of receptor-incorporated BLMs (DOPC/cholesterol and PC/cholesterol, for abbreviations see below) was achieved by the folding method in which a bilayer is formed from lipid phase containing a known fraction of an ionophore spread on the air-solution interface. However, it should be noted that the control of membrane resistance of receptor-incorporated BLMs is important for the development of quantitative potentiometric BLM sensors. Because of an impedance matching between the membrane resistance and the input impedance of the electrometer ($>10^{15}$ Ω), a membrane resistance below 60 Ω (an aperture of a Teflon film: ca. 200 μm in diameter) was required for potential measurements. The L-α-phosphatidylcholine (PC) or L-α-dioleoyl phosphatidylcholine (DOPC), together with cholesterol in different mole fractions, most often formed BLMs having large resistances (above 60 GΩ). Consequently, BLMs of membrane resistance below 60 GΩ are required for quantitative BLM sensors. The formation of such BLMs could successfully be achieved by incorporating added ionic sites, together with

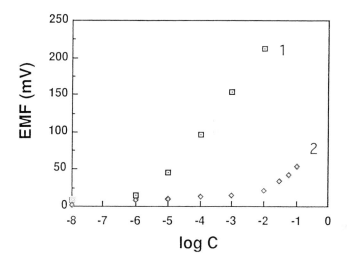

Figure 6. Potentiometric response curves for (1) K^+ and (2) Na^+ ions in 100 mM tetramethyl ammonium chloride containing 5 mM Tris/HCl buffer (pH 7.4) with a BLM containing valinomycin and NaTFPB (DOPC: valinomycin: cholesterol: NaTFPB = 4:1:2:0.007 (molar ratio).

receptors, into the BLMs. Simultaneous incorporation of valinomycin and an anionic site, sodium tetrakis (4-fluorophenyl) borate (NaTFPB), into the BLMs enabled formation of BLMs, the resistances of which are always lower than 60 GΩ. With such membrane (DOPC: valinomycin: cholesterol : NaTFPB = 4 : 1 : 2 : 0.007 in a molar ratio), all BLMs (n = 6) thus formed could be used for potentiometric measurements. Figure 6 shows the potentiometric response curves for K^+ and Na^+ ions of the valinomycin-incorporated BLM thus formed. The BLMs containing valinomycin and NaTFPB generated potentiometric Nernstian response to K^+ ions in the concentration range from 1.0×10^{-5} M to 1.0×10^{-2} M. This demonstrates that the incorporation of NaTFPB lowered the detection limit for K^+ ions as compared with that (1.0×10^{-4} M) of the valinomycin-based BLM without anionic sites. The present approach was also valid for other receptor-incorporated BLMs (H. Sato, M. Sugawara, Y. Umezawa, in preparation). Thus, the potentiometric responses of the receptor-based BLMs could be improved by incorporation of anionic sites.

Conclusion

New sensing membranes described above demonstrate that mimicking receptor signalings in biomembranes is useful for designing highly sensitive and selective sensing systems for ions and molecules. Although examples of such sensing membrane are limited, the results shown above strongly suggest the general applicability of the present approaches to other receptor signalings for the development of new sensing membranes, if purified preparation of receptors are obtained and membrane systems are appropriately designed.

Acknowledgements
This work was supported from the Grant-in-Aids for Scientific Research by the Ministry of Education, Science and Culture, Japan. The support from the Nissan Science Foundation (Tokyo, Japan) was also acknowledged.

References

Adachi, Y., Sugawara, M., Taniguchi, K. and Umezawa, Y. (1993) Na^+, K^+-ATPase-based bilayer lipid membrane sensor for adenosine-5' triphosphate. *Anal. Chim. Acta* 281 : 577–584.
Alberts, B., Bray, D., Lewis, J., Ratt, M., Roberts, K. and Watson, J.D. (1994) *Molecular Biology of the Cell*. Third Edition. Garland Publishing, Inc., New York & London.
Beaugé, L. and Berberián, G. (1984) Acetyl phosphate act as a substrate for Na^+ transport by $(Na^+ + K^+)$-ATPase. *Biochim. Biophys. Acta* 772 : 411–414.
Blair, T.L., Yang, S.-T., Smith-Palmer, T. and Bachas, L.G. (1994) Fiber optic sensor for Ca^{2+} based on an induced change in the conformation of the protein calmodulin. *Anal. Chem.* 66 : 300–302.
Campos, M., Berberián, G. and Beaugé, L, (1988) Some total and partial reactions of Na^+/K^+-ATPase using ATP and acetyl phosphate as a substrate. *Biochim, Biophys. Acta* 938, 7–16.

Ikeda, T.S., Hwang, S.S., Coady, M., Hirayama, B.A., Hediger, M.A. and Wright, E.M. (1989) Characterization of a Na⁺/glucose cotransporter cloned from rabbit small intestine. *J. Memb. Biol.* 110:87–95.

Minami, H., Sugawara M., Odashima, K., Umezawa, Y., Uto, M., Michaelis, E.K. and Kuwana, T. (1991a) Ion channel sensors for glutamic acid. *Anal. Chem.* 63:2878–2795.

Minami, H., Sugawara M., Odashima, K., Umezawa, Y., Uto, M., Michaelis, E.K. and Kuwana, T. (1991b) An evaluation of signal amplification by the ion channel sensor based on a glutamate receptor ion channel protein. *Anal. Sci.* 7 (supplement):1675–1676.

Minami, H., Sato, N., Sugawara M. and Umezawa, Y. (1991c) Comparative study on the potentiometric responses between a valinomycin-based bilayer lipid membrane and a solvent polymeric membrane. *Anal. Sci.* 7:853–862.

Morf, W.F., Seiler, K., Rusterholz, B. and Simon, W. (1992) Design of a calcium-selective optode membrane based on neutral ionophores. *Anal. Chem.* 62:738–742.

Nagase, S., Kataoka, M., Naganawa, R., Komatsu, R., Odashima, K. and Umezawa, Y. (1990) Voltammetric anion responsive sensors based on modulation of ion permeability through Langmuir-Blodgett films containing synthetic anion receptors. *Anal. Chem.* 62: 1252–1259.

Odashima, K., Sugawara M. and Umezawa Y. (1991) Biomembrane mimetic sensing chemistry. *Tren. Anal. Chem.* 10:207–215.

Odashima, K., Kotato, M., Sugawara M. and Umezawa Y. (1993) Voltammetric study on a condensed monolayer of a long alkyl cyclodextrin derivative as a channel mimetic sensing membrane. *Anal. Chem.* 65:927–936.

Odashima, K., Sugawara M. and Umezawa Y. (1994) *In*: T.E. Malouk and D.J. Harrison (eds): *Interfacial Design and Chemical Sensing* (ACS Symposium Series 561) American Chemical Society, Washington, D.C., Chapter 11, pp 123–134.

Ottenbacher, D., Jähnig, F. and Göpel, W. (1993) A prototype biosensor based on transport proteins: electrical transducers applied to lactose permease. *Sensor. Actuator. B* 13–14:173–175.

Ozawa, T., Kakuta, M., Sugawara M., Umezawa, Y. and Ikura, M. An optical calcium ion sensor based on a calmodulin mediated Ca²⁺ signaling pathway monitored by surface plasmon resonance; submitted.

Sugao, N., Sugawara M., Uto, M., Minami, H. and Umezawa, Y. (1993) Na⁺/D-glucose cotransporter based bilayer lipid membrane sensor for D-glucose. *Anal. Chem.* 65:363–369.

Sugawara M., Kojima, K., Sazawa, H. and Umezawa, Y. (1987) Ion-channel sensors. *Anal. Chem.* 59:2842–2846.

Sugawara M., Omoto, M., Yoshida, H. and Umezawa, Y. (1988) Enhancement of uphill transport by a double carrier membrane system. *Anal. Chem.* 60:2301–2303.

Sugawara M., Kataoka, M., Naganawa, R., Komatsu, R., Odashima, K. and Umezawa, Y. (1989) Biomimetic ion channel sensors based on host-guest molecular recognition in Langmuir-Blodgett membrane assemblies. *Thin Solid Films* 180:129–133.

Sugawara M., Yosida, H., Henmi, A. and Umezawa, Y. (1991) Enhancement of analyte concentration based on a feed/receiving volume effect in uphill transport membrane transport. *Anal. Sci.* 7:141.

Sugawara M., Sazawa, H. and Umezawa, Y. (1992) Effect of membrane surface charge on the host-guest complex of valinomycin in a synthetic lipid monolayer at the air-wate interface. *Langmuir* 8:609–612.

Sugawara M., Sugao, N., Umezawa, Y., Adachi, Y., Taniguchi, K., Minami, H., Uto, M., Odashima, K., Michaelis, E.K. and Kuwana, T. (1993) *In*: F.A. Shultz and I. Taniguchi (eds) *Redox Mechanisms and Interfacial Properties of Molecules of Biological Importance 1993*. Proceedings Vol. 93–11, The Electrochemical Society, Pennington, pp 268–279.

Sugawara M., Khoo, S.B., Yoshiyagawa, S., Yagi, K., Sato, H., Namba, M., Wakabayashi, M., Minami, H., Sazawa, H., Odashima, K. and Umezawa, Y. (1994) Factors affecting background permeabilities of ordered mono-, bi- and multilayers as channel mimetic sensing membranes. *Anal. Sci.* 10:343–347.

Sugawara M., Okude, K., Tohda, K. and Umezawa, Y. (1996a) Concentration of dopamine by proton-driven uphill transport using lasalocid A as the carrier. *Anal. Sci.* 12:331–335.

Sugawara, M., Hirano, A., Rehák, M., Nakanishi, J., Kawai, K., Sato, H. and Umezawa, Y. (1996b) Electrochemical evaluation of chemical selectivity of glutamate receptor ion channel proteins with a multi-channel sensor. *Biosensors and Bioelectronics;* submitted.

Umbach, J.A., Coady, M.J. and Witght, E.M. (1990) Intestinal Na^+/glucose cotransporter expressed in *Xenopus* oocytes is electrogenic. *Biophys. J.* 57:1217–1224.

Uto, M., Yoshida, H., Sugawara M. and Umezawa, Y. (1986) Uphill transport membrane electrodes. *Anal. Chem.* 58:1798–1803.

Uto, M., Michaelis, E.K., Hu, I.F., Umezawa, Y. and Kuwana, T. (1990) Biosensor development with a glutamate receptor ion-chemical reconstituted in a lipid bilayer. *Anal. Sci.* 6: 221–225.

Yagi, K., Khoo, S.B., Sugawara, M., Sakaki, T., Shinkai, S., Odashima, K. and Umezawa, Y. (1996) Channel mimetic sensing membranes for alkaline metal cations based on oriented monolayers of calixarene esters. *J. Electroanal. Chem.* 401:65–79.

Frontiers in Biosensorics I
Fundamental Aspects
ed. by F. W. Scheller, F. Schubert and J. Fedrowitz
© 1997 Birkhäuser Verlag Basel/Switzerland

Genetically modified *Escherichia coli* for colorimetic detection of inorganic and organic Hg compounds

J. Klein, J. Altenbuchner and R. Mattes

Institute of Industrial Genetics, University of Stuttgart, D-70569 Stuttgart, Germany.

Summary. A sensitive colorimetric bacterial system was developed for the detection of Hg(II) and organomercury compounds. The bioactive species, a recombinant *Escherichia coli*, produces proportionally elevated levels of the enzyme β-galactosidase with increasing amounts of Hg. This is due to a reporter plasmid which carries a Hg(II)-inducible promoter (*mer* promoter) from the Hg resistance transposon Tn*501* regulating the transcription of a promoterless *lacZ* gene. Additionally, a pMB1 origin of replication without the natural RNA polymerase start site is fused downstream of the *mer* promoter leading to a Hg(II)-inducible plasmid replication, which results in an improved signal-to-noise ratio. To enhance the sensitivity of this cellular biosensor, the transport proteins for Hg(II) uptake are constitutively produced by a helper plasmid. To enable the detection of organically bound Hg, the *Streptomyces lividans* organomercurial lyase, an enzyme which catalyses the cleavage of C-Hg-bonds of organomercurial compounds, is also provided by the helper pasmid. Hg(II) and phenylmercuric acetate (PMA) concentrations as low as 5×10^{-10} M (0.1 ppb) may be detected within a few minutes.

Introduction

The response of biological systems to compounds within the environment has been a conceptual basis for the development of bioassays. An example is the inhibition of growth due to toxic substances which allows an accurate biospecific determination of concentrations of e. g. antibiotics (minimal inhibitory concentration). But these methods are tedious and time consuming and often not specific, at least when mixtures of unknown composition have to be tested. In addition some biological systems have acquired resistance mechanisms to overcome life-threatening effects of toxic compounds. Four principal biological strategies are observed which are summarised in Figure 1. These include passive strategies, for example a general change in permeability which reduces uptake, or a mutation in the gene encoding the target component may render the cell resistant. In contrast, active strategies involve the acquistion of additional genes leading to the production of new enzymes, which metabolise and thus inactivate the toxic component or facilitate its export.

Generally the latter active mechanisms of resistance are highly specific and include the production of distinct proteins, mainly enzymes, which interact with the substance in question. Biochemical methods measuring

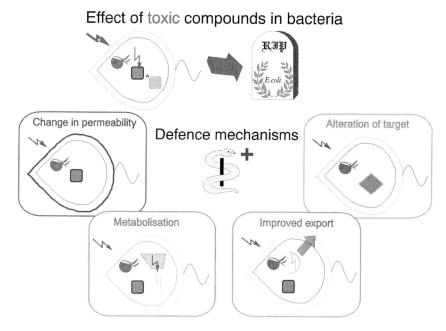

Figure 1. Schematic view of possible defence mechanisms for microbial resistance to toxic compounds.

the activity of such specific enzymes allowed the development of faster and more precise alternative tests rather than merely monitoring growth inhibition. In consequence it is also conceivable to quantify specific intracellular reactions which lead to growth retardation alternatively. These modern possibilities for qualitative or quantitative assays with microbial cells allow the development of biospecific microbial sensors.

When studying active resistance mechanisms of microbial cells it became apparent that the genetic determinants involved are usually regulated. The production of encoded specific proteins is induced in response to the toxic substance on a level which is usually concentration dependent. The application of reporter genes using gene fusions allowed to identify the regulatory components of gene expression, a method now widely used in such research. We attempted to redirect the aim of reporter gene analysis for the understanding of regulatory mechanisms into the modular development of specific biosensoric cells. For this purpose we evaluated the genetic modules necessary for accurate and concentration-dependent production of reporter enzymes with respect to the inducing agent.

As a model system the genetic components for resistance against mercury compounds were used. These are most important heavy metal environmental pollutants in form of inorganic and organic Hg compounds, released by a wide spectrum of industrial and agricultural processes or by the weathering of Hg-bearing rocks (Robinson and Tuovinen, 1984). Human exposure to even small amounts (ng to µg) of Hg compounds, especially the more toxic organomercurials, leads to nervous and physical disorders (Aschner and Aschner, 1990; De Flora et al., 1994). The need for analysis and monitoring environmental and food contamination by Hg compounds led to the development of sensitive and reliable physicochemical detection procedures like atomic absorption spectrophotometry (Omang, 1971), or more sensitive, cold vapour atomic fluorescence detection (Bloom and Fitzgerald, 1988). Although these detection methods are highly sensitive, they have some disadvantages: sample preparation, extraction and interpretation is laborious and involves expensive equipment and highly skilled staff. This makes it impossible to perform measurements without delay (*on line*) at the site of interest (*on site*). Simple biological systems based on proteins with high affinity to Hg compounds may contribute to fill this gap between time-consuming Hg work-up protocols and the desirable *on site* and *on line* measurement of Hg. One of these biological approaches is the use of specific antibodies in an ELISA-test system to detect Hg(II) (Wylie et al., 1991). A further way may be the development of specific cellular Hg biosensors (Klein et al., 1989; 1991; Selifonova et al., 1993; Tescione and Belfort, 1993) based on bacterial Hg resistance mechanisms.

Specific Hg detoxification systems are present in many Gram-negative and Gram-positive bacteria isolated from polluted areas (Silver et al., 1989; Silver and Walderhaug, 1992). Although they belong to very different taxonomic groups the microorganisms display similar resistance mechanisms encoded by different but homologous genes which always are organised in regulated operon entities (Fig. 2). The resistance mechanism is based on the effective import of Hg(II) into the bacterial cell by transport proteins (MerT and MerP) and immediate enzymatic detoxification by reduction of the ions via a mercuric reductase (MerA) to the less toxic and volatile Hg (Summers, 1986). Some organisms are even able to deal with a broad range of organomercurial compounds. The organomercurial lyase (MerB) cleaves the C-Hg bond of organomercurials and releases Hg(II), which is subsequently reduced by MerA (Fig. 3). The various *mer* genes are arranged in an operon, whereby transcription is controlled by a regulatory protein (MerR) which specifically recognises Hg(II). It may act as a positive activator of transcription in the presence of Hg(II) and as a repressor in their absence, as exemplified for Tn*501* (Lund and Brown, 1989; O'Halloran et al., 1989) (Fig. 4). In contrast, for the operon of *Streptomyces lividans* MerR was shown to act solely as repressor, whose action is relieved by the inducer compound Hg(II) (Brünker et al., 1996).

Concerning their efforts to elucidate the regulation mechanism of the narrow spectrum resistance determinants of the transposon Tn*501* or the similar Tn*21* (Silver et al., 1989), Lund and Brown (1989) or Ross et al. (1989), Park et al. (1992) and Condee and Summers (1992) have used

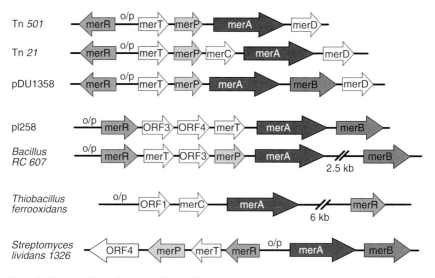

Figure 2. Organization of genes in bacterial mercury resistance operons.

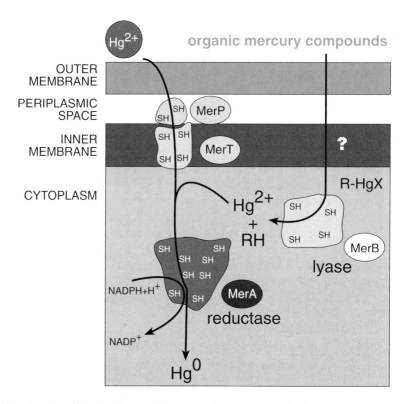

Figure 3. Detoxification of mercurial compounds in gram-negative bacteria.

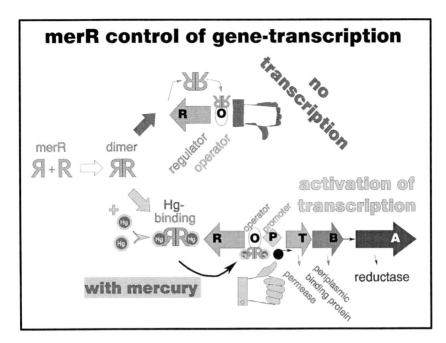

Figure 4. Transcriptional control of *mer* operon expression by the regulator protein MerR. Repression by MerR dimer but activation by Hg(II)-[MerR]$_2$-complex.

transcriptional fusions with reporter genes such as *lacZ* or *lux*. The transcriptional fusion of a reporter gene to a *mer* operon, for example to the well-known Tn*501* operon, creates a bacterial strain which is able to monitor the presence of Hg(II) by the production of quantitatively detectable reporter enzyme activities. In that way luminometric biosensors specific for Hg(II) were described by Tescione and Belfort (1993) or Selifonova et al. (1993). The investigations presented here aimed at the systematic improvement of a colorimetric Hg-specific bacterial biosensor.

Materials and methods

Bacterial strains, media and growth conditions

All cloning and induction experiments were carried out in *Escherichia coli*, strain JM109 (*recA1 supE44 endA1 hsdR17 gyrA96 relA1 thi* Δ (*lac-pro-AB*) F′[*traD36 proAB*$^+$ *lacI*^{q} *lacZ*ΔM15], Yanisch-Perron et al., 1985) using 2 YT medium (16 g/l tryptone, 10 g/l yeast extract, 5 g/l NaCl). The following antibiotics were used for selection and/or maintenance of plasmids in JM109: 100 μg/ml ampicillin, 25 μg/ml chloramphenicol, 50 μg/ml kanamycin or 20 μg/ml tetracycline.

DNA manipulations and cell transformation

All DNA manipulations were carried out as described elsewhere (Sambrook et al., 1989). All enzymes were purchased from Boehringer Mannheim/ Germany, except *Taq* DNA polymerase from Pharmacia LKB GmbH, Freiburg/Germany and used according to the manufacturers' instructions. *E. coli* transformations were performed using the TSS method, originally described by Chung et al. (1989).

Plasmids

All used plasmids are summarised in Table 1. pJOE115 is a pBR322-derivative carrying the Hg resistance transposon (Altenbuchner et al., 1981). A 2340 bp *Eco*RI fragment from pJOE115 containing *merR*, the regulatory region *op*, *merT* and *merP* (*merRopTP* cassette) was integrated into the *Eco*RI restriction site of plasmid pKO4 (McKenney et al., 1981) to obtain pJKS1. A promoterless 3811 bp *lacZ* gene of pRU676::Tn*1731* (Ubben and Schmitt, 1987) was transferred as a *Bam*HI fragment into the *Bam*HI-digested plasmid pJKS1 to create pJKS3. pJKS19 was derived from pUC19 (Vieira and Messing, 1982) by inserting the above-mentioned *merRopTP* cassette and *Bam*HI *lacZ* fragments into the appropriate restriction sites. pJKS68 is a derivative of pJKS3. The origin of replication of pBR322 lacking the appropriate RNA polymerase start site was first inserted into *Nru*I-digested pIC19 as a 434 bp *Hae*III fragment, and from there transferred as a 467 bp *Eco*RI-*Hin*dIII fragment into *Eco*RI-*Hin*dIII-cut pKO4. The *Eco*RI *merRopTP* cassette fragment (2340 bp) and the *Bam*HI *lacZ* fragment (3811 bp) were successively inserted into this plasmid resulting in the reporter plasmid pJKS68. A 686 bp *Eco*RI-*Ava*I fragment of pJOE115, which carried *merRopT'*, was isolated and inserted into a helper plasmid which provide *merRopT''*as *Eco*RI fragment (Klein, 1992). This 702 bp *Eco*RI fragment containing only a functional *merR* gene and the regulatory region was inserted into *Eco*RI-digested pKO4, followed by cloning of the *Bam*HI *lacZ* fragment (3811 bp) into this plasmid to give pJKS39. The genes *merT* and *merP* were obtained as a 993 bp *Bgl*I fragment of pJOE115. Sticky ends were filled in with Klenow polymerase and the fragment was blunt-end ligated into *Eco*RV-cut pACYC184 (Chang and Cohen, 1978) to yield pJKS13. The gene *merB* was isolated as a 655 bp *Bam*HI fragment from pJOE851-461 (Sedlmeier and Altenbuchner, 1992) via PCR and ligated to *Bam*HI-digested pJKS13 to obtain pJOE2004.

Hg-specific β-galactosidase induction in liquid cultures

E. coli strains bearing the constructed plasmids were grown overnight in 2YT medium at 37 °C with the appropriate antibiotics, diluted 1 : 100

Table 1. Plasmids used in this study

Plasmid	Resistance[a]	Relevant genotype and features[b]	Origin of replication	Source (reference)
pACY184	Cm^{r}Tc^r	–	p15A	Chang and Cohen, 1978
pACYC184::Tn*21*	Cm^{r}Tc^r	Tn*21* in pACYC184	p15A	De la Cruz and Grinstedt, 1982
pBR322	Ap^{r}Tc^r	–	pMB1	Bolivar et al., 1977
pIC19H	Ap^r	*rop*$^-$	pMB1	Marsh et al., 1984
pJOE115	Ap^r	Tn*501* in pJOE100	pMB1	Altenbucher, unpublished
pJOE851-461	Ap^r	*merRAB (S. lividans)*	pMB1	Sedlmeier and Altenbuchner, 1992
pKO4	Ap^r	*galK*	pMB1	McKenney et al., 1981
pRU6776::Tn*1731*	Km^r	*aphA lacZ* in Tn*1731*	pMB1	Ubben and Schmitt, 1987
pUC18	Ap^r	*rop*$^-$	pMB1	Vieira and Messing, 1982
pJKS3	Ap^r	*merRopTP lacZ*	pMB1	this study
pJKS13	Cm^r	*ptetA meropTP*	p15A	this study
pJKS19	Ap^r	*merRopTP lacZ*	pMB1	this study
pJKS68	Ap^r	*merRopTP lacZ ori*	pMB1	this study
pJKS81	Cm^r	*ptetA meropT-lacZ*	p15A	this study
pJOE2004	Cm^r	*ptetA meropTP B*	p15A	this study

[a] Ap^r: ampicillin resistance, Cm^r: chloramphenicol resistance, Km^r: kanamycin resistance, Tc^r: tetracycline resistance.
[b] *op*: operator/promoter region of Tn*501*, *TP*: *merTP* of Tn*501*, *T-lacZ*: translational fusion of MerT and LacZ, *ori*: origin of replication of pBR322, *B:* *merB* of *Streptomyces lividans,* inactivated or deleted genes are not mentioned.

in fresh medium and cultivated to the late exponential growth phase (A_{600} = 0.7). 2 ml of the growing cultures were transferred to 10-ml test tubes and different concentrations of Hg(II), phenylmercury acetate or other heavy metal salts were added for induction. The cultures were further incubated at 37 °C for fifteen minutes or as otherwise indicated. β-galacto-sidase (β-gal) assays were immediately carried out as described by Miller (1972). The activities are the mean values of three independent measurements with standard deviations of less than 10%. The detection threshold was defined as the Hg concentration which led to a twofold increase in activity compared to uninduced β-gal activity.

Results

Transcriptional fusion of lacZ as reporter gene to the mer genes of Tn501

The concentration of Hg(II) can be determined via a reporter enzyme activity whose production is dependent on a transcriptional fusion of the reporter gene to a Hg resistance operon (Klein et al., 1989; Tescione and Belfort, 1993; Selifonova, 1993). For convenience and to provide a broad application of the resulting Hg specific biosensor, there should be a rapid and simple colorimetric assay to measure the activity of the reporter enzyme. Furthermore, low concentrations of Hg(II), Hg(I) or organo-mercurials should not inhibit the enzyme. Both requirements were met by the β-galactosidase (β-gal) of *E. coli*. A series of chromogenic and even highly sensitive chemiluminescent substrates are available and, using purified β-gal, it was confirmed that enzyme activity is not influenced by Hg(II) concentrations as high as 5×10^{-6} M (1000 ppb) (data not shown).

For transcriptional fusion of the *lacZ* gene to the *mer* operon of Tn*501*, observations of Nakahara et al. (1979) and Lund and Brown (1987) were taken into account who showed that the presence of intact Hg transport and regulatory proteins without a functional reductase led to Hg(II) supersensitive cells. Without reduction to the volatile Hg(0), the *mer* operon inducing Hg(II) were efficiently imported into the cells and accumulated. Consequently, a promoterless reporter gene fused within *merA* downstream of *merR*, the regulatory region and *merT* and *merP* of Tn*501* was expected to be induced even at very low concentrations of Hg(II). A *merA* deleted 2340 pb *Eco*RI fragment of Tn*501* which contained the transport protein genes *merT* and *merP* and the regulatory elements was inserted upstream of a promoterless *lacZ* gene on the high copy plasmid pKO4 to create pJKS3 (Tab. 1). Translation stop codons in all three reading frames at the 5′ end of *lacZ* prevented translational fusions. To analyse induction of *E. coli* JM109 (pJKS3) by Hg(II), the strain was incubated for one hour with fairly low concentrations of HgCl$_2$ up to 6.25×10^{-8} M (12.5 ppb) and the β-gal activities were determined. As illustrated in Figure 5 A, a linear correlation of Hg(II) concentration and enzyme activity could be shown in the investigated concentration range. The detection threshold was about 10^{-8} M (2 ppb).

Figure 5. Genetic factors influencing the biosensor sensitivity. (A) Hg(II)-dependent β-gal synthesis of *E. coli* JM109 (pJKS3). (B) Influence of the transport protein level on the β-gal production after induction with 10^{-7} M (20 ppb) Hg(II). (C) Influence of the copy number of the reporter plasmid. All strains were cultivated to the exponential growth phase in 2YT and induced with the indicated Hg(II) concentrations. After different induction times [(A) 60 min, (B) 5–120 min, (C) 15 min], the β-gal activities were measured. The insets show the molecular organization of the test plasmids.

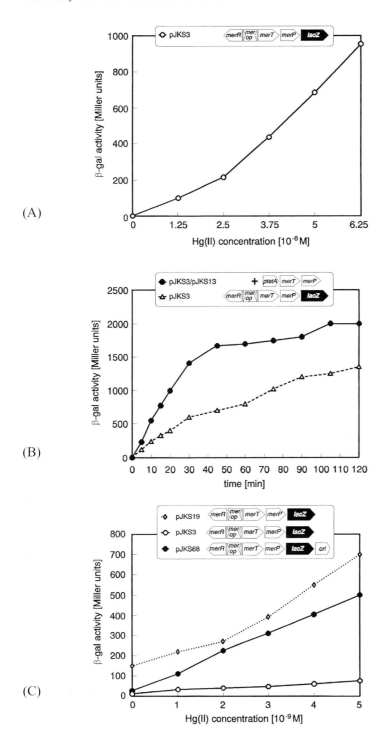

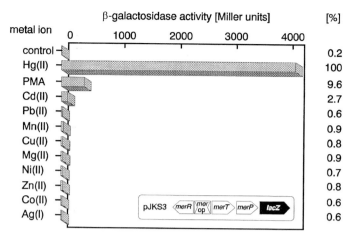

Figure 6. Specificity of the biosensor *E. coli* (pJKS3). Cultures were induced as described in materials and methods except 1 h induction time. The concentration of Hg(II) and PMA was 2×10^{-6} M, whereas the concentrations of the other metals were 10 times higher (2×10^{-5} M). The β-gal activity induced by Hg(II) (4169 Miller Units) was set to 100% and used as reference value.

The biosensor is specific for Hg(II)

A useful biosensor for Hg requires a high sensitivity as well as a high specificity towards the compounds to be monitored. *E. coli* JM109 (pJKS3) was incubated with different metal ions and phenylmercury acetate (PMA) and the correspondig β-gal activities were measured. $HgCl_2$, PMA, $CdSO_4$, $Pb(CH_3COO)_2$, $MnCl_2$, $Mg(CHCOO)_2$, $NiSO_4$, $ZnSO_4$, $CoCl_2$ and $AgNO_3$ were used in the test. According to the manufacturers' analyses, contaminations with other heavy metal ions, especially Hg(II), were negligible (0.002 to 0.005 %). Figure 6 summarises the results. The specificity was mainly restricted to Hg(II). Although the Tn*501* system is known to confer the narrow spectrum resistance, the organomercurial compound PMA induce β-gal synthesis, but only at a very high concentration. This might be due to Hg(II) contaminations within the organic compound. Of the heavy metal ions, only Cd(II) slightly interfered with the Hg(II)-specific system *in vivo*. This is in agreement with *in vitro* transcription assays using purified MerR (Ralston and O'Halloran, 1990). It interacted with MerR *in vivo* at ten times higher concentrations than Hg(II) and caused a forty times less transcriptional activation. In contrast to the *in vitro* results, Zn(II) and Ag(I) did not effect our biosensor E. Coli; JM109 (pJKS3).

Constitutively synthesised mer transport proteins considerably lower
the time of β-gal induction by Hg(II)

Hg(II) ions have to enter the bacterial cell to induce transcription of the *mer*
genes. Since the transport genes have to be induced first to import the inducer effectively, consitutively synthesised *mer* transport proteins were
expected to decrease the time of induction. A 993 bp fragment from Tn*501*
which encodes *merTP* and additionally contains the *mer* promoter and
operator was inserted downstream of the *tetA* promoter of pACYC184 to
give rise to pJKS13. The *tetA* promoter (*ptetA*) is a weak, constitutive promoter which should provide moderate transcription of *merTP*. To prove this
assumption β-gal was translationally fused to the N-terminus of MerT
(plasmid pJKS81, Tab. 1). As expected, the transcription signals of the *tetA*
promoter and the translation initiation signals of MerT led to an elevated
β-gal production (about 4000 Miller units). The plasmid pJKS13 was
transformed into *E. coli* JM109 (pJKS3). The resulting strain JM109 (pJKS3,
pJKS13) was incubated with Hg(II) for different times and the β-gal activities were measured. As illustrated in Figure 5B, the strain exhibited higher
β-gal amounts after shorter induction times compared to the original strain
JM109 (pJKS3). Constitutive expression of *merT* and *merP* seem to provide
an improved Hg(II) uptake, which resulted in accelerated and higher β-gal
production. Induction of the strain JM109 (pJKS3, pJKS13) by 10^{-7} M
Hg(II) (20 ppb) for 45 minutes was sufficient to produce nearly 90% of the
maximal β-gal activity obtained after 120 minutes of induction. By using
an induction time of 15 minutes, which is less than an *E. coli* generation time, the sensitivity of the biosensor was about 0.5 to 1 ppb (2.5 to
5×10^{-9} M). In comparison to the original strain JM109 (pJKS3) with a
detection threshold of about 2 ppb after 60 minutes of induction, this
means a two- to fourfold increase in sensitivity and an fourfold decrease in
induction time.

Hg(II)-induced plasmid replication improves the signal-to-noise ratio

To further lower the threshold value of Hg detection, the difference between
Hg(II)-induced β-gal activity and uninduced β-gal basal level, the so-called
signal-to-noise ratio, had to be increased. To find out if a change in the plasmid copy number had an influence on this ratio the same *mer-lacZ* fusion as
in pJKS3, was introduced into plasmid pUC19 (pJKS19, Tab. 1) which has
a higher copy number than pKO4 (McKenney et al., 1981). This increase in
the copy number (pJKS19) was accompanied by an increase in the appropriate Hg(II)-induced β-gal activities, but, additionally, by an increase in the
basal level (Tab. 2, Fig. 5C). The signal-to-noise ratio was not influenced
since basal and induced activities changed proportionally (Tab. 2). Thus this
ratio could only be influenced by alteration of the reporter plasmid copy

Table 2. Signal-to-noise ratio fo *E. coli* JM109 strains carrying different reporter plasmids at 5×10^{-10} M (0.1 ppb) and 5×10^{-9} M (1 ppb) Hg(II). The helper plasmid pJKS13 is present in all strains.

Plasmid	β-gal Activity (Miller units)		Signal-to-noise ratio
	uninduced	induced	
0.1 ppb			
pJKS3	15	20	1.3
pJKS19	150	180	1.2
pJKS68	26	78	3
1 ppb			
pJKS3	15	75	5
pJKS19	150	700	4.7
pJKS68	26	500	19.2

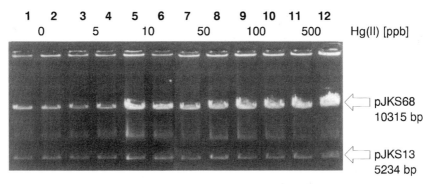

Figure 7. Hg(II)-dependent plasmid amplification of pJKS68. *E. coli* JM109 (pJKS68, pJKS13) was induced for 1.5 hours with different Hg(II) concentrations (in duplicate incubations). Plasmid DNA was isolated, cut with *Bam*HI and aliquots (according to 0.5 A_{600}) were loaded on a 1% agarose horizontal electrophoresis gel (lanes 1 to 12). The sizes of the linearised plasmids in base pairs are shown on the right.

number. We attempted hence a controlled increase of reporter plasmid copy number by Hg(II) MerR activated transcription. Therefore, an additional pMB1 origin of replication (*ori*) of pBR322 (Bolivar et al., 1977) missing its indigenous RNAII promoter was inserted as a 434 bp *Hae*III fragment at the 3′ end of *lacZ* resulting in pJKS68. This additional *ori* is placed in pJKS68 downstream of the Hg(II)-inducible *mer* promoter, analogous to work of Panayotatos (1984) who used the *lac*UV5 promoter.

The new reporter plasmid pJKS68 and the *merTP* helper plasmid pJKS13 were cotransformed into *E. coli* JM109. As shown in Figure 5C, JM109 (pJKS68, pJKS13) produced comparable high β-gal amounts as JM109 (pJKS3, pJKS13) by induction with Hg(II). However, the basal enzyme activity in crude extracts was only 26 Miller units (Tab. 2), giving a sixfold decrease compared to JM109 (pJKS19, pJKS13). Using strain

JM109 pJKS68, pJKS13) the detection threshold could be decreased tenfold from 5×10^{-9} M (1 ppb) with JM109 (pJKS3, pJKS13) to 5×10^{-10} M (0.1 ppb) Hg(II). This is accompanied by a two- to threefold increase of the signal-to-noise ratio at 0.1 ppb (Tab. 2). Proof of Hg(II)-mediated plasmid amplification was given by incubating *E. coli* JM109 (pJKS68, pJKS13) with different Hg(II) concentrations and concomitant determination of the amount of reporter plasmid DNA. Whole plasmid DNA was purified by a quick lysis procedure (Sambrook et al., 1989). After digestion with *Bam*HI, identical aliquots of the linearised plasmids were loaded on a 1 % horizontal agarose gel. As illustrated in Figure 7 the band intensity of the reporter plasmid appeared to expand proportionally to the increase in Hg(II) concentration whereas the signal of the linearised fragment of the reference plasmid did not change.

Detection of organomercurial compounds by using merB

The gene *merB* encodes the organomercurial lyase, which is able to cleave C-Hg bonds of organomercurial compounds. As seen in Figure 2 this gene is missing in Tn*501* but appears in other *mer* operons isolated from microorganisms displaying a so-called broad spectrum resistance. We speculated that this enzyme should release Hg(II) from organomer-curials inside the biosensing bacterial cell and, by this way, induced β-gal synthesis. The promoterless *merB* gene was amplified from *Streptomyces lividans* (Sedlmeier and Altenbuchner, 1992) as a 655 bp *Bam*HI PCR fragment. This fragment was inserted into *Bam*HI-digested *merTP* helper plasmid pJKS13 downstrem of the *tetA* promoter and *merTP* to get pJOE2004. The *merTPB* helper plasmid pJOE2004 and the reporter plasmid pJKS68 were cotransformed into *E. coli* JM109 to verify the supposed induction response for Hg(II) and PMA. As shown in Figure 8 B the strain appeared to produce β-gal in proportional relation to the added PMA concentration whereas JM109 (pJKS68, pJKS13), lacking *merB*, did not respond to organic PMA. The β-gal activity linearly increased up to a concentration of about 1.5×10^{-8} M. The detection threshold for PMA was 3×10^{-10} M PMA (0.1 ppb, data not shown). As a control, both strains were compared under induction with different Hg(II) concentrations and both showed significant and comparable β-gal production over the tested range, even at 5×10^{-10} M (Fig. 8 A, Fig. 5 C). Both strains produced β-gal in linear correlation to the added Hg(II) up to 6.25×10^{-8} M. In JM109 (pJKS68, pJOE2004) lower PMA concentrations led to a maximum β-gal response in comparison to Hg(II). This might probably be due to the res-pective intracellular availability of the inducer. Hg(II) is specifically transported into the cell via MerT and MerP whereas organic Hg com-pounds are supposed to enter the bacterial cell efficiently via passive diffusion.

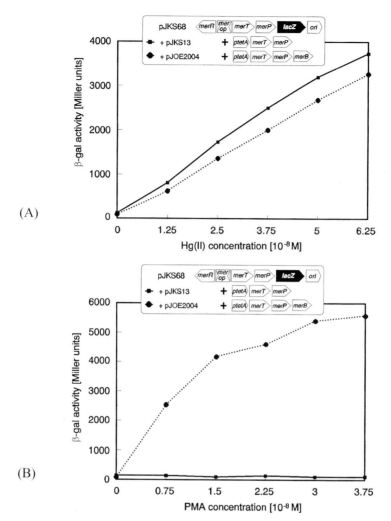

Figure 8. β-gal induction behaviour of *E. coli* JM109 (pJKS68, pJKS13) and JM109 (pJKS68, pJOE2004) at various concentrations of Hg(II) (A) and PMA (B) after 15 minutes induction time.

Development of an agar plate assay

A future perspective of the Hg(II) biosensor may be the experimental solution of a simple "Hg detection stick" which would resemble colour-based pH indicator sticks. Such a development may be the basis of a broad application of such biosensors. It should be possible to immobilise the bacterial cells on solid media or supports without affecting their induci-bility. A promising step into this direction was achieved by the develop-ment of a Hg-specific plate assay on solid agar medium. In principle, this test is analogous to the histochemical detection of β-gal (Messer and Viel-

(A)

E.coli JM109 (pJKS68, pJKS13)

(B)

E.coli JM109 (pJKS68, pJOE2004)

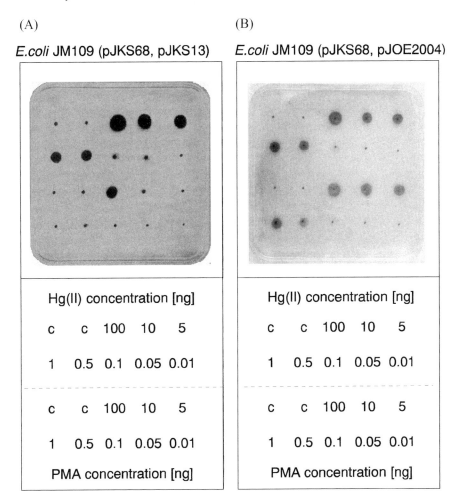

Hg(II) concentration [ng]				
c	c	100	10	5
1	0.5	0.1	0.05	0.01
c	c	100	10	5
1	0.5	0.1	0.05	0.01
PMA concentration [ng]				

Hg(II) concentration [ng]				
c	c	100	10	5
1	0.5	0.1	0.05	0.01
c	c	100	10	5
1	0.5	0.1	0.05	0.01
PMA concentration [ng]				

Figure 9. Agar plate assay with *E. coli* JM109 (pJKS68, pJKS13) (A) or JM109 (pJKS68, pJOE2004) (B). 10 µl of sterile Millipore water were used as a negative control (c).

metter, 1965). 100 µl of an overnight culture were spread on M9 minimal agar plates which were supplemented with 0.2% casamino acids, and the cells were grown for about 5 hours at 37 °C. Serial dilutions of Hg-solutions were spotted in a 10-µl volume, and the plates were incubated for further 30 minutes at 37 °C. Hg-dependent production of β-gal was detected by a soft agar overlay which contained 1 mg/ml X-gal as β-gal substrate. After 30 minutes of incubation at room temperature, blue colour development indicated β-gal activity and thus added Hg which appeared as Hg concentration dependent. A minimum of 0.1 ng Hg(II) or PMA in a 10-µl test volume (5×10^{-8} M, 3×10^{-8} M) could be detected, as seen in Figure 9.

Discussion

Transcriptional fusions of the regulatory elements of the Tn*501* specified Hg resistance and *lacZ* were used for the development of a sensitive microbial based Hg sensor. The use of LacZ as a reporter enzyme offered the possibility to rapidly analyse the signal by eye, by spectrophotometrical measurement or by chemiluminescence. Using AMPGD [3-(2′-spiroadamantane)-4-methoxy-4-(3′-b-D-galactopyranosyl-oxy)-phenyl-1,2-dioxetane] (Galacto-Light assay, Tropix, Bedford) and a luminometer, very low amounts of β-gal can be detected. With our microbial biosensor, the number of *E. coli* cells per assay could be reduced from about 10^9 cells with β-ONPG to 10^6 cells with AMPGD as substrate (data not shown). But with AMPGD the same concentration of Hg(II) was necessary to induce expression of *lacZ* and to significantly increase the β-gal activity above noninduced levels. Regarding the need of instrumental equipment, luminometer, and the cost of substrate AMPGD, the chromogenic assays may be preferable.

In the original construction, *lacZ* was fused with the N-terminal region of *merA* downstream of *merRopTP* which inactivated the Hg reductase gene. It was concluded that a functional Hg reductase would lower the inducer concentration by reduction of Hg(II). Indeed, if a functional Hg reductase, provided by the Hg resistance transposon Tn*21* on the plasmid pACYC184 (Tab. 1, De la Cruz and Grinstedt, 1982) was introduced into JM109 (pJKS3), about a tenfold higher concentration of HgCl$_2$ was needed to induce a detectable β-gal activity (data not shown).

Two other changes caused a dramatic improvement in the sensitivity of the assay and the time needed to perform *lacZ* induction. The first one was the addition of constitutively synthesised transport proteins MerT and MerP. The importance of the Mer transport proteins in activation of the *mer* operon transcription was shown before (Lund and Brown, 1987; Nakahara et al., 1979). Hg(II)-independent, constitutively produced Mer transport proteins, provided by a helper plasmid, reduced the time needed for Hg(II)-mediated induction of *lacZ* in pJKS3 from more than 120 min for JM109 (pJKS3) to 45 min for JM109 (pJKS3, pJKS13). In addition, the concentration of HgCl$_2$ needed for significant β-gal induction was reduced from about 1×10^{-8} M after 60 minutes of induction for JM109 (pJKS3) to 5×10^{-9} M after 15 minutes induction time using JM109 (pJKS3, pJKS13). The second increase in sensitivity from 1 to 0.1 ppb Hg(II) resulted from coupling the reporter plasmid replication and the transcription of the *mer* promoter. The reporter plasmid pJKS68 carries two replication origins. The original pMB1 origin of replication provides the normal pKO4 replication and the second, *mer* promoter dependent origin of replication leads to a Hg(II) inducible plasmid replication. As illustrated in Figure 7 the copy number increased about three- to fourfold at 10 ppb (5×10^{-8} M) and tenfold at 500 ppb (2.5×10^{-5} M). Expression of *lacZ* was enhanced by the induction of transcription and the concomitant amplification in the copy number of the reporter plasmid.

A unique feature of the microbial biosensor presented in this paper is the expansion of its specificity from inorganic Hg(II) to organomercurials like PMA by providing the organomercurial lyase gene (*merB*) from the broad spectrum resistance of *Streptomyces lividans* on a helper plasmid. The specific activity of MerB in cells carrying the plasmid pJOE2004 was not determined, however, it is high enough to ascertain PMA concentrations with a comparable sensitivity (3×10^{-10} M) as HgCl$_2$ (5×10^{-10} M).

The maximal sensitivity of the described biosensor is about 3 to 5×10^{-10} M for a specific Hg compound after 15 minutes Hg(II) induction. Tescione and Belfort (1993) and Condee and Summers (1992) constructed *lux*-based reporter plasmid bearing *E. coli* strains. They described significantly higher detection thresholds of 2×10^{-8} M (30 min induction) or 1×10^{-8} M (2 to 3 minutes induction). The need for *mer* specific transport proteins to lower the detection threshold was reported by Selifonova et al. (1993) who described a bioluminescent sensor *E. coli* strain using fusions of Tn*21 mer* genes to the *Vibrio fisheri luxCDABE* operon. In contrast to their reported similar sensitivity (5×10^{-10} M) after 40 minutes induction time they described a very slow growth of the sensor strain with only an increase of 0.2 units in the optical density in 24 hours combined with the need for an adaptational growth over three days before the start of the Hg test procedure. In contrast to this poor growth performance our sensor strain constructs show normal cultivational behaviour. From this it is obvious that the lowest detection threshold with the minimal growth and induction time is performed by the strains described in this report. They provide a practical approach to differentiate organic from inorganic Hg compounds and measure the important bioavailable Hg which is not tightly bound to macromolecules in soil or sewage (Selifonova et al., 1993). By using these two biosensors, which recognise inorganic and organic Hg, it would be possible to differentiate Hg contaminations. Risk assessment could be facilitated focusing on organomercurial contaminations, which are much more toxic than Hg(II). The investigations reported here focused on Hg salt solutions, because only rare data are available referring to the behaviour of Hg compounds in samples containing high amounts of organic substances. Too little is known about the mobility and the availability of biologically active Hg in complex organic material. Using the bacterial biosensors, it may be possible to shed more light on these important questions.

Besides Hg, regarded as the most toxic heavy metal, microorganisms developed a variety of resistance mechanisms against other heavy metals such as Cr(VI) or Ni(II) (Silver et al., 1989; Silver and Walderhaug, 1992). If the transcription of these resistance genes is specifically regulated, the same experimental solutions as described in this report may be used to provide other heavy metal biosensors with altered specificity.

Instead of these *in vivo* assays, it may also be possible to make use of the specific binding of a heavy metal to the regulatory protein of the appropriate resistance system for the development of an advanced *in vitro* test. The Hg-sensing protein component of the Hg resistance system is the regulatory protein MerR. It has a nanomolar sensitivity ($K_D < 10^{-8}$ mol $\times$ l^{-1}) and is highly selective for Hg (Ralston and O'Halloran, 1990). The affinity constant of MerR for Hg(II) ($K > 10^8$ l$\times$ mol^{-1}) is comparable to the affinity (association) constant of some antibodies for their antigen ($K = 10^8$ l-10^{11} l$\times$ mol^{-1}). These properties of MerR make this protein the best candidate as sensor component in the development of a Hg-specific *in vitro* assay. Such a test could be developed as sandwich test similar to the conventional enzyme-linked immuno sorbant assay (ELISA) using antibody enzyme conjugates for signal production. MerR should be coupled to a reporter enzyme, similar to an antibody-enzyme conjugate, which may be tested by gene fusion of *merR* to a reporter gene of interest.

References

Altenbuchner, J., Choi, C.L., Grinstedt, J., Schmitt, R. and Richmond, M.H. (1981) The transposon Tn*501* (Hg) and Tn*1721* (Tc) are related. *Genet. Res. Camb.* 37:285–289.

Aschner, M. and Aschner, J.L. (1990) Mercury neurotoxicity: mechanisms of blood-brain barrier transport. *Neurosci. Biobehav. Rev.* 14:169–174.

Bloom, N. and Fitzgerald, W.F. (1988) Determination of volatile merury species at the picogramm level by low-temperature gas chromatography with cold-vapour atomic fluorescence detection. *Anal. Chim. Acta* 208:151–161.

Bolivar, F., Rodriguez, R.L., Greene, P.J., Betlach, M.C., Heyneker, H.L., Boyer, H.W., Crosa, J.H. and Falkow, S. (1977) Construction and characterization of new cloning vehicles. II. A multipurpose cloning system. *Gene* 2:95–113.

Brünker, P., Rother, D., Sedlmeier, R., Klein, J., Mattes, R. and Altenbuchner, J. (1996) Regulation of the operon responsible for broad-spectrum resistance in *Streptomyces lividans* 1326. *Mol. Gen. Genet.* 251: (in press).

Chang, A.C.Y. and Cohen, S.N. (1978) Construction and characterization of amplifiable multicopy DNA cloning vehicles derived from the p15A cryptic miniplasmid. *J. Bacteriol.* 134:1141–1156.

Chung, C.T., Niemela, S.L. and Miller, R.H. (1989) One-step preparation of competent *Escherichia coli*: transformation and storage of bacterial cells in the same solution. *Proc. Natl. Acad. Sci. USA* 86:2172–2175

Condee, C.W. and Summers, A.O. (1992) A *mer-lux* transcriptional fusion for real-time examination of *in vitro* gene expression kinetics and promoter response to altered superhelicity. *J. Bacteriol.* 174:8094–8101.

De Flora, S., Benicelli, C. and Bagnasco, M. (1994) Genotoxicity of mercury compounds – A review. *Mut. Res.* 317:57–79.

De la Cruz, F. and Grinstedt, J. (1982) Genetic and molecular characterization of Tn*21*, a multiple resistance transposon from R100.1. *J. Bacteriol.* 151:222–228.

Klein, J. (1992) *Quecksilbernachweis über Reporterenzyme – Biosonde und Enzymsensor.* Ph.D. thesis, University of Stuttgart, Germany.

Klein, J., Altenbuchner, J. and Mattes, R. (1989) Mercury detection with transcriptional fusions in Tn*501*. 13th Workshop on Procaryotic genetics, Disentis, CH (13–16. September, 1989).

Klein, J., Altenbuchner, J. and Mattes, R. (1991) A new method to detect mercury using bacteria as biosensor. *In*: M. Reuss, H. Chmiel, E.-D. Gilles and H.-J. Knackmuss (eds): *Biochemical Engineering – Stuttgart.* Gustav Fischer, Stuttgart, New York, pp 323–326.

Lund, P.A. and Brown, N.L. (1987) Role of the *merT* and *merP* gene products of transposon Tn*501* in the induction and expression of resistance to mercuric ions. *Gene* 52:207–214.

Lund, P.A. and Brown, N.L. (1989) Regulation of transcription in *Escherichia coli* from the *mer* and *merR* promoters in the transposon Tn*501*. *J. Mol. Biol.* 205:343–353.

Marsh, J.L., Erfle, M. and Wykes, E.J. (1984) The pIC plasmid and phage vectors with versatile cloning sites for recombinant selection by insertional inactivation. *Gene* 32: 481–485.

McKenney, K., Shimatake, H., Court, D., Schmeissner, U., Brady, C. and Rosenberg, M. (1981) A system to study promoter and terminator signals recognized by *Escherichia coli* RNA polymerase *In*: J.G. Chirikjan and T.S. Papas (eds): *Gene amplification and analysis, Vol. II: Analysis of nucleic acids by enzymatic methods.* Elsevier-North Holland Press, Amsterdam, pp 383–415.

Messer, W. and Vielmetter, W. (1965) High resolution colony staining for the detection of bacterial growth requirement mutants using naphthol a_{20}-dye techniques. *Biochem. Biophys. Res. Commun.* 21:182–186.

Miller, J.H. (1972) *Experiments in molecular genetics.* Cold Spring Harbor Laboratory, Cold Spring Harbor, N.Y.

Nakahara, H., Silver, S., Miki, T. and Rownd, R.H. (1979) Hypersensitivity to Hg(II) and hyperbinding activity associated with cloned fragments of the mercurial resistance operon of plasmid NR1. *J. Bacteriol.* 140:161–166.

O'Halloran, T.V., Frantz, B., Shin, M.K., Ralston, D.M. and Wright, J.G. (1989) The MerR heavy metal receptor mediates positive interaction in a topologically novel transcription complex. *Cell* 56:110–129.

Omang, S.H. (1971) Determination of mercury in natural waters and effluents by flameless atomic absorption spectrophotometry. *Anal. Chim. Acta* 53:415–420.

Panayotatos, N. (1984) DNA replication regulated by the priming promoter. *Nucl. Acids Res.* 6:2641–2648.

Park, S.-J., Wireman, J. and Summers, A.O. (1992) Genetic analysis of the Tn*21 mer* operator-promoter. *J. Bacteriol.* 174:2160–2171.

Ralston, D.M. and O'Halloran, T.V. (1990) Ultrasensitivity and heavy metal selectivity of the allosterically modulated MerR transcription complex. *Proc. Natl. Acad. Sci. USA* 87:3846–3850.

Robinson, J.B. and Tuovinen, O.H. (1984) Mechanisms of microbial resistance and detoxification of mercury and organomercury compounds: Physiological, biochemical and genetic analyses. *Microbiol. Rev.* 48:95–124.

Ross, W., Park, S.-J. and Summers, A.O. (1989) Genetic analysis of transcriptional activation and repression in the Tn*21 mer* operon. *J. Bacteriol.* 171:4009–4018.

Sambrook, J., Fritsch, E.F. and Maniatis, T. (1989) *Molecular cloning: a laboratory manual.* 2nd edn. Cold Spring Harbor Laboratory Press, New York.

Sedlmeier, R. and Altenbuchner, J. (1992) Cloning and DNA sequence analysis of the mercury resistance genes of *Streptomyces lividans. Mol. Gen. Genet.* 236:76–85.

Selifonova, O., Burlage, R. and Barkay, T. (1993) Bioluminescent sensors for detection of bioavailable Hg(II) in the environment. *Appl. Environ. Microbiol.* 59:3083–3090.

Silver, A. and Walderhaug, M. (1992) Gene regulation of plasmid- and chromosome-determined inorganic ion transport in bacteria. *Microbiol. Rev.* 56:195–228.

Silver, S., Misra, T.K. and Laddaga, R.A. (1989) DNA sequence analysis of bacterial toxic heavy metal resistances. *Biol. Trace Elem. Res.* 21:145–163.

Summers, A.O. (1986) Organization, expression and evolution of genes for mercury resistance. *Annu. Rev. Microbiol.* 40:607–634.

Tescione, L. and Belfort, G. (1993) Construction and evaluation of a metal ion biosensor. *Biotechnol. Bioeng.* 42:945–952.

Ubben, D. and Schmitt, R. (1987) A transposable promoter and transposable promoter probes derived from Tn *1721. Gene* 53:127–134.

Vieira, J. and Messing, J. (1982) The pUC plasmids and M13mp7-derived system for insertion mutagenesis and sequencing with synthetic universal primers. *Gene* 19:259–268.

Wylie, D.E., Carlson, L.D., Carlson, R., Wagner, F.W. and Schuster, S.M. (1991) Detection of mercuric ions in water by ELISA with a mercury-specific antibody. *Anal. Biochem.* 194:381–387.

Yanisch-Perron, C., Vieira, J. and Messing, J. (1985) Improved M13 phage-cloning vectors and host strains: nucleotide sequences of the M13mp18 and pUC vectors. *Gene* 33: 103–119.

Thin layers/Interfaces

Frontiers in Biosensorics I
Fundamental Aspects
ed. by F. W. Scheller, F. Schubert and J. Fedrowitz
© 1997 Birkhäuser Verlag Basel/Switzerland

Studies on pyroelectric response of polymers modified with azobenzene moieties

L. Brehmer[1], Y. Kaminorz[1], R. Dietel[1], G. Grasnick[2] and G. Herkner[2]

[1] Institute of Solid State Physics, University of Potsdam, D-14469 Potsdam, Germany;
[2] Institute of Thin Film Technology and Microsensorics r.s., D-14513 Teltow, Germany

Summary. In this review we present investigations on new materials and arrangements for new pyroelectric sensor devices. We applied pyroelectrical measurements on thin films based upon poly(vinyl alcohol)s and poly(siloxane)s with azobenzene side chains and discuss the relaxation behaviour and stability of poled pyroelectric polymers. Results on special poly(vinyl alcohol)s with side chains consisting of azobenzene unit and aliphatic head group are comparable with those achieved by poly(vinylidene fluoride) (PVDF). The new radiation-sensitive material seemed to be suitable as a detector system for the *in situ* determination of glucose in blood. Further developments of a complex detector system are now under investigation.

Introduction

In recent years, electrically poled polymers containing pyroelectrically active groups with large dipole moments have been successfully developed for pyroelectric devices, although the market is still dominated by inorganic materials like lithium tantalate ($LiTaO_3$), triglycine sulphate (TGS) and lead zirconium titanate (PZT). Pyroelectric sensors made from electrically poled polymer films such as poly(vinylidene fluoride) (PVDF) and copolymer poly-(vinylidene fluoride-co-trifluor ethylene) (P(VDF/TrFE)) containing pyro-electrically active groups have been established on the market in the last 3 years. As Dorozhkin et al. (1983), Mader and Meixner (1990), Ploss et al. (1990) and Robin, Broussoux and Dubois (1991) reported PVDF and its copolymers are suitable for pyroelectric applications. High-electric-field poling is used to generate dipole orientations of the pyroelectrically active units in the polymer matrices. This poling is performed around the glass transition temperatures of these materials and they are cooled to room temperature with the electric field constant. For device applications, the materials should show both large pyroelectric coefficient and long-term stability.

Compared to pyroelectric crystals, poled polymers have advantageous thermal properties and a good processability into thin films. However, such materials (expect PVDF and P(VDF/TrFE)) often have the disadvantage of insufficient long-term stability of their pyroelectric properties. According to our investigations it seems that polymers with electrical dipoles incorporated into side chains have the advantage of better long-term stability in comparison to host-guest systems consisting of host polymer and dipole-containing molecules as guests.

Polymers with flexible and dipole-rich side chains seemed to be suitable for construction of pyroelectric sensitive devices. One of the requirements was the processability of chosen materials into thin films. Therefore we prepared polymers with azobenzene side chains based upon poly(vinyl alcohol) and poly(siloxane) and investigated the relaxation behaviour, the temperature dependence and long-term stability of pyroelectric response of the film polymers after poling. Poly(vinyl alcohol) and poly(siloxane) have been chosen as polymeric backbones for their known film-forming properties and formation of side-chain polymer.

Experimental part

Materials and experimental setup

Poly(vinyl alcohol) with azobenzene side chains (PVA-Az)
Poly[vinyl-4′-n-dodecyloxyazobenzene-4-carboxylate)-co-(vinyl alcohol)] (PVA-Az) was prepared by Einhorn esterification of poly(vinyl alcohol) with 4′-n-dodecyloxyazobenzene-4-carboxylic acid as Janietz and Bauer (1991) reported. The poly(vinyl alcohol) used was a commercially availabe product (Fluka, molecular weight 72 000 g · mol⁻¹, degree of polymerization 1600). 4′-n-dodecyloxyazobenzene-4-carboxylic acid was obtained by coupling p-aminobenzoic acid with phenol followed by the reaction with n-dodecyl bromide. Synthesis of copolymer with statistical distribution of side chains is carried out by reacting the acid chloride of 4′-n-dodecyloxyazobenzene-4-carboxylic acid with poly(vinyl alcohol) in dry pyridine.

The degree of esterification (D.E.) of the obtained optimum polymer is calculated from elemental analysis, D.E. = 85%. The dipole moment P calculated (MOPAC, AM1) for E- and Z-conformation of side-chain molecule was 4.87 D and 2.44 D, resp.

Infrared transmission spectra were recorded on a Nicolet Magna 550 FTIR spectrometer at a resolution of 4 cm⁻¹. KBr pellets were prepared from poly(vinyl alcohol) compounds.

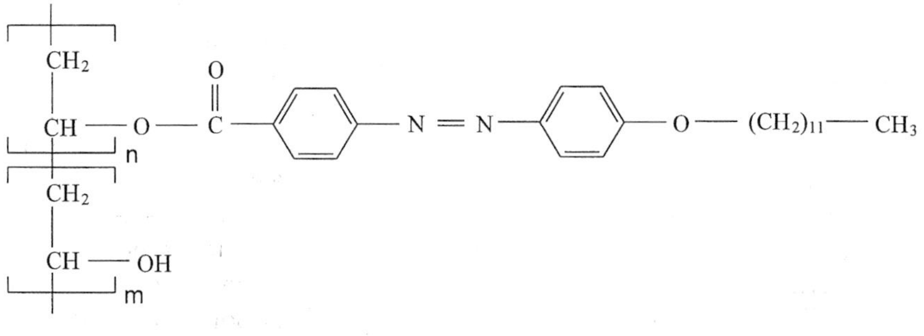

Figure 1. Formulae PVA-Az, statistical copolymer, ratio n:m = 5.66:1.

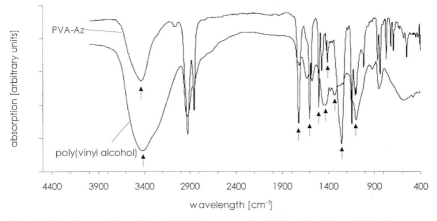

Figure 2. Fourier-transform infrared (FITR) spectra of poly(vinyl alcohol) and PVA-Az.

The spectrum of poly(vinyl alcohol) shows an intensive absorption band at wave number 3420 cm^{-1} which corresponds to stretching vibrations of hydroxyl groups despite taking water in the KBr-matrix into account. Two broad absorption bands at wave number 1430 cm^{-1} and 1330 cm^{-1} are assigned to a combination of OH- and CH$_2$-vibrations and another one at wave number 1095 cm^{-1} belongs to the (C-O)-stretching vibration.

A decreasing intensity of the OH-stretching vibration at wave number 3420 cm^{-1} is observed for PVA-Az due to the coupling of the azobenzene side chains. Several additional absorption bands indicate characteristic groups of side chains. Two intensive absorption bands at wave number 1720 cm^{-1} and 1255 cm^{-1} belong to the ester group and the bands at 1600 cm^{-1}, 1580 cm^{-1}, 1500 cm^{-1} and 1405 cm^{-1} indicate vibrations of the aromatic ring system.

Poly(siloxane) with azobenzene sice chains (PS-Az)
A copoly(siloxane) with azo dye in the side chain has been synthesized from commercially available linear poly(methyl-H-co-dimethyl-siloxane) (Chemiewerk Nünchritz GmbH, molecular weight 4370 g·mol^{-1} (GPC), degree of polymerization: 65). The incorporation of an azo group with high dipole moment into the polymer involves a high degree of polarizability of aromatic system in the side chains. The side chain molecules were prepared as Grasnick 1991 reported by esterification of the azo dye Disperse Red 1 (Aldrich) with 10-undecenoic acid and then coupled to the siloxane backbone via hydrosilation with hexachloroplatinic acid as catalyst giving a copolymer with statistical distribution of side chains.

The dipole moment P calculated (MOPAC, AM1) for E- and Z-conformation of side chain molecule was 6.53 D and 4.18 D, resp.

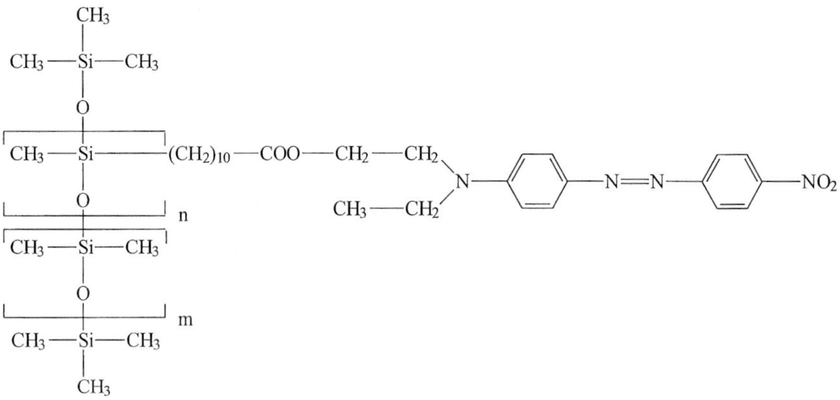

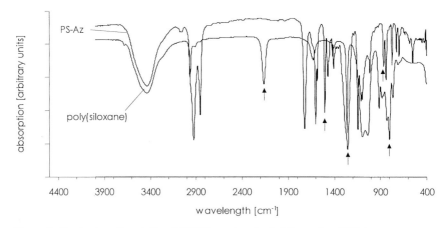

Figure 3. Formulae PS-Az, statistical copolymer, ratio n:m = 0.97:1.

Figure 4. Fourier-transform infrared (FITR) spectra of poly(siloxane) and PS-Az.

Infrared transmission spectra were recorded on a Nicolet Magna 550 FITR spectrometer at a resolution of 4 cm^{-1}. Poly(siloxane) samples were measured as capillary layers between 2 KBr-pellets with only the side chain polymer resolved in toluene.

Figure 4 shows the absorption bands of poly(siloxane) compound which are characteristic for structural elements.

The (Si-H)-stretching vibration appears at wave number 2160 cm^{-1}. Other bands at wave number 1260 cm^{-1} and 805 cm^{-1} belong to (CH$_3$)-bending and (Si-C)-stretching vibration, resp.

The spectrum of PS-Az is mainly dominated by toluene absorptions caused by the presence of toluene as solvent. The absorption bands at wave number 1260 cm^{-1} and 805 cm^{-1} (see above) correspond to the main chain,

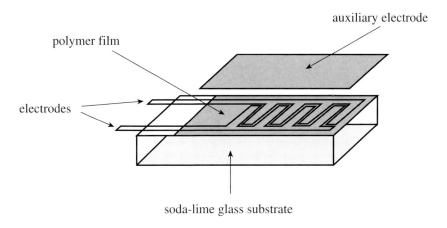

Figure 5. Experimental arrangement on interdigitated structure.

whereas characteristic absorption bands at 1520 cm^{-1}, 1340 cm^{-1} and 860 cm^{-1} belong to the nitro group. The absence of (Si-H)-stretching vibration at wave number 2160 cm^{-1} indicates the presence of side chain at the polymer. Spectrum of PS-Az does not show double bond absorption at wave number 1640 and 915 cm^{-1} indicating an almost complete grafting of vinyl monomer onto poly(siloxane) backbone and the absence of free monomer.

Sample preparation and experimental procedure
Perfect films for pyroelectric measurements have to be both thin and homogeneous. Therefore samples have been prepared on evaporated interdigitated structures (shown schematically in Fig. 5) and in sandwich arrangement (between 2 evaporated electrodes) by spin-coating or dipping procedures from chloroform (PVA-Az) and toluene (PS-Az) solutions. Film thicknesses were approximately 1 μm and have been measured by ellipsometry and interference microscopy. For the homogenization of the electric field in vertical direction on interdigitated structures the evaporation of an electrically non-contacted auxiliary electrode floating onto thin polymer layer proved to be very effective.

The poling of polymer films is performed for PVA-Az at T = 175 ± 5 °C while applying an voltage of 360 V and for PS-Az at T = 135 ± 5 °C and 100 V, resp., followed by cooling to room temperature with the field applied.

Pyroelectric measurements were performed by dynamic method with an amplitude-modulated laser diode (λ = 785 nm, 50 mW). The experimental arrangement was similar to what Bauer (1994) reported giving pyroelectric voltage response data as shown in Figure 6.

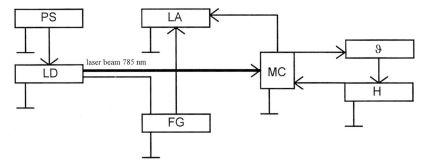

Figure 6. Experimental setup for dynamic measurements.

History: PS   power supply DIGI 35 CPU
                LD   laser diode HL25/MI-0785-50
                LA   lock-in amplifier EG & G 5210
                FG   function generator TOE 7720
                MC   measuring chamber
                H   heating Peltier element
                ϑ temperature indicator

Results and discussion

Pyroelectric response and long-term stability

In the following, we present systematical studies of pyroelectric voltage response. Theoretical value of pyroelectric voltage response of a thin layer determined by dynamic method can be calculated.

In the case of homogeneously poled pyroelectric layer the pyroelectric current I_- is given by equation (1):

$$I_- = p \cdot \frac{A}{d} \cdot \frac{\delta T}{\delta t} \qquad (1)$$

with

A   – area

p   – pyroelectric coefficient

d   – thickness of pyroelectric layer

$\dfrac{\delta T}{\delta t}$ – time derivative of temperature

For thin films and low modulation frequencies the temperature on the top and at the bottom of the film should be considered to be equal.

In this case the pyroelectric current is given by the following equation as reported by Bauer (1994)

$$|I_-| = k \cdot p \cdot A \cdot \sqrt{\omega} \tag{2}$$

where k is a constant that depends on the intensity of the laser beam light, the absorption coefficient of the sensor layer and the thermal dates of the substrate.

Measurable pyroelectric voltage at given modulation frequency ω depends on the total electric capacity C and the input resistance R of the lock-in amplifier or an impedance converter.

$$|U_{pe}| = \frac{\sqrt{R} \cdot |I_-|}{\sqrt{(R \cdot C \cdot \omega)^2 + 1}} \; ; \; \tau = R \cdot C \tag{3}$$

with limiting values

$$\omega \cdot \tau \ll 1 \; |U_{pe}| = R \cdot |I_-| = k_1 \cdot p \cdot A \cdot \sqrt{\omega} \tag{4}$$

$$\omega \cdot \tau \gg 1 \; |U_{pe}| = \frac{|I_-|}{C \cdot \omega} = k_2 \cdot \frac{p \cdot A}{\sqrt{\omega}} \tag{5}$$

That means, for low modulation frequencies ($\omega \cdot \ll 1$) the pyroelectric voltage increases proportionally with $\sqrt{\omega}$ and for higher frequencies it decreases proportionally with $1/\sqrt{\omega}$. The maximum value is determined by the relation $R = |1/(C \cdot \omega)|$.

Figures 7 and 8 show the pyroelectric voltage response of the poled polymers as measured by dynamic method. The magnitude of pyroelectric voltage of PVA-Az has been found to be more than two orders of magnitude higher than the corresponding value of PS-Az even if the calculated dipole moments of the PS-Az units are higher than for PVA-Az (see materials section). The flexibility of the poly(siloxane) backbone seemed to be the reason for low level of pyroelectric response and poor stability achieved with PS-Az samples. The electric capacity of sample and the impedance matching field effect transistor incorporated into the measuring system cause a decrease of the signal at high frequencies. As Figure 9 shows, an excellent long-term stability of pyroelectric response could be achieved for PVA-Az. A decay of the pyroelectric effect could not be observed, even after storing the samples at room temperature for more than a year. Compared to poled PVA-Az the pyroelectric response of polymer PVA-Az is not constant in time. The relatively fast decay $\tau \approx 100$ min at room temperature is shown in Figure 10.

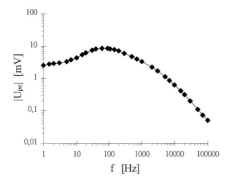

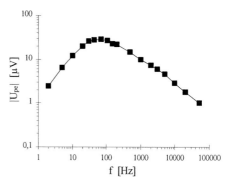

Figure 7. Pyroelectric voltage response $|U_{pe}|$ [mV] vers. modulation frequency f [Hz] of PVA-Az.

Figure 8. Pyroelectric voltage response $|U_{pe}|$ [μV] vers. modulation frequency f [Hz] of PS-Az.

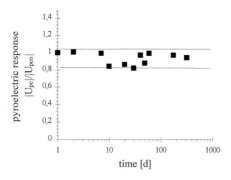

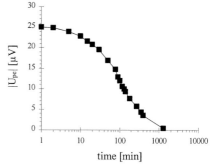

Figure 9. Long-term stability of pyroelectric response, PVA-Az, storage at RT, f = 30 Hz.

Figure 10. Decrease of pyroelectric response, PS-Az, storage at RT, f = 30 Hz.

Dielectric measurements

The dielectric behaviour was measured by an impedance-analyzer Schlumberger SI 1260. Dielectric loss of poled and unpoled PVA-Az films was determined at different temperatures and the frequency of the maximum dielectric loss (f_{max}) was estimated. It was found that the frequency f_{max} of the maximum dielectric loss increases with increasing temperature. As shown in Figure 11, the behaviour of f_{max} of the poled and unpoled sample plotted versus inverse temperature can be fitted by an Arrhenius equation. An activation energy E_a of 0.45 eV was calculated for the unpoled PVA-Az. The activation energy of the poled sample was determined to be 0.8 eV. The higher activation energy might be the reason for the better long-term stability of the poled PVA-Az sample.

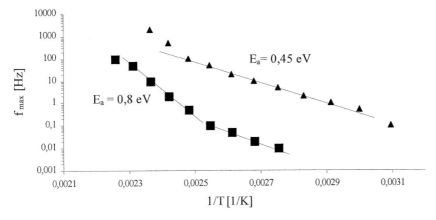

Figure 11. Activation energy diagram of poled and unpoled PVA-Az.

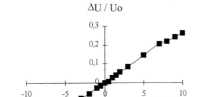

Figure 12. Voltage control of pyroelectric response, PVA-Az.

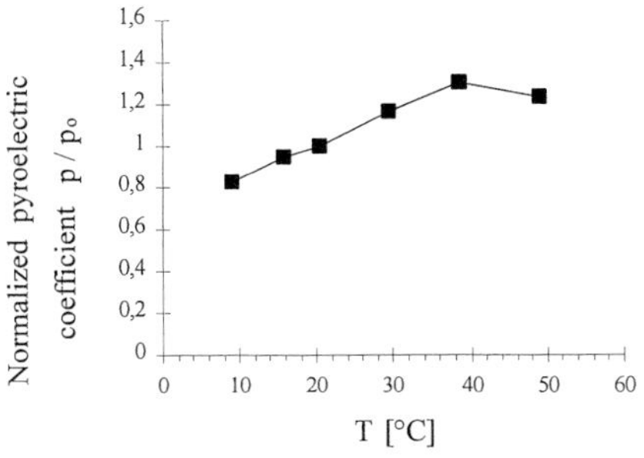

Figure 13. Temperature dependence of pyro-electric response, PVA-Az.

Control of pyroelectric response and temperature dependence

The control of maximum response and sensitivity is a very important property of a pyroelectric sensor device and therefore the aim of our investigations. The pyroelectric voltage response of the sample can be changed by a dc-voltage connected to the electrodes of the pyroelectric sensor device. When applying dc-voltage in the same direction as the poling voltage the pyroelectric response increases and in the opposite direction the response decreases with a slope of 2.7%/V (Fig. 12).

The temperature dependence of the pyroelectric response has been determined in a temperature range from 10 to 50°C. The relative pyroelectric coefficient shows in Figure 13 a nearly linear increase up to 35°C with a slope of 0.016 K^{-1}. A small decrease has been observed at higher temperatures.

Conclusions

It has been shown that the new polymer material PVA-Az has pyroelectric properties comparable to other commercial pyroelectric materials. Table 1 lists several pysical values of pyroelectric materials in comparison to PVA-Az. Long-term stability is sufficient for the use in pyroelectric sensor devices. Processability into thin films allows the fabrication of pyroelectric thin films as basis for sensitive sensor devices. The special method of arranging electrodes in lateral configuration is suitable for this process.

Conventional pyroelectric detectors consist of pyroelectric material, covered by 2 metal electrodes, a radiation absorbing layer on the top and a substrate on the back of the sensor device. The electrode configuration with auxiliary electrode allows the separation of absorbing and active electrodes as Brehmer et al. (1995) reported. The non-contacted auxiliary electrode can be used as radiation absorbing layer and covered for optimum thermal absorption. It provides a noticable enhancement of electrical polarization effect due to the mobilization of additional detection areas directly on the electrodes.

The found radiation-sensitive material and its device configuration seemed to be suitable for interesting applications in biosensorics, e.g. as new detector system for the in-situ determination of glucose in blood. In this case a change of pyroelectric signal caused by well-known enzymatic reactions could be directly correlated to glucose concentration in blood. The development of a multi-layered detector system is now under investigation.

Table 1. Pyroelectric materials in comparison to new PVA-Az (values of commercial materials as Bauer and Ploss (1990) reported)

Material	$LiTaO_3$	TGS	PZT	PVDF	PVA-Az
$p \ [\mu C \cdot m^{-2} \cdot K^{-1}]$	230	280	380	20	20
ε	47	38	290	9	4
$\tan \delta$	0.005	0.01	0.003	0.03	0.006
$F_I \ [10^{-10} \cdot mV^{-1}]$	0.72	1.2	1.5	0.09	0.08
$F_V \ [V \cdot m^2 \cdot J^{-1}]$	0.17	0.36	0.06	0.11	0.23
$F_D \ [10^{-5} \cdot (m^3 \cdot J^{-1})^{1/2}]$	4.9	6.6	5.8	0.56	1.8

Figures of merit: $F_I = p/(c \cdot \rho)$; $F_V = p/(\varepsilon_0 \cdot \varepsilon \cdot c \cdot \rho)$; $F_D = p/(c \cdot \rho \cdot (\varepsilon_0 \cdot \varepsilon \cdot \tan \delta)^{1/2})$

Acknowledgements
Thanks are due to Dr. D. Janietz for his side chain poly(vinyl alcohol)s, Dr. G. Knochenhauer for calculation of dipole moments and Dr. P. Frübing (all University of Potsdam, Institute of Solid State Physics) for many stimulating discussions. Two of the authors (G.G. and G.H.) thank the Ministerium für Wirtschaft, Mittelstand und Technologie of Land Brandenburg for financial support.

References

Bauer, S. (1994) Pyroelectrical investigation of charged and poled nonlinear optical polymers. *J. Appl. Phys.* 75:5306–15.

Bauer, S. and Ploss, B. (1990) A method for the measurement of the thermal, dielectric and pyroelectric properties of thin pyroelectric films and their applications for integrated heat sensors. *J. Appl. Phys.* 68:6361–6367.

Brehmer, L., Grasnick, G., Herkner, G. and Janietz, D. (1995). *Pyroelektrisches Dünnschichtsensorelement,* Pat. submitted to Deutsches Patentamt, Az. P 195 13 499.0-33.

Dorozhkin, L.M., Lazarev, V.V.; Pleskov, G.M., Chayanov, B.A., Nabiey, Sh. Sh, Nikiforov, S.M., Khokhlov, E.M., Chikov, V.A., Shigorin, V.D. and Shipulo, G.P. (1983) Thin-film pyroelectric detector made of an organic compound and used to measure parameters of pulsed laser radiation. *Soviet J. Quant. Electronics* 13(6):707–711.

Grasnick, G. (1991) *Siliconmodifizierung von Terephthalsäurehomo- und -copolyestern,* PhD thesis, Technical University of Dresden.

Janietz, D. and Bauer, M. (1991) Chromophoric poly(vinyl alcohol) derivatives, *1 Makromol. Chem.* 192:2635–2640.

Mader, G. and Meixner, H. (1990) Pyroelectric infrared sensor arrays based on the polymer PVDF. *Sens. Actuators, A* A22 (1–3):503–507.

Ploss, B., Lehmann, P., Schopf, H., Lessle, T., Bauer, S. and Thiemann, U. (1990) Integrated pyroelectric detector arrays with the sensor material PVDF. *Ferroelectrics* 109:223–228.

Robin, P., Broussoux, D., Dubois, J.C. and Thomson, C.S.F. (1991) Infrared detector comprising pyroelectric materials, EP 406053.

Frontiers in Biosensorics I
Fundamental Aspects
ed. by F. W. Scheller, F. Schubert and J. Fedrowitz
© 1997 Birkhäuser Verlag Basel/Switzerland

Polyelectrolyte layer systems

H. Möhwald and R. v. Klitzing[1]

Max-Planck-Insitut für Kolloid- und Grenzflächenforschung, D-12489-Berlin, Germany;
[1]*University of Mainz, D-55099 Mainz, Germany*

Summary. In this short contribution we demonstrate the feasibility of a simple technique to prepare an ultrathin film with enzymes entrapped that may be suitable for sensor applications. The films are impenetrable to macromolecules having a size of some nanometers but penetrable to small molecules. Quantitative data on diffusion coefficients and potential profile within these films are derived via optical probe techniques.

General approach

This contribution does not consider any potentially sensing biomolecule but considers some general aspects along a specific strategy to build a biosensor. This is not necessarily restricted to a specific sensor but may be more generally applied in various contexts. The structure that we envision (Fig. 1) is an enzyme in a hydrogel. The gel shall be permeable to substrate and product and near an electrode or an optical wave guide to measure a change followed by the enzymatic reaction. This calls for a thin film geometry, and for fast response we prefer an ultrathin film. Stability of the device suggests to prepare the matrix with polymers, and reproducibility demands to prepare a film with well-controlled thickness and density. It has been shown predominantly by the group of Decher, that stable, well-defined polymeric films can be prepared by consecutive alternating adsorption of oppositely charged polyelectrolytes (Decher et al., 1994). Into these films charged particles can be embedded likewise (Lvov et al., 1994), and along this line the groups of Kunitake and Lvov have incorporated nearly 20 different proteins (Lvov et al., 1995). It was shown that even functional cascades of electron transfer proteins can be built up.

These experiments have demonstrated the feasibility of an ultrathin biosensor with an enzyme embedded in a polyelectrolyte matrix, and we now have to turn to the quantitative aspects. This obviously requires more detailed experiments, and control also requires understanding. Therefore we shall concentrate in this work on the "enzyme free" polyelectrolyte film and ask the two questions:

(1) How are pH and electric field distribution inside the film related to those in the external medium?
(2) How penetrable are these films for ions and molecules and how does the diffusion depend on molecular details?

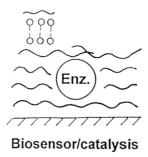

Biosensor/catalysis

Figure 1. Schematics of a design goal: An enzyme embedded, protected but accessible in an ultrathin polymerfilm.

Experimental

For the experiments described in detail elsewhere (v. Klitzing and Möhwald, 1995) we prepared a polyelectrolyte film with a fluorescent dye probe attached to a polymer in a defined depth below the surface. The film thickness is determined by X-ray reflectometry, and from neutron reflectometry we can infer that the dye is located with precision better than 1 nm (Schmitt et al., 1993) (Fig. 2). The film on a glass substrate is excited by a totally externally reflected light, and the dye emission is detected via a monochromator and an optical multichannel analyzer. The film on the glass slide is part of a liquid cell that enables exchange of the aqueous environment during the optical measurement.

The fluorescence of the dye used, fluorescein isothiocyanate (FITC), depends on pH, and we can therefore measure the local pH and electric field at the position of the dye (v. Klitzing and Möhwald, 1995). By adding quenchers to the outside medium and measuring time-dependent fluorescence we can determine the penetration of the matrix by these molecules (v. Klitzing and Möhwald, 1996a). We used two types of quenchers or quenching mechanisms. For rhodamin B (RhB) addition FITC fluorescence is reduced due to radiationless energy transfer to RhB. In this case RhB fluorescence increase can be measured accompanying FITC fluorescence decrease, and this gives more reliability to the analysis. The energy transfer occurs over a characteristic distance of 5 nm and depends on the sixth power of the distance. We did not intend to deconvolute this distance dependence, but considered the transfer an "all or nothing" process to estimate the diffusion coefficient. Such a process is essentially given for the second type of quencher, a paramagnetic molecule (2,2,6,6,-Tetramethyl-4-piperidinol-1-oxide, TEMPOL) penetrating and, on collision, increasing the rate of radiationless transitions to the ground state.

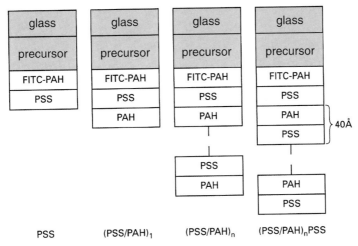

Figure 2. Preparation of the polyelectrolyte films by consecutive adsorption of polycation polyallylhydrochloride (PAH) and polyanion polystyrenesulfonate (PSS) from aqueous solutions. Only the structure of the decisive film covering the dye layer (FITC-PAH) is given in detail.

Local pH measurements

Figure 3 shows measurements of the FITC fluorescence as a function of pH in the outside medium for different dye depths within the film. One realizes that the pH in the film responds to the outside pH indicating that the film is proton permeable and that there is no local buffering capacity. One also observes a shift of the titration curves, depending on the coating thickness. This can be understood as a gradient in the local electric field ψ from the surface into the bulk of the film, since the midpoint of the titration curve $pK(x)$ and that in the absence of a field $pK(\infty)$ are related by the Henderson-Hasselbach equation

$$pK(x) = pK(\infty) - \frac{F}{RT} \cdot \psi.$$

From the measurement of $pK(x)$ we then derived $\psi(x)$ as given in Figure 4. The dependence is qualitatively as expected for a negatively charged surface, and in this case the outer layer is negatively charged. (For a positively charged surface we observed almost no $\psi(x)$ dependence (v. Klitzing and Möhwald, 1995)). Within the accuracy of the measurement $\psi(x)$ can be described by an exponential dependence as expected within the Gouy-Chapman theory (curve in Fig. 4).

The screening length of 3 nm deduced from Figure 4 is a factor of about 3 smaller than in the water phase at these conditions. This indicates either different ionic content or different dielectric constant of the polyelectrolyte film.

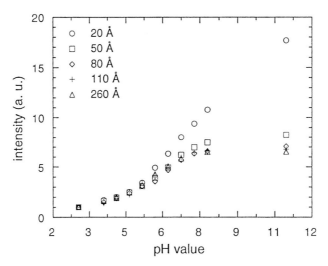

Figure 3. Integrated fluorescence normalized to display equal intensities at pH = 3.0 in dependence on the pH value of the external buffer solution. The parameter is the thickness of the polyelectrolyte film on the FITC-PAH layer.

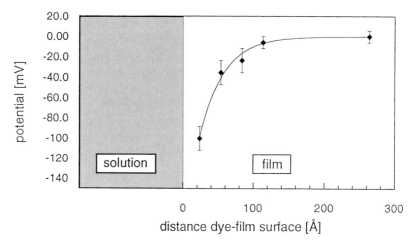

Figure 4. Potential profile in a polyelectrolyte film with a PSS surface, calculated by using the pK's determined by the data curves shown in Figure 3. The solid line corresponds to the fit using a simple exponential function.

Film penetration

Figure 5 shows the fluorescence intensity as a function of time after adding a RhB solution to the outside medium for different film thickness. For a very thin film one observes a very fast decay to a steady value. This corresponds to a fast dye adsorption at the film surface. From the reduction in intensity one may estimate a surface density of the dye slightly above

$$\frac{1}{\pi \cdot (5 \text{ nm})^2} \approx \frac{1}{100 \text{ nm}^2} \text{ which is a few percent of a monolayer.}$$

For medium thickness the time dependence can be analyzed in more detail, thus yielding the diffusion coefficient. We verified that the data can be described by a Stern-Volmer plot, and thus could convert fluorescence intensities into local concentrations near the dye.

Measurements at different times and coverages can thus be converted into concentrations at different depths and for different times (Fig. 6). The profile thus derived can be described by an error function that would be expected from a diffusion model (v. Klitzing and Möhwald, 1996 b). From a fit to the data one derives diffusion coefficients between 10^{-16} and 10^{-14} cm²/sec, typical for a polymeric glass. It is remarkable that the data are best described with a model considering a much larger diffusion coefficient for the outer 100 Å of the film compared to the inner regions. This should be ascribed to a looser packing and hence easier penetration of the outer parts of the film.

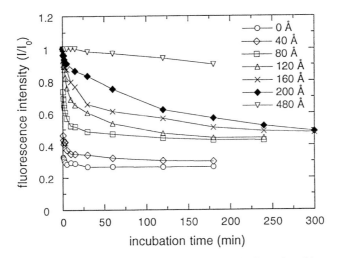

Figure 5. Integrated fluorescence normalized with respect to the intensity without any quencher in dependence of time after flushing the Rhodamine solution into the cell. The parameter is the thickness of the polyelectrolyte film on the FITC-PAH layer.

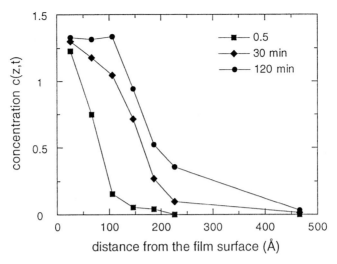

Figure 6. Concentration profile of Rhodamine molecules inside the polyelectrolyte film determined by the distance from the film surface. The data are calculated by the results shown in Figure 5 for different times after flushing the Rhodamine solution into the cell.

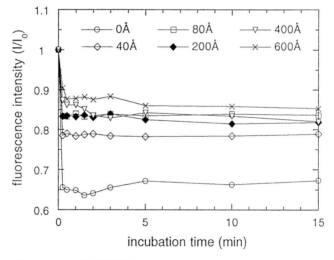

Figure 7. As Figure 5, but with TEMPOL as quencher.

Beyond the discussion of a proper model the experiments yield direct information on penetration of a species into the film and thus give guidelines on manufacturing parameters. A future task will surely be to prepare an easily swellable and hence penetrable matrix, and we expect vastly varying diffusion coefficients between the above mentioned value and the one of pure water which is *ten* orders of magnitude larger.

Although we cannot conclusively describe the transport mechanism we can already state that the process drastically depends on the tpye of diffusant. To prove this Figure 7 gives the fluorescence intensity as a function of time after adding the smaller quencher molecule TEMPOL. The decay can hardly be resolved, indicating a transport process which is at least a factor of 200 faster than for RhB. Hence it appears as if the transport drastically depends on molecular size. Still we note that there may be also other factors influencing permeability like charge, hydrophilicity or solubility in the polymer matrix which we have not yet assessed.

Conclusions

We have shown that the polyelectrolyte adsorption technique provides well-defined ultrathin films that entrap large molecules like enzymes, but that are diffusible in a selective way to low molecular weight compounds. Local electric and permeability properties can be measured in situ by optical techniques. The system and methodology can thus be extended to develop a sensor based on a biological recognition principle.

Acknowledgements
This work was supported by the Deutsche Forschungsgemeinschaft, the Stiftung Volkswagenwerk, the Fonds der Chemischen Industrie and the EU Human Capital and Mobility Programm. We thank G. Decher for many helpful comments and F. Eßler for help with the preparation.

References

Decher, G., Lehr, B., Lowack, K., Lvov, Yu. and Schmitt, J. (1994) New nanocomposite films for biosensors: layer-by-layer adsorbed films of polyelectrolytes, proteins or DNA. *Biosens. Bioelectron.* 9:677–684.

Lvov, Yu., Haas, H., Decher, G., Möhwald, H., Mikhailov, A., Mtschedlishvily, B., Morgunova, E. and Vainshtein, B. (1994) Successive deposition of alternate layers of polyelectrolytes and a charged virus. *Langmuir* 10:4232–4236.

Lvov, Yu., Ariga, K., Ichinose, I. and Kunitake, T. (1995) Assembly of multicomponent protein films by means of electrostatic layer-by-layer adsoprtion. *J. Amer. Chem. Soc.* 117: 6117–6123.

Schmitt, J., Grünewald, T., Decher, G., Pershan, P.S., Kjaer, K. and Lösche, M. (1993) Internal Structure of layer-by-layer adsorbed polyelectrolyte films: A Neutron and X-ray reflectivity study. *Macromolecules* 26:7058–7063.

v. Klitzing, R. and Möhwald, H. (1995) Proton concentration profile in ultrathin polyelectrolyte films. *Langmuir* 11:3554–3559.

v. Klitzing, R. and Möhwald, H. (1996a) Transport through ultrathin polyelectrolyte films. *Thin Solid Films*, in press.

v. Klitzing, R. and Möhwald, H. (1996b) A realistic diffusion model for ultrathin polyelectrolyte films. *Macromolecules*, in press.

Frontiers in Biosensorics I
Fundamental Aspects
ed. by F.W. Scheller, F. Schubert and J. Fedrowitz
© 1997 Birkhäuser Verlag Basel/Switzerland

Förster energy transfer in ultrathin polymer layers as a basis for biosensors

M. Völker[1] and H.-U. Siegmund[2]

[1]*Bayer AG, Zentrale Forschung und Entwicklung Uerdingen, D-47812 Krefeld, Germany;*
[2]*Bayer Corp., Diagnostics Division, Elkhart, IN 46515, USA*

Summary. A method of detecting the binding of analyte molecules to biospecific receptors, like antibodies, is described. Förster energy transfer is used in connection with monomolecular organic films. The films are built up from pre-polymerized materials using the Langmuir-Blodgett or self-assembly techniques. Fluorescent dyes (as energy transfer donors) as well as reactive groups for covalent immobilization of protein receptors are integrated into the polymers. Several different methods for immobilizing biomolecules are described, including the use of protein A and the biotin/streptavidin couple. The studies suggest that the combination of Förster transfer and ultrathin organic films can be used for the construction of biosensors working either by displacement or by competitive assays. The mannose/concanavalin A, digoxin/antibody, and mouse immunoglobulin G/antibody systems are investigated. Further, a simplified meter optimized for measuring the fluorescence ratio at two wavelengths is described.

Introduction

Langmuir-Blodgett (LB) and Self Assembly (SA) films have been a subject of academic studies for a long time, and investigations have been carried out to find practical applications in industry and technology. It has been frequently suggested that these ultrathin organic films could be used in the field of biological detecting and sensing (Fuchs et al., 1991). The sensitive layers of biosensors are often technical membranes, where considerable expertise is required in loading them with biomolecules, e.g. from the blotting techniques and solid phase binding assays. Since monomolecular films are much thinner than conventional membranes, a gain in detection speed can be expected in those types of sensors where molecules, ions, or electrons have to penetrate the sensor's surface (Kauffmann et al., 1994). However, their sometimes poor stability and the more complex way of preparing them may have been the reason for their lack of a commercial success. On the other hand, monomolecular films can have several other advantages over conventional materials: a well-defined and extremely thin structure, in the case of LB films a resemblance to biological membranes (Ahlers et al., 1990), and their capability of taking up various guest molecules, a rather interesting property for biosensor development. Since their stability can be improved by using pre-polymerized materials (Tippmann-Krayer et al., 1991; Lvov et al., 1993) and since new methods have been developed for scaling up film production (Embs et al., 1993, and references cited within), it seemed worth designing a sensitive and versatile biosensing concept based on LB or SA films. This paper describes the development of polymeric materials, biochemical reagents, and instrument equipment necessary for the detection of a biospecific binding.

Förster energy transfer as a sensing principle

A physical effect that works very efficiently in monomolecular organic films is the Förster energy transfer. It is well known that two fluorophores, one of them emitting light at a wavelength where the second one absorbs, show the phenomenon of a radiationless Förster energy transfer by dipole-dipole interaction. This energy transfer is highly dependent on the distance between the dyes, showing an r^{-6} dependence for individual molecules and, within well-defined limits, an r^{-4} dependence for a planar dye arrangement (Kuhn, 1970). Though deviations from the ideal theoretical behavior have been demonstrated (Fromherz and Reinbold, 1988), the critical distance, where deactivation of the donor via energy transfer is equally probable as by all other processes, has been generally found to be around 5 to 10 nm. These dimensions can be adjusted very well with the geometry of monomolecular organic films, where the typical layer thickness is in the order of 1 to 3 nm. The difficulty of a penetration of the dye molecules into the monolayer systems was overcome by using amphiphilic cyanine dyes incorporated into appropriate matrix monolayers, and extensive studies dealing with fluorescence energy transfer in LB films have been described by Kuhn et al. (1972).

In basic biochemical investigations, Förster energy transfer has also been applied as a tool to determine distances in macromolecules and to detect biological binding. A review of immunological test based on this idea can be found in a review by Stryer (1978), and additional interesting applications have been occasionally described (Thompson and Patchan, 1995; see also references cited in Siegmund and Becker, 1993). However, they all require that different kinds of biomolecules have to be labeled with different chromophores. The energy transfer is used to detect binding between receptor and analyte in the presence of fluorescence-labeled, but non-bound molecules. In chemical sensing, the fluorescence energy transfer technique has initiated some investigation as well (Gabor et al., 1995).

Our goal was to show the feasibility of a concept that shifts the double-labeling of proteins, as necessary in a homogeneous energy transfer immunoassay, to an immobilization of proteins onto a fluorescence-labeled monomolecular organic film. We were further interested in a higher degree of variation of the fluorescence donor dye (F_1 in Fig. 1), lessening restrictions regarding compatibility with the labeled protein. The basic concept of a sensing principle developed in our group has been described previously (Hugl et al., 1989; Siegmund and Becker, 1993), and it is illustrated in Figure 1.

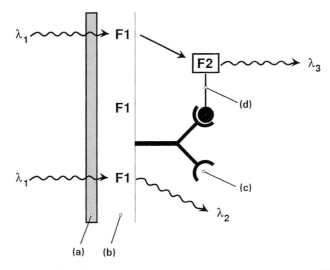

Figure 1. Biosensor Principle. On a glass or polymer substrate (a), a monomolecular (LB or SA) organic film (b), containing a donor dye F_1, is equipped with immobilized biological receptor molecules (c), preferably a monoclonal antibody. To this receptor a derivative of the analyte molecule (d) labeled with the acceptor dye F_2 is bound, its presence can be detected by the fluorescence λ_3 when F_1 is excited at λ_1. This device is brought into contact with a solution of unlabeled analyte. In a competitive way, the labeled analyte (d) can then be (partially) replaced, which causes a reduction in the fluorescence λ_3 of F_2, whereas the fluorescence λ_2 of F_1 will be restored. These signal changes should therefore be dependent on the concentration of the analyte.

Polymer materials for ultrathin films containing fluorophores

A critical point in the concept is the selection of fluorophores to obtain an optimum in sensitivity. Not only do the wavelengths have to match, but the concentration of donor dye in the film has to be optimized to yield maximum energy transfer and minimum self-quenching. Preformed polymers for the formation of LB films carrying covalently bound fluorophores were optimized and extensively characterized for the application in a Förster energy transfer sensor (Siegmund et al., 1991; general information and further literature on the preparation of LB films can be found, e.g., in Fuchs et al., 1991). It was found that amphiphilic polymethacrylates synthesized by statistical radicalic copolymerization from octadecyl-methacrylate and 4-(2,3-dimethyldioxolanyl)-methylenemethacrylate were good matrices for introducing additional comonomers without sacrificing their excellent film-forming properties. Coumarin dyes were chosen as fluorophores for their wide variability in emission wavelengths. Due to their generally large Stokes shifts, the detection apparatus can be kept fairly simple. They were converted to their methacryl amides or methacryl esters and then copolymerized with the above-mentioned monomers with varying dye content (Fig. 2, compounds **1** to **5**). LB films prepared from these

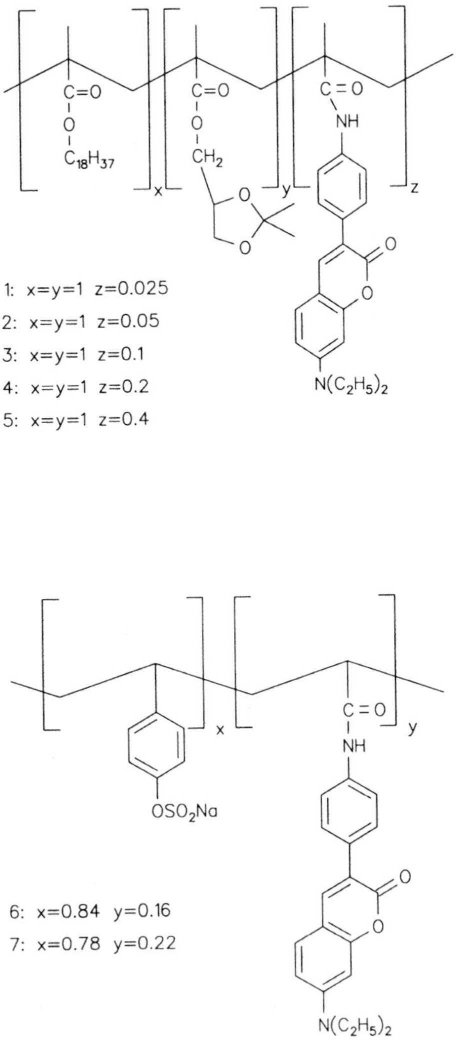
C=O
O
C18H37
x
C=O
O
CH2
O
O
y
C=O
NH
O
O
N(C2H5)2
z
1: x=y=1 z=0.025
2: x=y=1 z=0.05
3: x=y=1 z=0.1
4: x=y=1 z=0.2
5: x=y=1 z=0.4
x
OSO2Na
C=O
NH
y
O
O
N(C2H5)2
6: x=0.84 y=0.16
7: x=0.78 y=0.22

Figure 2. Polymers used to form monomolecular organic films carrying fluorescent groups.

materials were characterized by fluorescence spectroscopy under various conditions, with and without Förster energy transfer, and fluorescence microscopy. They showed sufficient physical and chemical stability to survive conditions experienced at the exposed part of a biosensor, like variations in pH range, ion strength, temperature, and enzyme activities present in sample solutions (Siegmund and Becker, 1993).

Self-assembled mono- and multilayers were prepared using the method of Decher and Hong (1991), where alternating layers of polyanions and polycations are absorbed from aqueous solutions to a solid substrate bearing surface charges. Polylysine hydrochloride was generally used as the positively charged intermediate layer. In a similar synthetic approach as for the LB films, statistical copolymers were prepared that consisted of fluorophores and of negatively charged groups necessary for the layer formation (Siegmund et al., 1992 a). Compounds 6 and 7 of Figure 2 are some typical examples of this category. However, since the chemical environment for the fluorophores is much different within polyionic multilayers compared to amphiphilic LB multilayers, the system needed to be re-optimized. It is generally known and has been observed by us as well that fluorescence dyes in high local densities are prone to self-quenching, both as monomers dissolved and polymer-bound. When arylsulfonates were used as anionic components in conjunction with coumarin monomers, monolayers prepared from these polymers showed a fluorescence comparable with LB monolayers containing the same fluorophore. Though we did not investigate their behavior further, an explanation for this may be that the fluorophores are kept spatially separated within the polymer film by the repelling forces of the charged groups. On the other hand, the density of coumarin fluorophores can be brought to somewhat higher values with the lighter ionic groups compared to using the heavier amphiphilic side chains. Whereas the weight percentage for the coumarin-containing monomer in compound 5 is 22%, the values are 25% or 33% for compounds 6 and 7, respectively.

The high physico-chemical stability of this kind of SA layers has been demonstrated by Lvov et al. (1993). We were able to confirm their results for our functionalized derivatives, varying pH, temperature, ion strength in the same way as it was done for LB films. Ellipsometry and reflectometry show a regular increase in film thickness, depending on the number of layers. They survive the conditions used for antibody-binding studies, like incubation and washing steps. In addition, fluorescence microscopy studies performed on layers made from fluorophore-containing materials, like compounds 6 or 7, revealed that a homogeneity comparable to LB films can be obtained.

The substrate materials carrying the films were selected to be free of intrinsic fluorescence, to avoid quenching of the desired fluorescence, to provide a surface suitable for the application of monomolecular films, and to be translucent for having the highest possible flexibility to perform a fluorescence assay. Typically, floatglass (microscopic slides) or pieces of macroscopic polymer films of several hundred micrometer thickness were chosen, like polycarbonate (Makrofol®). When started with a polylysine layer, self assembly worked excellently on vinylidene chloride surface-activated poly(ethylene terephthalate) which is used as substrate for manufacturing photographic films (Agfa AG). These substrates can be covered with another piece of glass or film, thus forming a cuvette-type arrangement for applying and measuring liquid samples (Siegmund and Becker, 1993).

Rhodamine derivatives were generally chosen as the best-matching Förster acceptor dyes. Some other fluorescent dyes, like fluorescein derivatives, Lucifer yellow, and other coumarin derivatives, were tried with limited success. Reactive rhodamine derivatives ready to react with primary amino groups, like isothiocyanates, are commercially available from different sources. Proteins, like antibodies, as well as low molecular weight analyte molecules carrying at least one amino group can be labeled in this way using standard suppliers' procedures. Tetramethyl rhodamine B isothiocyanate (TRITC) has an excitation wavelength of 552 nm and emits at 575 nm, thus is capable of being a Förster transfer acceptor for the coumarin fluorophores of the compounds **1** to **7** (Fig. 2) which are excited at 405 nm and emit at 497 nm. Since this emission peak is rather broad (on the order of 100 nm, see spectra below), there is enough spectral overlap between the two fluorophores.

Immobilizing biomolecules on ultrathin films

We have investigated two major scenarios: either a low molecular weight analyte or a receptor protein, like an antibody, is immobilized to an ultrathin film that covers the sensor surface. There are several practicable methods for both of these scenarios. An easy and rather flexible method for immobilizing either type of biomolecules is the use of hydrophobic "anchors" in connection with LB films. This has been done for the mannose/concanavalin A system (Siegmund and Becker, 1993) as well as for digoxin/antibody, using N-hydroxysuccinimidyl stearate as the "anchor". However, the physico-chemical stability of this immobilization method proved to be insufficient, particularly when trying to immobilize proteins.

Another way is to incorporate reactive groups into the polymers that from the organic films. As described above for the fluorophores, a number of functional groups known to react with proteins can be introduced – either by copolymerization or by a polymer analog reaction. This technology has been used in the past for various purposes, e.g., for the modification of macroscopic membranes or for chromatography column materials. A review by Scouten (1987) gives a rather broad overview on this subject and describes the basic strategies of immobilization. To these reactive polymers, biomolecules can be coupled either before or after the films are cast. Carrying out the reaction in the bulk phase before the film is cast is usually only practicable for the immobilization of low molecular weight analyte analogs, like mannose, digoxin or biotin. However, for immobilizing antibodies, as commonly known, it is necessary to not only bind them to the sensor surface, but to do so with the right orientation. Antibodies are Y-shaped molecules, with the bindings sites at both top ends of the "arms" of the "Y". Thus, a desired orientation is to immobilize them at the "stem" of the "Y" with the "arms" pointing up. The following methods turned out to be successful in our hands: (a) the immobilization of protein A, an antibody-binding protein, in a first step, then followed by the application of antibodies; (b) the activation of the antibodes by a periodate cleavage reaction, then binding them to hydrazide group bearing surfaces; (c) an initial immobilization of streptavidine, followed by the application of biotinylated receptors. Binding studies observed by Förster energy transfer using either of the above methods are described below.

Figure 3 shows some example of polymers used for protein immobilization on LB and SA layers; a typical procedure for their synthesis has been described by Heiliger and Siegmund (1993). The "essentials" they all consist of are a reactive group and a part needed for the film formation. For LB films, as described previously by Embs et al. (1991), the latter is a close to equal ratio of hydrophilic and long-chain hydrophobic, preferably stearyl, side groups. In the case of polyionic SA films, the film-forming part has to be ionically charged. Such polymers can carry several different functionalities, like fluorophores and reactive groups, at the same time. We made successful attempts to synthesize such species, like compounds **8** to **11** (Fig. 3) for use in later experiments. For initial evaluation purposes, however, we found it to be more advantageous to have the flexibility of quickly combining various types of reactive groups with different fluorophores without having to go through a series of syntheses each time. For LB films only, this can be done by mixing the corresponding polymeric amphiphiles in solution before spreading them on a film balance to prepare them, in analogy to the method described by Kuhn et al. (1972) for monomeric amphiphiles. For LB and SA films, we frequently prepared "sandwich" assemblies, where a fluorophore-bearing layer is positioned closely underneath the top layer, which itself carries the reactive groups (see below, e. g., Fig. 5).

For the direct immobilization of underivatized proteins, we preferred to use N-hydroxysuccinimidyl, nitrophenyl ester, isocyanate, and isothiocyanate groups (Heiliger and Siegmund, 1993). They react with free amino groups of the proteins, assumed to be usually those of lysine side chains. The use of longer spacer groups turned out to increase the activity of the layers. The polymers **10** and **11** are examples carrying semicarbazide groups for the immobilization of periodate-treated proteins. This treatment cleaves glycosyl side chains at the "stem" part of the antibody-forming aldehyde and ketone groups, which then form semicarbazones (Scouten, 1987) with the exposed active groups of a monomolecular layer.

We performed several experiments to examine whether the immobilization of antibodies an ultrathin layers was covalent or non-covalent. Although various surface-sensitive physical analytical methods (TOF-SIMS, ESCA, AFM, SAXS, SPR, reflectometry, and electrokinetic measurements) indicated that substrates were covered with antibodies, we needed conclusive proof that immobilization really was covalent. Applying immunoanalytical methods, we eventually succeeded in verifying the covalent immobilization of periodate-oxidized mouse monoclonal antibodies that had been previously coupled to semicarbazide groups of the reactive polymer **11**.

8

9

10

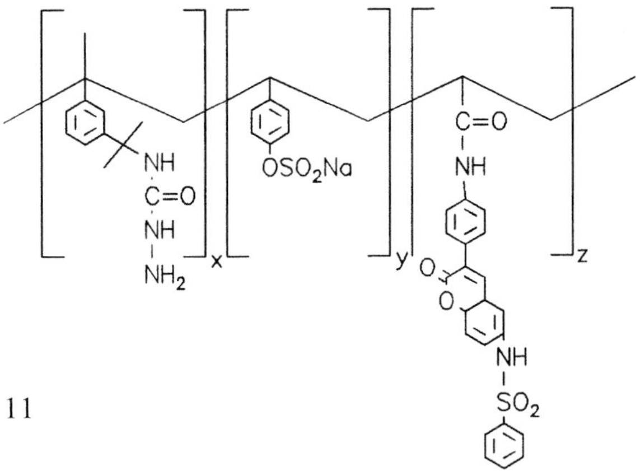

Figure 3. Polymers used to form monomolecular organic films carrying reactive groups to immobilize biomolecules. w = 1; x = 0,8–1.1 (for 8, 10) or x = 1 (for 9, 11); y = 0.1–0.4; z = 0.05 to 0.25.

First, we used an enzyme-linked immunosorbent assay (ELISA) to detect the immobilized antibody by a second peroxidase-labeled anti-mouse antibody. This result was then confirmed by capillary electrophoresis of antibody polymer conjugates prepared in solution. Unspecific binding between polymer and protein was excluded by the addition of sodium dodecyl sulfate (SDS) before capillary electrophoresis.

The antibody density on a substrate covered with an ultrathin polymer film is limited by the amount of reactive groups available in the copolymer. Due to synthetic restraints, however, it can become difficult to increase the fraction of reactive groups beyond a certain level, depending on the solubility of the reactive co-monomer and on its reactivity in the polymerization reaction. To avoid these problems, use of the system biotin/streptavidin can lead to a higher antibody density on the surface. Streptavidin is a protein which comprises four identical subunits, each capable of binding one biotin molecule. The binding constant for the couple is on the order of 10^{15} L · mol^{-1}, which is close to the strength of a covalent bond, thus rendering the connection between biotin and streptavidin virtually irreversible. The specific interaction between biotin and streptavidin on monolayers at the gas/water interface has been examined in detail by Ahlers et al. (1990), and this group also described the use of biotinylated antibodies docked to streptavidin for immunosensoric purposes (Müller et al., 1993). Self-assembled monolayers with biotinylated polylysine, however, have been previously described by Decher (1993). Immobilization of streptavidin on a biotinylated polylysine layer produces a well-organized two-dimensional streptavidin monolayer. Each streptavidin in this protein layer has two free binding sites, thus forming a bioreactive docking matrix (Fig. 4).

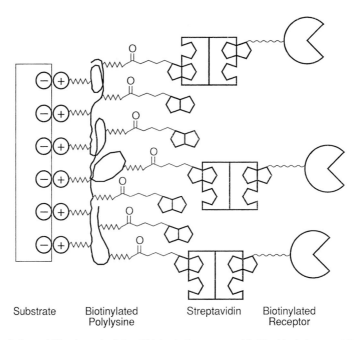

Substrate   Biotinylated   Streptavidin   Biotinylated
             Polylysine   Receptor

Figure 4. Immobilization principle of biological receptors aided by biotin/streptavidin.

Energy transfer experiments to monitor biospecific binding

For an initial evaluation of the system using biospecific binding studies, we
chose the lectin concanavalin A (Con A) as a receptor for mannosides
(Siegmund et al., 1992b). Con A was immobilized as a receptor onto a
mixed LB film containing the amphiphilic polymer **4** and N-hydroxysuccinimidyl stearate as the reactive group. A TRITC-labeled mannoside was
used as a ligand, causing a Förster energy transfer between the coumarin in
polymer **4** and the TRITC when it binds to Con A. Figure 5 shows the
fluorescence spectra obtained compared with the control experiment, where
Con A was substituted by bovine serum albumin (BSA) which does not
possess specific binding properties against mannosides. It can be clearly
seen that the TRITC band at 580 nm appears only in a significant intensity
when the specific receptor molecule is present. An addition of α-methyl
mannoside caused the acceptor fluorescence to disappear, due to a dissociation of the labeled mannoside from the lectin. A competitive replacement like this could be used for the determination of unlabeled analyte concentration.

   To extend the principle to an immunosensor, we chose the detection of
digoxin as a model system (Siegmund et al., 1992a). The steroid digoxin is
a drug used for the treatment of congestive heart failure and other related

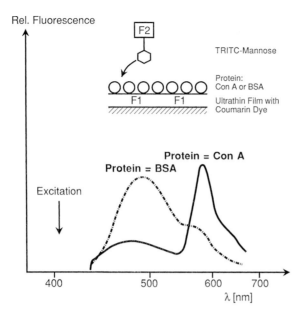

Figure 5. Binding of TRITC mannoside to Con A immobilized to an LB film, detected by Förster energy transfer (—). Coumarin donor (F_1) fluorescence at 495 nm, rhodamine acceptor (F_2) fluorescence at 577 nm. Fluorescence excitation was at 405 nm. Control experiment (– · –) with BSA instead of Con A. For details see Siegmund et al. (1992 b).

diseases. Its concentration in serum during therapy is critical (0.8 to 2 ng/ml), therefore a fast and easy monitoring will improve its effectiveness and avoid toxicity (Sommer et al., 1990). First, protein A was immobilized on a polymer substrate which was coated by a reactive organic polymer film (LB or SA) that also contained coumarin. To this surface, monoclonal digoxin antibodies were bound with the F_c-part resulting in an end-on orientation (Fig. 6, insert). Digitoxigenin, which is the aglycon of the glycoside digitoxin and easy to derivatize at its primary hydroxyl group at position 3, was labeled with TRITC to yield compound **12** (Fig. 7). It shows a close resemblance to digoxin and can be recognized by the antibody. After excitation at 405 nm, the fluorescence of the acceptor dye (TRITC) appears at 580 nm in addition to the fluorescence of the donor dye (F_1, coumarin) at 495 nm (Fig. 6). The incomplete quenching of the donor dye fluorescence is likely to be caused by the relatively large distance between the dyes. In a control experiment, BSA was immobilized instead of protein A, and no significant acceptor fluorescence appears in this case.

Advancing from this simple design, we did time course experiments in a flow cell cuvette. Two SA layers were transferred to the inner surface of a cuvette (Helma 136-QS, 0.1 mm path length). The first layer was the polycation polylysine hydrobromide, and the second layer was made of the polyanion **11**. The digoxin antibody was immobilized with protein A

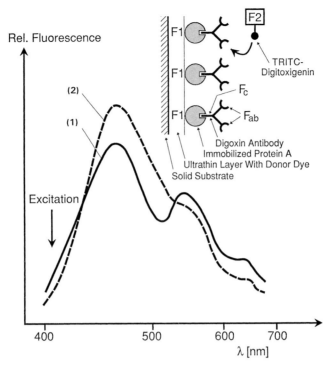

Figure 6. Förster energy transfer of anti-digoxin/TRITC-digitoxigenin. (1) Digoxin antibody, or (2) BSA were immobilized on polymer **11**. TRITC digitoxigenin **12** was then added in both cases.

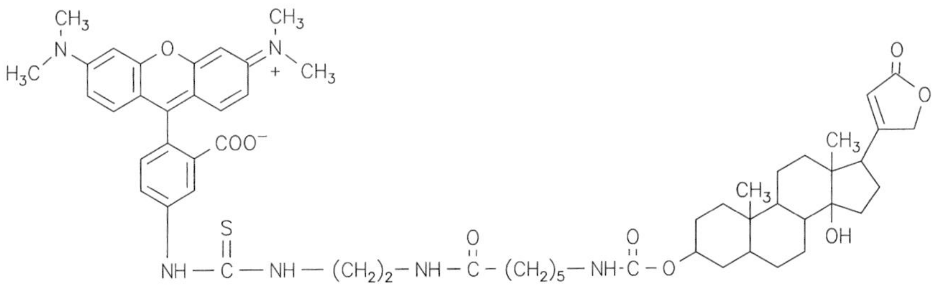

12

Figure 7. TRITC-labeled digitoxigenin derivative.

in the same way as described above and shown in Figure 6 (insert). Subsequently, carbonate buffer, a solution of TRITC-labeled digitoxigenin **12** in the same buffer, again buffer, and a second time a solution of **12** were pumped through the cuvette. As can be seen in Figure 8, the Förster ratio (i.e., the quotient of the fluorescence intensities of the acceptor dye F_2 and the donor dye F_1) increased when the labeled analyte **12** flowed through the cuvette. The signal increased again when **12** flowed a second time through the cuvette, indicating that the antibody was not yet saturated. However, considerable unspecific binding manifested by the peaks was also observed. Washing with buffer removed excess TRITC digitoxigenin, leaving the specific signal. In control experiments, either the non-specific antibody anti human serum albumin (HSA) or no antibody were used. In both cases, only unspecificity was observed, leaving no specific signal after being washed with buffer. However, the system proved to be difficult to use for a quantification of the analyte, since the unspecific background was generally too high, excessive washing steps could disrupt the film structure, and the immobilized receptor density appeared to be too low.

A step forward towards a quantitative sensor was achieved, however, when the system streptavidin/biotin was applied for the Förster energy transfer biosensor (Diederich, 1996). In an extention of the concept described above and shown in Figure 4, an immunosensor matrix was built up on a substrate primed with a polyionic SA multilayer, then coated

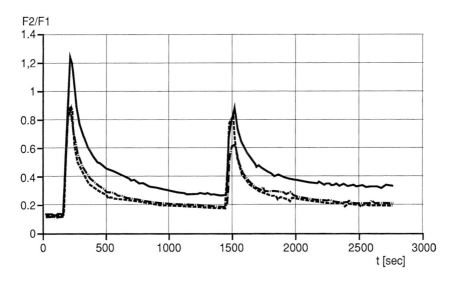

Figure 8. Flow cell experiment with TRITC digitoxigenin **12**. (——) anti-digoxin, (– –) no antibody, (–·–·) anti-HSA immobilized. Flow rate: 16.7 µl/sec of 10 µg/ml compound **12** in 0.01 mol/L carbonate buffer pH 8.

with biotinylated polylysine, streptavidin, and finally with biotinylated protein A to bind antibodies. To perform Förster transfer experiments with this assembly, one of the proteins had to be labeled with a donor dye F_1, otherwise the distance between donor and acceptor dyes F_2 would be too big. Fluorescein isothiocyanate (FITC) labeled protein A was used to immobilize anti mouse immunoglobulin G (IgG), and rhodamine isothiocyanate (RITC) labeled mouse IgG was detected. After excitation at 470 nm, the quotient of the fluorescence of the acceptor dye RITC at 577 nm and the donor dye FITC at 530 nm were measured. In a control experiment, no anti mouse IgG was used. The results of both measurements are shon in Figure 9. Förster ratios for the control are much lower, demonstrating the applicability of the system for Förster energy transfer biosensing.

A compact fluorescence detector

In order to use the above system in a biosensor or a benchtop instrument, a simple and inexpensive apparatus was developed; a block diagram is depicted in Figure 10 (Siegmund and Becker, 1993). One excitation and two detection channels are used. After passing through a heat-absorbing glass, the light of a 75 Watt xenon lamp is coupled into a waveguide and then focused on the sample through an interference filter which is chosen to get optimum excitation of the donor dye at λ_1 (see Fig. 1 for wavelength definition). The fluorescence is detected with two

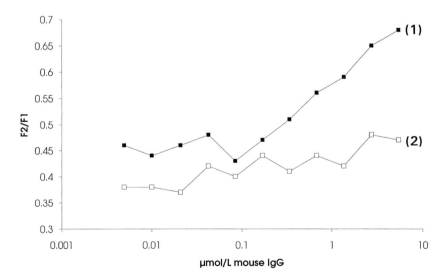

Figure 9. Förster transfer ratio versus concentration of mouse IgG. (1) with anti mouse IgG, (2) control without anti mouse IgG.

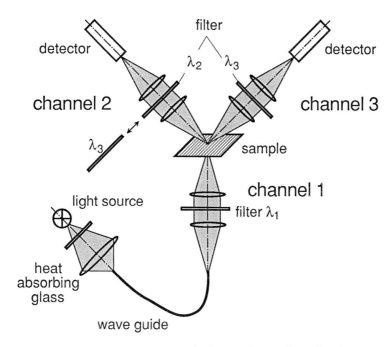

Figure 10. Block diagram of the Förster transfer detector. See text for explanations.

channels, where the wavelength is again selected by narrow band pass filters with maximum transmission at λ_2 and λ_3, the maximum emissions of the donor and the acceptor dye, respectively. The fluorescence is focused on photomultiplier tubes to record the photocurrent. An alternative arrangement is the use of only one detection channel, where the filters with maximum transmission at λ_2 and λ_3 are switched mechanically. The advantage of this is that only one channel has to be calibrated and the size of the instrument can be reduced. Switching mechanically, however, is slower than the simultaneous detection of both fluorescence intensities, prolonging the time needed for a single measurement. For flow cell measurements, like in Figure 8, two independent channels were preferred.

Perspectives

The above studies show that the combination of radiationless Förster energy transfer and monomolecular polymer films provide a promising way to detect the presence of biomolecules for which a specific receptor can be obtained. Though more work is still required in order to get reproducible and precise dose response, our current experimental state suggests that this is feasible as well. As it is generally true for immunosensors, the advances in monoclonal antibody preparation open up a wide field of potential applications for this technology, since receptors for a variety of analytes can be obtained. The frequently observed tendency of biological systems to self-organization, as seen in the biotin/streptavidin couple, can be of further help in sensor construction. An instrument based on this technology may be employed in a physician's office or for low volume routine laboratory testing.

Acknowledgements
The work presented was the result of a larger research team cooperation, and a considerable number of persons have contributed to it. The authors explicitly wish to thank Drs. A. Diederich and M. Lösche for cooperation concerning the system biotin/streptavidin and Drs. A. Becker, L. Heiliger, H. Hugl and D. Möbius for the helpful discussions and other contributions, as well as M. Averdick, A. Koreik and D. Riesebeck for their expert technical assistance. This work was funded in part by grants of the Bundesministerium für Forschung und Technologie (BMFT) within the project "Ultradünne Polymerschichten".

References

Ahlers, M., Müller, W., Reichert, A., Ringsdorf, H. and Venzmer, J. (1990) Spezifische Wechselwirkung von Proteinen mit funktionellen Lipidmonoschichten – Wege zur Simulation von Biomembranprozessen. *Angew. Chem.* 102:1310–1327; *Angew. Chem. Int. Ed. Engl.* 29: 1269–1285.

Decher, G. (1993) Supramolekulare Chemie: Ultradünne Schichten aus Polyelektrolyten. *Nachr. Chem. Tech. Lab.* 41:793–800.

Decher, G. and Hong, J.D. (1991) Buildup of Ultrathin Multilayer Films by a Self-Assembly Process: II. Consecutive Adsorption of Anionic and Cationic Bipolar Amphiphiles and Polyelectrolytes on Charged Surfaces. *Ber. Bunsenges. Phys. Chem.* 95: 1430–1434.

Diederich, A. (1996) *Dissertation*, University of Mainz, Strukturelle und funktionelle Untersuchungen von Proteinmomo- und -multischichten an technischen Grenzflächen.

Embs, F., Funhoff, D., Laschewski, A., Licht, U., Ohst, H., Prass, W., Ringsdorf, H., Wegner, G. and Wehrmann, R. (1991) Preformed Polymers for Langmuir-Blodgett Films – Molecular Concepts. *Adv. Mater.* 3:25–31.

Embs, F.W., Winter, H.H. and Wegner, G. (1993) Langmuir-Blodgett multilayer assembly by a continuous process using a steadily flowing subphase. *Langmuir* 9:1618–1621.

Fromherz, P. and Reinbold, G. (1988) Energy transfer between fluorescent dyes spaced by multilayers of cadmium salts of fatty acids. *Thin Solid Films* 160:347–353.

Fuchs, H., Ohst, H. and Prass, W. (1991) Ultrathin Organic Films: Molecular Architectures for Advanced Optical, Electronic and Bio-Related Systems. *Adv. Mater.* 3:10–18.

Gabor, G., Chadha, S. and Walt, D.R. (1995) Sensitivity enhancement of fluorescent pH indicators using pH-dependent energy transfer. *Anal. Chim. Acta* 313:131–137.

Heiliger, L. and Siegmund, H.-U. (1993) Beschichtete Träger, Verfahren zu ihrer Herstellung und ihre Verwendung zur Immobilisierung von Biomolekülen an Oberflächen von Festkörpern. *German Patent Application* (Bayer AG) DE 43 19 037.

Hugl, H., Kuckert, E., Möbius, D., Ohst, H., Rolf, M., Rosenkranz, H.J., Schopper, H.C., Siegmund, H.-U., Sommer, K. and Wehrmann, R., Optischer Biosensor. *German Patent Applications* (Bayer AG) DE 39 38 598 (1989), DE 40 13 713 (1990).

Kauffmann, F., Hoffmann, B., Erbach, R., Heiliger, L., Siegmund, H.-U. and Völker, M. (1994) Ca^{2+} sensor with amphiphilic Langmuir-Blodgett membranes. *Sensors Actuators B* 18–19: 60–64.

Kuhn, H. (1970) Classical aspects of energy transfer in molecular systems. *J. Phys. Chem.* 53:101–108.

Kuhn, H., Möbius, D. and Bücher, H. (1972) Spectroscopy of Monolayer Assemblies. *In*: A. Weissberger and B. Rossiter (eds): *Physical Methods of Chemistry*, Vol. 1, Part 3B. John Wiley, New York, pp 577–702.

Lvov, Y., Decher, G. and Möhwald, H. (1993) Assembly, Structural Characterization, and Thermal Behavior of Layer-by-Layer Deposited Ultrathin Films of Poly(vinyl sulfate) and Poly(allylamine). *Langmuir* 9:481–486.

Müller, H., Ringsdorf, H., Rump, E., Wildburg, G., Zhang, X., Angermaier, L., Knoll, W., Liley, M. and Spinke, J. (1993) Attempts to Mimic Docking Processes of the Immune System: Recognition-Induced Formation of Protein Multilayers. *Science* 262:1706–1708.

Scouten, W.H. (1987) A Survey of Enzyme Coupling Techniques. *In*: K. Mosbach (ed): *Methods in Enzymology*, Vol. 135. Academic Press, New York, pp 30–65.

Siegmund, H.-U., Becker, A. and Möbius, D. (1991) Characterization and Optimization of Langmuir-Blodgett Films Containing Polymer-Bound Fluorophores. *Adv. Mater.* 3: 605–608.

Siegmund, H.-U., Heiliger, L., van Lent, B. and Becker, A. (1992a) Optischer Festphasenbiosensor auf Basis fluoreszenzfarbstoffmarkierter polyionischer Schichten. *German Patent Application* (Bayer AG) DE 42 08 645.

Siegmund, H.-U., Becker, A., Ohst, H. and Sommer, K. (1992b) An optical biosensor principle based on fluorescence energy transfer. *Thin Solid Films* 210/211:480–483.

Siegmund, H.-U. and Becker, A. (1993) A new way of biosensing using fluorescence energy transfer and Langmuir-Blodgett films. *Sensors Actuators B* 11:103–108.

Sommer, R.G., Belchak, T.L., Bloczynski, M.L., Boguslawski, S.J., Clay, D.L., Corey, P.F., Folz, M.M., Fredrickson, R.A., Halmo, B.L., Johnson, R.D., Marfurt, K.L., Runzheimer, H.-V. and Morris, D.L. (1990) A Unitzed Enzyme-Labeled Immunometric Digoxin Assay Suitable for Rapid Testing. *Clin. Chem.* 36:201–206.

Stryer, L. (1978) Fluorescence energy transfer as a spectroscopic ruler. *Annu. Rev. Biochem.* 47:819–846.

Thompson, R.B. and Patchman, M.W. (1995) Lifetime-Based Fluorescence Energy Transfer Biosensing of Zinc. *Anal. Biochem.* 227:123–128.

Tippmann-Krayer, P., Riegler, H., Paudler, M., Möhwald, H., Siegmund, H.-U., Eickmans, J., Scheunemann, U., Licht, U. and Schrepp, W. (1991) Thermostability of Polymeric Langmuir-Blodgett Films. *Adv. Mater.* 3:46–51.

Frontiers in Biosensorics I
Fundamental Aspects
ed. by F.W. Scheller, F. Schubert and J. Fedrowitz
© 1997 Birkhäuser Verlag Basel/Switzerland

Coupling of enzyme reactions to the charge transfer at the interface of two immiscible solvents

M. Senda

Department of Bioscience, Fukui Prefectural University, Matsuoka-cho, Fukui 910-11, Japan

Summary. Electrochemical principle of ion-selective electrodes (ISEs) based on the ion-transfer reactions across a polarizable organic or oil/aqueous or water interface is described; the amperopmetric ISE and the potentiometric ISE are addressed. Electrochemical sensors and biosensors based on amperometric ISEs are discussed in some details. Amperometric sensors for monitoring ammonia (and other volatile amines) can be constructed on the basis of amperometric ammonium-ISE, where the pulse amperometric technique can successfully be employed. Urea and creatinine biosensors can also be fabricated by immobilizing urease or creatinine deiminase, respectively, on the surface of the amperometric ammonia sensor. Somme favored characteristics of amperometric ISE-based sensors and biosensors are discussed with reference to the sensors and biosensors described.

Introduction

The potentiometric enzyme electrode, or sensor, a combination of a poten-tiometric ion-selective electrode or a semiconductor field-effect device with an immobilized enzyme, has been extensively studied and has found applications in environmental, biological and clinical analysis (for a review: Kuan and Guilbault, 1987; Campanella and Tomassetti, 1990; Winquist and Danielsson, 1990). In contrast, this review is concerned with the *amperometric* enzyme electrode, or sensor, that is an amperometric ion-selective electrode-based sensor and biosensor.

Recent electrochemical studies on ion-transfer reactions across the inter-face between two immiscilbe electrolyte solutions, or, in short, the oil/water (O/W) interface, like the nitrobenzene/water or 1,2-dichloroethane/water interface, have shown that this interface can be electrochemically polarized and that the transfer of ions that takes place across the O/W inter-face within the polarizable potential range, or through what is called the potential window, can be studied using voltammetric or polarographic tech-niques. Thus, the polarizable O/W interface can function as an electrode interface which responds voltammetrically to a specified ion (or ions) that is (or are) transferable across the interface throught the potential window. In other words, we have obtained an ion-selective electrode based on a pola-rizable O/W interface. According to the theory of voltammetry, there are two available types of ion-selective electrodes; an amperometric ion-selec-tive electrode and a potentiometric ion-selective electrode. The former gives a current response which is proportional to the concentration of the analyte ion, whereas the latter gives a potential response which changes linearly with the logarithm of the concentration (strictly speaking, the activity) of the analyte ion (for a review, Senda et al., 1989, 1991; Senda, 1995; Girault, 1993). In this review, the electrochemical principle of ion-selective electrodes based on the ion-transfer reactions across a polarizable O/W interface is briefly described, and electrochemical sensors and biosensors based on amperometric ion-selective electrodes are discussed in some detail.

Electrochemical principle of an ion-selective electrode (ISE). Amperometric ISE vs potentiometric ISE

We first consider an electrochemical cell,

$$R_1 \mid B_1^+, A_1^- \ (O) \mid B_2^+, A_2^- \ (W) \mid R_2, \tag{I}$$

where (O) represents the organic or oil phase and (W) the aqueous or water phase, and the interface indicated by * is the O/W interface to be polairzed. B_1^+ and B_2^+ and A_1^- and A_2^- are the supporting electrolyte cations and anions, respectively, and R_1 and R_2 are the reference electrodes. An electric fields is applied across the O/W interface by the two reference electrodes. The cell potential, E, is defined as the terminal potential of the right-hand reference electrode (R_2), referred to that of the left-hand one (R_1), and is related to the potential difference across the O/W interface, $\Delta\phi \ (= \Delta_o^w\phi = \phi^w - \phi^o$, ϕ^α being the Galvani potential of the α-phase ($\alpha = o, w$)) according to

$$E = \Delta\phi - \Delta E_{ref}. \tag{1}$$

Here, ΔE_{ref} is determined by the compositon of the reference-electrode system of the electrochemical cell, here R_1 and R_2 in cell I. When ions B_1^+ and A_1^- are extremely hydrophobic, like tetrabutylammonium (TBA$^+$) and tetraphenylborate (TPB$^-$) ions, while ions B_2^+ and A_2^- are extremely hydrophilic, like Li$^+$ and Cl$^-$ ions, and as long as the potential difference across the O/W interface does not exceed a certain range of positive and negative magnitudes, the transfer of B_1^+ and A_1^- from O to W and that of B_2^+ and A_2^- in the opposite direction should be negligible; hence, no current will flow across the interface (Fig. 1 A, where the positive (or anodic) current stands for the transfer of cations from W to O and of anions from O to W and the negative (or cathodic) current for the transfer of ions in the opposite direction). Thus, the O/W interface behaves as an ideal-polarized interface in this range of potential difference, which we call the polarizable potential range, or the potential window of the O/W interface.

We consider the second case when a moderately hydrophilic or semi-hydrophilic (that is, moderately hydrophobic or semi-hydrophobic) ion j is present in one of the two phases, e. g., a semi-hydrophilic cation B_3^+, like tetramethylammonium (TMA$^+$) ion, in the aqueous phase (W),

$$R_1 \mid B_1^+, A_1^- \ (O) \mid B_3^+, B_2^{+,} A_2^- \ (W) \mid R_2. \tag{II}$$

*

(amperometric ISE) (test solution)

Then, the transfer of the j-ion, here the B_3^+ ion, across the interface takes place at an intermediate potential within the polarizable potential range of cell I; the current, i, associated with the transfer of the j-ion is observed in this polarizable potential range. The kinetics of the ion transfer at the interface can be expressed, in analogy with ordinary electrode kinetics, by

$$i = z_j FA \left(k_f c_j^{w,s} - k_b c_j^{o,s} \right), \tag{2}$$

where z_j is the number, including the sign, of the charge of the j-ion, F the Faraday constant, A the interface area, $c_j^{\alpha,s}$ the surface concentration of the j-ion in the α-phase ($\alpha = o, w$), and k_f and k_b are the rate constants for the transfer of j-ion from W to O and from O to W, respectively. The rate constants are functions of $\Delta\phi$, and may be expressed by a Butler-Volmer type equation, as given by

$$k_f = k_s \exp \left[(\alpha z_j F/RT)(\Delta\phi - \Delta\phi^0) \right]$$

and $\hspace{10cm}$ (3)

$$k_b = k_s \exp \left[(-\beta z_j F/RT)(\Delta\phi - \Delta\phi^0) \right]$$

where k_s is the standard rate constant, that is, the rate constant at $\Delta\phi = \Delta\phi_j^0$, $\Delta\phi_j^0$ being the standard (or, strictly speaking, formal) potential of the j-ion transfer at the O/W interface, α and β ($\alpha + \beta = 1$) are the transfer coefficients, and R and T have their usual meanings. The standard potential is related to the standard Gibbs energy of transfer of the j-ion from O to W, $\Delta_o^w G_{tr,j}^0$, by

$$\Delta\phi_j^0 = - \Delta_o^w G_{tr,j}^0 / z_j F. \tag{4}$$

Thus, the ion-transfer current across the O/W interface is observed at about the standard potential or the half-wave potential within the polarizable potential range (Fig. 1 B, where the ion transfer current vs potential ($i - \Delta\phi$) curve is represented by a polarographic wave with the half-wave potential, $\Delta\phi_{1/2}$, and the limiting current, i_1). In cell II, the left-hand half-cell $R_1 \mid B_1^+$, A_1^- (O) $\mid$ constitutes an amperometric ISE which is immersed in an external

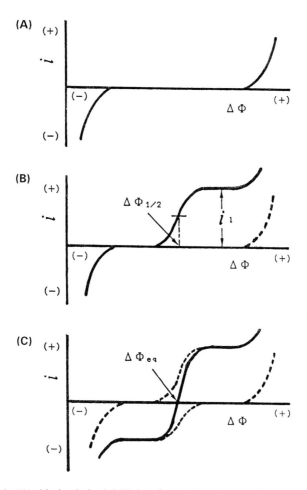

Figure 1. (A): Ideal-polarized O/W interface; (B): Voltammetric (polarographic) current vs potential curve; (C): Composite positive-negative (or anodic-cathodic) current vs potential curve and equilibrium potential at zero current (Senda and Yamamoto, 1996, after modification, with permission).

or test solution containing an analyte ion $|B_3^+, B_2^+, A_2^- (W)|$ with a reference electrode $|R_2$. The voltammetric current, usually observed as the limiting current in polarography or the peak current in potential sweep voltammetry etc., is proportional to the concentration of the analyte ion.

We consider the third case when a semi-hydrophilic (or $-$ hydrophobic) j-ion, here B_3^+ ion, is present in both phases,

$$R_1 \mid B_1^+, A_1^-, B_3^+ \, (O) \mid B_3^+, B_2^+, A_2^- \, (W) \mid R_2. \tag{III}$$

(potentiometric ISE)   (test solution)

We then observe the composite positive-negative current wave of the transfer of j-ion from W to O and from O to W (Fig. 1C); the zero current potential or equilibrium potential, $\Delta\phi_{j,eq}$ (at $i = 0$ in Eq. 2), should be given by Nernst equation,

$$\Delta\phi_{j,eq} = \Delta\phi_j^0 + (RT/z_jF)\ln[c_j^o/c_j^w]. \tag{5}$$

Therefore, the *emf*, that is, the cell potential for cell III at $i = 0$, E_{eq}, is given by $E_{eq} = \Delta\phi_{j,eq} + \Delta E_{ref}$ (see Eq. 1). In cell III, the left-hand half-cell $R_1|B_1^+, A_1^-, B_3^+ (O)|$ constitutes a potentiometric ISE which is immersed in an external or test soluton containing an analyte ion $|B_3^+, B_2^+, A_2^- (W)|$ with a reference electrode $|R_2$. Thus, the potentiometric ISE gives the potential response at the zero-current control, which changes linearly with the logarithm of the concentration (activity) of the analyte ion (for a given, constant concentration of the analyte ion, here B_3^+ ion, in the organic phase of the ISE). In this chapter, for the sake of simplicity, the activity of ions is equated to their concentration, unless otherwise stated.

In many cases, in order to make the O/W interface selectively responsive to a specified ion M, a hydrophobic ionophore L which associated selectively with the specified ion M to form a hydrophobic complex LM (and, in some cases, L_nM ($n = 2...$) etc.) is added in the organic phase. Then, the transfer of the M-ion from W to O is facilitated by the formation of the complex LM in O,

$$M(W) + L(O) = ML(O). \tag{6}$$

Usually, the (formal) formation constant of the complex, K^o, defined by $K^o = c_{LM}^o/c_L^o c_M^o$, is set to be much larger than unity so that the concentration of the free M-ion in the O-phase is negligibly small compared with that of the complex ML, that is, $c_{ML}^o \gg c_M^o$. Then, the standard potential of the transfer of the M-ion at the O(containing L)/W interface should be replaced when $c_L^o \gg c_M^o$ (and usually also $c_L^o \gg c_M^w$) by

$$\Delta\phi_M^0 = \Delta\phi_M^0 + (RT/z_MF)\ln[1/(c_L^o K^o)]. \tag{7}$$

Consequently, the standard potential (and hence the half-wave potential etc.) of the M-ion transfer at the O(containing L)/W interface is selectively shifted to a less positive (when $z_M > 0$, or less negative when $z_M < 0$) potential within the potential window to make the interface selectively responsive to the M-ion present in the W-phase.

As stated above, amperometric ISE gives a current response, I, which is proportional to the concentration of the analyte ion(s) in a test solution, c_M. The current response is measured under the control of the potential applied to the electrode interface, E_{app}, which may be conventionally represented by the half-wave potential, $E_{1/2}$, in polarography or the peak potential, E_p, in potential sweep voltammetry and so on,

$$I = k_{amp} c_M \quad \text{at} \quad E = E_{app} \text{ (or } E_{1/2}, E_p). \tag{8}$$

where k_{amp} is a constant which depends on the experimental conditions, such as the mode of the potential control and current measurement, the design of the electrode and cell, etc. On the other hand, the potentiometric ISE gives a potential response, E, which changes linearly with the logarithm of the concentration (activity) of the analyte ion in the test solution, c_M, where the current following across the electrode interface, I, is usually controlled at zero,

$$E = k_{pot} \log [c_M] + b \quad \text{at} \quad I = 0, \tag{9}$$

where k_{pot} and b are the constants and $k_{pot} = 2.303 \, RT/z_M F$ for a nernstian response.

Porentiometric ion-selective electrodes and their analytical applications have been extensively studied and well developed, while amperometric ion-selective electrodes based on a polarizable O/W interface are rather new types of ISE; their analytical applications have only started recently. Some of the important properties of sensors and biosensors based on the amperometric ISE are discussed in the following.

Amperometric ISE sensors

The left-hand half-cell of cell II represents an amperometric ISE that consists of an organic phase (O) and can be used for monitoring an analyte, here B_3^+ in the aqueous phase (W). Likewise, an amperometric ISE that consists of an aqueous phase (W) for monitoring an analyte in the organic phase (O) can be constructed (Osakai et al., 1984). The former is occasionally called an organic-solvent electrode, like a nitrobenzene electrode, and the latter a water electrode. For practical purposes, organic solvents of lower vapor pressure, such as o-nitrophenyloctylether, o-nitrophenylphenylether, and 2-fluoro-2′-nitrodiphenylether (Sawada et al., 1990), can be used in place of nitrobenzene, though usually at the cost of lowering the conductivity of the organic phase of the electrode by about one tenth at the same supporting electrolyte concentration. Also, it is common to stabilize the liquid electrode by gelation; a nitrobenzene poly(vinylchloride) gel electrode and a water agar-gel electrode have been studied (Osakai et al., 1984). Also, stabilization of the liquid electrode interface by placing a thin, hydrophilic membrane at the interface has proved to be very useful in constructing amperometric ISE sensors (Hundhammer et al., 1987; Yamamoto et al., 1989, 1990).

The fabrication of an amperometric ultramicro ISE sensor is feasible by constructing a polarizable O/W interface at the tip (a few tens μm in diameter) of a micro glass-pipette (Senda et al., 1987; Ohkouchi et al., 1991). Unfortunately, in vivo application of an amperometric ultramicro ISE sensor, for exmaple, for acethylcholine, was unsuccessful, mainly because of the lack of sufficient sensitivity (Miyazaki, unpublished data). Amperometric ultramicro-hole ISE sensors also appear to be promising (Osborne et al., 1994).

Some important properties of ion sensors based on amperometric ISE are discussed in the following by taking a laboratory-made ammonium ion sensor as an example.

Ammonium ion sensor

An amperometric ammonium ion-selective electrode (NH_4^+-ISE) can be constructed by adding an ionophore of ammonium ion in the organic phase (Osakai et al., 1987; Osborne and Girault, 1995 a). In this study, an iono-phore dibenzo-18-crown-6 ether (DB18C6) was used. Thus, the electro-chemical cell for an amperometric determination of ammonium ion or ammonia gas with this NH_4^+-ISE is represented by

$$*$$

Ag | AgCl | 0.1 M TBACl | 0.1 M TBATPB | | 0.05 M MgCl$_2$ | | Test Soln.

                              0.02 M DB18C6   0.05 M L-Lys   (pH 9.0)

          (W)   (NB)   HSM ↑   (W)   GPM

                           Ag | AgCl-   (IV)

← (GPM-covered amperometric ammonium ion sensor) →

(Yamamoto et al., 1989) where TBACl and TBATPB are tetrabutylammonium chloride and tetrabutylammonium tetraphenylborate, respectively. The polarizable O/W interface (indicated by *) is stabilized by placing a hydrophilic semipermeable membrane (HSM, a dialysis membrane of 20 μm thickness) at the interface. The NB (containing DB18C6)/W interface is also responsive to both potassium and sodium ions ($E_{1/2}(K^+) = -132$ mV and $E_{1/2}(Na^+) = -14$ mV whereas $E_{1/2}(NH_4^+, pH 7.3) = -21$ mV vs $E_{1/2}((CH_3)_4N^+)$, at $c_i^o = 0.05$ mM). Therefore, in order to eliminate any interference caused by soidum, potassium, or other ions, the NH_4^+-ISE surface is covered by a gas-permeable membrane (GPM, a Teflon membrane 50 μm in thickness, Sumitomo Denko FP-200) with an inner solution of 0.05 M MgCl$_2$, 0.05 M L-lysine (pH 8.5) between the GPM and the polarized NB/W(HSM) interface. A counter reference electrode made of Ag/AgCl(W) is connected to the cell through the inner solution. The GPM-covered sensor is immersed in a test solution, usually of pH 9.0, into which an aliquot of the sample solution is added. The pulse amperometric technique (see below) can be used to record the current response of the sensor. A laboratory-made GPM-covered ammonium-ion sensor gave linear current response to the concentration of ammonium ion in the test solution (Yamamoto et al., 1989).

Pulse amperometry

In amperometric ISE the flow of the response current across the interface results in a change in the state of the electrolyte distribution at and near the electrode interface, which is a disadvantage of amperometric ISE sensors, expecially when the sensors are used for a long period. This can, however, be practically eliminated by means of a pulse amperometric technique (Fig. 2); the electrode potential is controlled first at the initial potential E_i at which a negligible ion-transfer current flows. After a fixed, relatively long waiting time T (e. g. 5 s) the potential is abruptly changed to E_{app} (= ΔE + E_i) at which an ion-transfer current flows for a short period τ (e.g. 100 ms). The potential pulse is ended by a return to the initial potential E_i and the electrode is kept at E_i until the next potential pulse is applied after a fixed waiting time and so on. Since the ion-transfer reaction at the O/W interface is generally reversible, the electrode interface returns to its original state at the end of the waiting time; thus, a highly reproducible current response can be obtained for a long term of measurement. The current is usually sampled at a time near to the end of the pulse, and a signal proportional to this sampled value is recorded. Thus, the recorded current vs potential

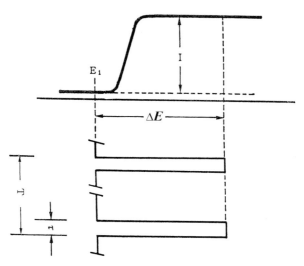

Figure 2. Pulse amperometry (Senda and Yamamoto, 1996, after modification, with permission).

curve is equivalent to a normal-pulse polarogram in polarography with a dropping mercury electrode. When the applied potential E_{app} is large enough to give a limiting current, the current signal is directly proportional to the concentration of the analyte ion in the test solution (W), or the surface concentration of the analyte ion at the W-side of the O/W (HSM) interface when the interface is stabilized by placing an HSM.

Current sensitivity and response time

The amperometric ISE sensors with the above-stated ammonium ion sensor usually gave linear calibration curves up to 0.4 mM (the upper limit of the concentration of the linearity should be improved by increasing the concentration of the ionophore in the organic phase). The relative standard deviation was usually a few % or less, and the detection limit was $1-2 \cdot 10 \, \mu M$ (Yamamoto et al., 1989, 1990).

The response time of the amperometric ISE sensor, in which the ion-selective O/W interface is stabilized by placing an HSM at the interface, is determined by the diffusion process of ions within the HSM and the possible stagnant layer on the membrane-covered electrode surface; it was typically about 20 s with the pulse amperometric technique. The response time of the above-stated laboratory-made ammonium ion sensor was typically about 60 s, which should be attributable to the diffusion process across the GPM and the inner solution layer.

When two or more ion components which give the current response at one and the same amperometric ISE are present in a test solution, the ISE sensor gives the current responses of the components each independently at their half-wave potentials. Therefore, if their half-wave potentials are reasonably separated, the concentration of these components can be simultaneously determined with one and the same ISE sensor (see below) (Yamamoto et al., 1990).

Volatile amine sensors

The laboratory-made ammonium-ion sensor discussed above also gives a current response to the ammonium ions of other volatile amines, like trimethylamine or trimethylammonium ion. Therefore, the GPM-covered volatile amine sensor can be used to quantify the volatile amine content in foods. Since trimethylammonium ion also gives an ion-transfer current at an NB(without ionophore)/W interface, a separte determination of ammonia and trimethylamine in foods can be made by using two sensors: one based on the NB(with ionophore)/W interface and the other on the NB(without ionophore)/W interface. The former gives the sum of the current responses of two amines, and the latter that of only trimethylamine (Yamamoto et al., 1989; Senda and Yamamoto, 1993b). The additivity of the response in amperometric ISEs is also advantageous in other applications of ampero-metric ISE sensors (see below).

Amperometric ISE biosensors

Enzyme sensors or biosensors based on amperometric ISE can be designed in nearly the same way as biosensors based on other electrochemical devices that have been extensively developed in recent years (Turner et al., 1987; Wise, 1990). These applications involve enzymatic reactions, which usually take place in an immobilized-enzyme layer on an electrode surface. The progress of the reaction is typically monitored by measuring the rate of formation of a product or the disappearance of a reactant. If the product or the reactant is electroactive (that is, giving response on the amperometric ISE), its concentration may be directly monitored. In the following, some important features of biosensors based on amperometric ISE are discussed taking a laboratory-made urea biosensor as an example.

Urea biosensor

The principle of a laboratory-made urea biosensor is shown in Figure 3 (Osakai et al., 1988; Senda and Yamamoto, 1993a, b; Yamamoto and Senda, 1993). Urease is immobilized on the surface of a GPM-covered ammonium ion sensor (cell V, below). Urea is transported by diffusion through the immobilized-enzyme layer where it is enzymatically hydrolyzed by

$$NH_2CONH_2 + H_2O \xrightarrow{\text{urease}} CO_2 + 2NH_3. \tag{10}$$

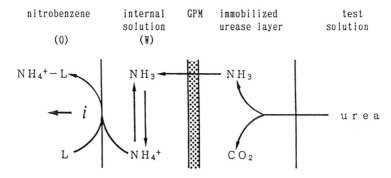

Figure 3. Principle of the amperometric urea biosensor (Senda and Yamamoto, 1996, with permission).

The product NH_3 is in reversible equilibrium with its dissociated ionic form, NH_4^+; these decomposition products are also transported by diffusion through the immobilized-enzyme layer while they are produced. The rate of the enzymatic reaction may be expressed by the Michaelis-Menten equation; when the substrate concentration, c_{urea}, is much lower than K_M, K_M being the Michaelis constant, the rate is proportional to the substrate concentration. Then, as a result of the diffusion process associated with the enzymatic reaction, it can be shown (Senda and Yamamoto, 1993a) that the (total) concentration of the products (that is, $NH_3 + NH_4^+$) on the GPM surface of the ammonium ion sensor is proportional to the concentration of urea on the surface of the immobilized-urease layer facing the test solution. Consequently, the current response of the urease-immobilized GPM-covered ammonium ion sensor is proportional to the concentration of urea.

The electrochemical cell for the amperometric determination of urea with the laboratory-made urea biosensor immersed in a test solution is represented by

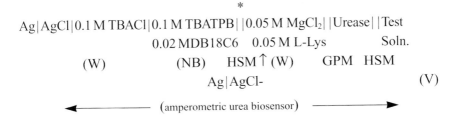

Jack bean urease, usually 100 U, was immobilized by covering the GPM surface of the ammonium ion sensor by a hydrophilic semipermeable membrane (HSM, a dialysis membrane 20 μm in thickness) with a urease solution (0.1 M tris-HCl, pH 8.5, 15% bovine serum albumin) layer between the GPM and HSM. A schematic illustration of the urea biosensor is shown in Figure 4. The pulse amperometric technique, as stated above, was used to record the current response. The representative current-response curve of the laboratory-made uera biosensor is shown in Figure 5. The biosensor gave a current response linear to the concentration of urea in the test solution up to 1.0 mM. Beyond this limit of concentration a downward deviation from the linearity was observed, which may be attributable to the parabolic dependence of the enzymatic reaction rate on the substrate concentration (Michaelis-Menten equation), though a certain disagreement between the experimental results and the theoretical prediction is left unsolved. The relative standard deviation of the current response was 3.8 % (n = 5, at 20 μM), and the lifetime was more than 20 days. The laboratory-made urea biosensor was successfully applied to the determination of urea in biological fluids (Yamamoto and Senda, 1993). Application of the film

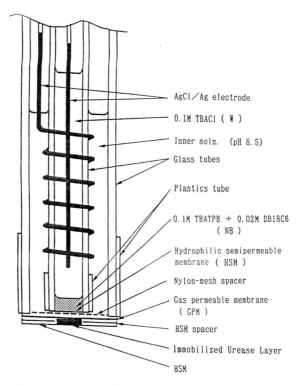

AgCl/Ag electrode

0.1M TBACl (W)

Inner soln.   (pH 8.5)

Glass tubes

Plastics tube

0.1M TBATPB + 0.02M DB18C6
          (NB)

Hydrophilic semipermeable
membrane (HSM)

Nylon-mesh spacer

Gas permeable membrane
(GPM)

HSM spacer

Immobilized Urease Layer

HSM

Figure 4. Schematic cross-section of a laboratory-made amperometric urea biosensor. (Senda and Yamamoto, 1996, with permission).

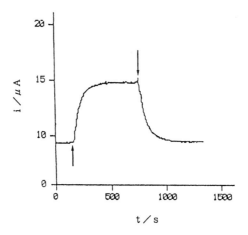

Figure 5. Pulse amperometric current vs time curve obtained with a laboratory-made urea biosensor immersed in a test solution (see cell B). A 0.2-mM urea solution was added to the test solution at the time indicated by an upward arrow followed by washing the sensor by the base solution at the time indicated by a downward arrow. $E_i = 0.25$ V, $\Delta E = 100$ mV, $\tau = 100$ ms, T = 5 s. (Senda and Yamamoto, 1993a, with permission).

technology with UV excimer laser photoablation techniques to fabricate a microelectrode strip sensor for urea as well as ammonia is interesting (Osborne and Girault, 1995b).

A creatinine biosensor can also be fabricated in much the same way as the urease biosensor, but using creatinine deiminase (50 U) in place of urease. The laboratory-made creatinine biosensor showed similar characteristics to those of the urea biosensor, and gave a current response proportional to the concentration of creatinine in a test solution in the range of 20 μM to 0.20 mM (Yamamoto, 1991; Yamamoto and Senda, 1993) or as high as 1 mM (Osborne and Girault, 1995c). The design and fabrication of amperometric biosensors using other enzyme(s) based on the ammonia or ammonium ion-ISE sensor could be feasible.

Correction for residual current

An advantage of the amperometric urea biosensor mentioned above is that the correction for the residual current due to the resicual ammonium ion that may be present in test solution can be relatively easily achieved with the amperometric ISE biosensor, compared with the potentiometric ISE biosensor. For this purpose, a (ammonium ion) sensor of the same structure as the urea biosensor, but without urease in the immobilized-enzyme layer, was fabricated. This urease free urea sensor gave a current response proportional to the concentration of ammonium ion present in the test solution, but no current response to urea. The (normal) urea biosensor gives the sum of the current responses each proportional to their concentrations when both urea and ammonium ion are present in the test solution. The current sensitivities to ammonium ions of the two sensors coincided with each other within an experimental error of less than 5%. Therefore, the concentration of urea corrected for the residual ammonium ion can be computed from the difference between the current response of the (urease-immobilized) urea biosensor and that of the urease-removed urea sensor (Senda and Yamamoto, 1993a, b).

Finally, it is noted that recent developments in electron-transfer voltammetry at the polarizable O/W interface also appear to be interesting for further studies on the coupling of enzyme reactions to the charge transfer at the interface of two immiscible electrolyte solutions.

References

Campanella, L. and Tomassetti, M. (1990) The membrane sensors in the environmental, biological and clinical analysis. *In*: D.L. Wise (ed.): *Bioinstrumentation: research, developments and applications.* Butterworths, Boston, pp 1369–1428.

Girault, H. (1993) Charge transfer across liquid/liquid interface. *In*: J.O.M. Bockris, B.E. Conway and R.F. White (eds): *Modern Aspects of Electrochemistry.* Butterworth, London, pp 1–62.

Hundhammer, B., Dhawa, S.K., Bekele, A. and Seidlitz, H.J. (1987) Investigation of ion transfer across the membrane-stabilized interface of two immisible electrolyte solutions. *J. electroanal. Chem.* 217:253.

Kuan, S.S. and Guilbault, G.G. (1987) Ion-selective electrodes and biosensors based on ISEs.
In: A.P.F. Turner, I. Karube and G.S. Wilson (eds): *Biosensors: fundamentals and applications.* Oxford Univ. Press, Oxford, pp 135–152.

Ohkouchi, T. Kakutani, T., Osakai, T. and Senda, M. (1991) Voltammetry with an ion-selective
microelectrode based on polarizable oil/water interface. *Anal. Sci.* 7:371–376.

Osakai, T., Kakutani, T. and Senda, M. (1984) Ion transfer voltammetry with the interface between polymer-electrolyte gel and electrolyte solution. *Bunseki Kagaku* 33:
E371–377.

Osakai, T., Kakutani, T. and Senda, M. (1987) A novel amperometric ammonia sensor. *Anal. Sci.*
3:521–526.

Osakai, T., Kakutani, T. and Senda, M. (1988) A novel amperometric urea sensor. *Anal. Sci.*
4:529–530.

Osakai, T., Nuno, T., Yamamoto, Y. Saito, A. and Senda, M. (1989) A microcomputer-control-
led system for ion-transfer voltammetry. *Bunseki Kagaku* 38:479–485.

Osborne, M.D. and Girault, H.H. (1995a) Amperometric detection of the ammonium ion by
facilitated ion transfer across the interface between two immiscible electrolyte solutions.
Electroanalysis 7:425–434.

Osborne, M.D. and Girault, H.H. (1995b) The liquid-liquid micro-interface for amperometric
detection of urea. *Electroanalysis* 7:714–721.

Osborne, M.D. and Girault, H.H. (1995c) The micro water/1,2-dichloroethane interface as a
transducer for creatinine assay. *Mikrochim. Acta* 117:175–185.

Osborne, M.D., Shao, Y., Pereira, C.M. and Girault, H.H. (1994) Micro-hole interface for the
amperometric determination of ionic species in aqueous solutions. *J. electroanal. Chem.*
364:155–161.

Sawada, S., Osakai, T. and Senda, M. (1990) Polarizability of o-nitrophenyl ethers/water interfaces and its applicability to ion-transfer voltammetry. *Bunseki Kagaku* 39:539–549.

Senda, M. (1991) Ion-selective electrodes based on polarizable electrolyte/electrolyte solutions
interface. *Anal. Sci.* 7(S):585–590.

Senda, M. (1995) Electroanalytical chemistry at liquid/liquid interfaces. *Denki Kagaku* 63:
368–372.

Senda, M. and Yamamoto, Y. (1993a) Urea biosensor based on amperometric ammonium ion
electrode. *Electroanalysis* 5:775–779.

Senda, M. and Yamamoto, Y. (1993b) Amperometric ion sensors and their applications in
food chemistry and clinical chemistry. *In*: E. Pungor (ed): *Proceedings of the 2nd Bioelectroanalysis Symposium,* Matrafured, Hungary, 1992: Akademiai Kiado, Budapest,
pp 139–160.

Senda, M. and Yamamoto, Y. (1996) Amperometric ion-selective electrode sensors. *In*: A.G.
Volkov and D.W. Deamer (eds): *Liquid-Liquid Interface: Theory and Methods.* CRC Press,
Inc., Boca Raton, FL, pp 277–293.

Senda, M., Kakutani, T., Osakai, T. and Ohkouchi, T. (1987) A new amperometric acetylcholine
ion-selective microelectrode. *In*: E. Pungor (ed.): *Proceedings of the 1st Bioelectroanalysis
Symposium,* Matrafured, Hungary, 1986. Akademiai Kiado, Budapest, pp 353–364.

Senda, M., Kakiuchi, T., Osakai, T., Nuno, T. and Kakutani, T. (1989) Theory of ion-selective
electrodes; amperometric ISE and potentiometric ISE. *In*: E. Pungor (ed.): *Proceedings of the
5th Symposium on Ion-Selective Electrodes,* Matrafured, Hungary, 1988. Akademiai Kiado,
Budapest, pp 559–568.

Senda, M., Kakiuchi, T. and Osakai, T. (1991) Electrochemistry at the interface between two
immescible electrolyte solutions. *Electrochim. Acta* 36:253–262.

Turner, A.P.F., Karube, I. and Wilson, G.S. (eds) (1987) *Biosensor, fundamentals and applications.* Oxford Univ. Press, Oxford.

Winquist, F. and Danielsson, B. (1990) Semiconductor field effect devices. *In*: A.E.G. Cass (ed.):
Biosensors: a practical appraoch. IRL Press, Oxford University Press, Oxford, pp 171–190.

Wise, D.L. (ed.) (1990) *Bioinstrumentation: research, development and applications.* Butterworths, Boston.

Yamamoto, Y. (1991) *Study on amperometric ion-selective electrode sensors,* PhD Thesis. Kyoto
University.

Yamamoto, Y. and Senda, M. (1993) Amperometric ammonium ion sensor and its application to biosensors. *Sensor. Actuator. B* 13 – 14:57 – 60.

Yamamoto, Y., Nuno, T., Osakai, T. and Senda, M. (1989) A volatile amine sensor based on the amperometric ion-selective electrode. *Bunseki Kagaku* 38:589 – 595.

Yamamoto, Y., Osakai, T. and Senda, M. (1990) Potassium and sodium ion sensor based on amperometric ion-selective electrode. *Bunseki Kagaku* 39:655 – 660.

Frontiers in Biosensorics I
Fundamental Aspects
ed. by F.W. Scheller, F. Schubert and J. Fedrowitz
© 1997 Birkhäuser Verlag Basel/Switzerland

Electrochemistry of heme proteins on organic molecule modified electrodes

S. Dong and T. Chen

Laboratory or Electroanalytical Chemistry, Changchun Institute of Applied Chemistry, Chinese Academy of Sciences, Changchun 130022, China

Summary. Electrochemistry of heme proteins on various organic functional electrodes is discussed. A bi-directional electrocatalytic mechanism for proteins on organic dye chemically modified electrodes (CMEs) is proposed. It depends on the relationship between formal potentials of the protein and the mediators.

After long adsorption time for preparing the CMEs, modifiers of a single functional group can act as efficient promoters for the direct electron transfer of cytochrome c. It was also found that the absorption behavior of cytochrome c on 4-pyridyl derivative CMEs is related to the concentration of the supporting electrolyte.

A successful immobilizing material for the enzymes, cryo-hydrogel membrane, was proven to be compatible with heme proteins. The immobilized proteins show good and stable electrochemical response on the electrode surface.

Introduction

Heme proteins are widely distributed in nature. They have important functions in respiratory processes, i.e., cytochrome c is an important electron carrier in the electron transfer process of mitochondrial respiratory chain, and hemoglobin and myoglobin take part in the transport and storage of dioxygen. The prosthetic group of these proteins is iron porphyrin called heme. The physiological functions of heme proteins are usually associated with the gain or loss of electrons in their active centers. This provides electrochemists with a chance to study the physiological reactions by electrochemical methods. Also knowledge of how these proteins work in physiological environment will undoubtedly promote their application in molecular electronic devices.

Early electrochemical studies of heme proteins showed the irreversible reduction of cytochrome c at mercury electrodes (Betso et al., 1972). High irreversibility of protein electron transfer at bare metal electrodes is due to several reasons: (i) proteins are generally strongly adsorbed at the electrode-soluton interface accompanied by denaturation. The denatured proteins undergo irreversible electrochemical reaction and hinder the electron transfer of free molecules from solution. (ii) The electroactive groups of most proteins are buried by the isolating protein fabric and are unable to contact the electrode. (iii) Highly ionic characteristic and asymmetric surface charge distribution of proteins also affect the reversibility of electron transfer reactions.

How to realize rapid electron transfer reactions of proteins continues to be a stimulating question for electrochemists. In the 1970's, mediators were added to solution to tirate proteins by the homogeneous catalytic reaction (Heineman et al., 1975, 1979). With the development of theories and applications of chemically modified electrodes (CME), various electrodes, made to function by small organic (Eddowes and Hill, 1977; Taniguchi et al., 1982) or inorganic molecules (Ohtani and Ikeda, 1993), electrode modification by polymers (Oliver et al., 1988; Hahn et al., 1990), conducting polymers (Bartlett and Farington, 1989; Caselli et al., 1991) and lipid membranes (Salamon and Tollin, 1991; Veyama et al., 1993), were employed to achieve the reversible and quasi-reversible electron transfer reaction of proteins. Here we introduce some of our recent work in this area.

Electron transfer of heme proteins in organic dye CMEs

Chemically modified electrodes have been widely used in electrochemical research of proteins (Hill, 1987; Armstrong et al., 1988). They improve the interfacial compatibility between proteins and electrode surface and facilitate the electron transfer of proteins. Modifiers are generally divided into two categories: promoters and mediators. Promoters are electroinactive in the potential window studied and proteins directly convey electrons with the electrode. On the other hand, mediators are electroactive in the potential window studied. They first take part in the electrode reaction and then convey electrons with proteins.

Recently, many works have reported quasi-reversible electron transfer reactions of heme proteins on dye-modified electrodes (Dong, 1991; Dong et al., 1995; Chen et al., 1995a). Dye-modified electrodes combined with thin layer spectroelectrochemistry is an effective method in determining formal potentials, electron transfer numbers and other thermodynamic and kinetic parameters of proteins (Dong et al., 1995).

Detection method – thin layer spectroelectrochemistry

Because of their large molecular weight, proteins usually exist in low concentration levels in solution, and the produced catalytic current is so low that it is often suppressed by charging current, residual current or the faraday current resulting from the mediators. Spectroelectrochemistry provides a suitable detection method. By choosing a proper detection wavelength, the background absorbance can be almost completely eliminated. One important feature of thin layer cells in bulk electrolysis (Bard and Faulkner, 1980), so the steady state can be quickly achieved. Three thin layer spectroelectrochemical techniques may be used in the study of electron transfer reactions of proteins (Dong et al., 1995). One is the spectropotentiostatic technique, in which equilibrium spectra are recorded under different applied electrode potentials. The second is potential step chronoabsorptometry, recording the absorbance variation with time after a step potential is applied. The third is cyclic voltabsorptometry, recording the absorbance variations under a certain detecting wavelength during the process of potential sweep. Using the spectropotentiostatic technique, several thermodynamic parameters, such as formal potential E^O and electron number n, may be determined (Dong et al., 1995; Heineman et al., 1984). Kinetic information can be obtained using cyclic voltabsorptometry. From difference between two inflexion points on the cyclic voltabsorptometric curve, ΔEp is determined (Zak et al., 1983) and then the standard heterogeneous rate constant k^O is obtained (Laviron, 1979). Potential step chronoabsorptometric curves can also be used to get kinetic information. On the occasion of electrocatalytic reaction occurring on chemically modified electrodes, the electrode and modifier should be considered as a whole, and the substrates react with k^O on this "new" electrode surface. Therefore, in this case, the determined k^O is an observed rate constant for the whole catalytic reaction.

Mediators – organic dyes

The backbone of various phenazine, phenothiazine and phenoxazine dyes has the following general structure (Gorton et al., 1991):

$$X = N, O, S$$

The large conjugation system is not only advantageous for dye molecules to be adsorbed strongly on the electrode surface, but also results in good electroactivity.

Using spectroelectrochemical methods, we systematically studied electron transfer reactions of several heme proteins on various phenazine, phenothiazine and phenoxazine modified platinum gauze electrodes, including methylene blue (Song and Dong, 1988; Song et al., 1990; Zhan et al., 1990; Zhang et al., 1990, Dong and Song, 1991), methylene green (Zhu and Dong, 1990a), janus green (Zhu and Dong, 1990b; 1992), toluidine blue (Dong and Chu, 1993b), brilliant cresyl blue (Dong and Zhu, 1989 and 1990, Dong et al., 1989), and azure A (Dong and Chu, 1992 and 1993a).

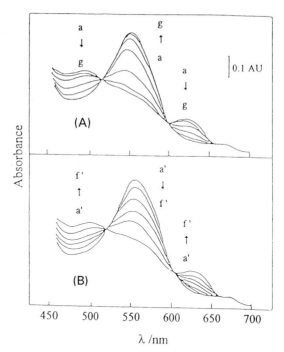

Figure 1. Spectra of 1.2 mg/ml hemoglobin on methylene blue modified Pt gauze electrode in optical transparent thin layer cell. Spectra of hemoglobin during (A): reduction at potentials (V): 0, − 0.1, − 0.14, − 0.16, − 0.17, − 0.18, − 0.2 (from a to g); (B): reoxidation at potentials (V): − 0.2, − 0.1, − 0.05, 0.0, 0.5, 0.1 (from a′ to f′). 2 min maintenance after each potential applied.

Figure 1 shows the thin layer spectropotentiostatic spectra of 1.2 mg/ml hemoglobin (Hb) on the methylene blue modified platinum gauze electrode (Song and Dong, 1988). With the applied potential changing from 0 V to − 0.2 V, the two absorption peaks at 490 nm and 625 nm of Hb(III) gradually decrease and eventually vanish, and at the same time the absorption peak at 550 nm of Hb(II) increases gradually. It indicates that Hb(III) is reduced during this process. The formal potential and electron number are − 0.154 V (vs. Ag/AgCl) and 4, respectively. The reaction of hemoglobin on methylene blue modified platinum electrode is a quasi-reversible process.

Figure 2 shows the cyclic voltabsorptometric curves of hemoglobin and myoglobin on methylene blue modified platinum electrode, respectively (Dong and Song, 1991). The reduction reaction proceeds much faster than the oxidation. On the other hand the oxidation reaction of cytochrome c on janus green modified platinum electrode (Zhu and Dong, 1992) proceeds much faster than the reduction. This difference results from the relationship between the formal potential of proteins and that of methylene blue. It will be discussed later.

Both, the reduction and the oxidation reactions of proteins can be fast at bi-mediator modified electrodes. Figure 3 shows the influence of bi-

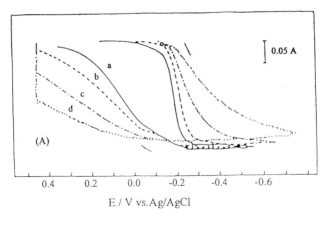

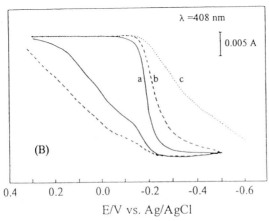

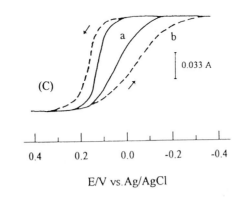

Figure 2. Cyclic voltabsorptometric curves of several heme proteins on modified Pt gauze electrode in optical transparent thin layer cell. 0.1 M phosphate buffer, 0.2 M KNO₃, pH 7.0; (A) 1.0 mg/ml hemoglobin, methylene blue/Pt electrode; sweep rate: a = 1, b = 2, c = 5, d = 10 mV/S. (B) 0.45 mg/ml myoglobin, methylene blue/Pt electrode; sweep rate: (a) 1, (b) 2, (c) 5 mV/s. (C) 1.0 mM cytochrome c, janus green/Pt electrode; (a) 1st cycle, (b) 11th cycle.

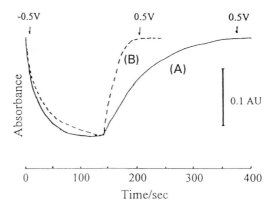

Figure 3. Double potential step chronoabsorptometry of 2 mg/ml hemoglobin. Detecting wavelength 405 nm; 0.10 M phosphate buffer, pH 7.0, 0.2 M KNO_3. Potential Step between -0.5 and $+0.5$ V. (A) At an adsorbed brilliant cresyl blue modified Pt gauze electrode; (B) at a bilayer modified Pt gauze electrode with brilliant cresyl blue adsorbed as the outer and bromopyrogal red as the inner layer.

mediator modification on the double potential step chronoabsorptometric curves of hemoglobin (Zhu and Dong, 1991). For example, on a bromo-pyrogal red (BPR) modified Pt electrode, no redox reaction of hemoglobin occurs. When the Pt electrode is modified by brillant cresyl blue (BCB), the times required for complete reduction and complete oxidation are 50 s and 200 s, respectively, (Fig. 3(A)). If in the electrode a second mediator, BPR, is introduced to construct a BPR/BCB/Pt bi-mediator modified electrode, the time required for complete reduction is nearly the same, but the time required for complete oxidation is shortened to 50 s (Fig. 3.(B)).

Mechanism – bi-directional catalytic reaction

We found that the shapes of cyclic voltabsorptometric curves are related to the difference between the formal potentials of the protein and the mediator (Fig. 4). Dyes are usually electroactive organic compounds. Their electrode reactions occur near the formal potential of redox proteins, such as hemoglobin and myoglobin, so a mediating mechanism can be proposed:

$$M_O + ne = M_R \tag{1}$$

$$M_R + P_O \underset{k_b}{\overset{k_f}{\rightleftharpoons}} M_O + P_R \tag{2}$$

where M_O and M_R represent the oxidized and reduced forms of the mediator, P_O and P_R repesent the protein dissolved in solution, k_f and k_b are the

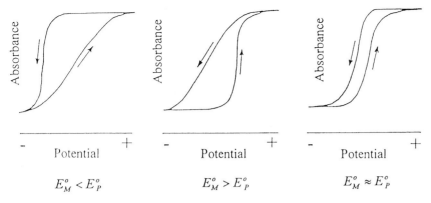

$$E_M^o < E_P^o \qquad\qquad E_M^o > E_P^o \qquad\qquad E_M^o \approx E_P^o$$

Figure 4. Relationship between the shape of cyclic voltabsorptometric curves and formal potentials of the protein E_p^o and the mediator E_M^o.

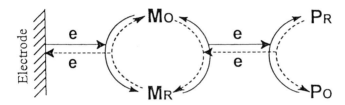

Figure 5. Schematic illustration of bi-directional catalytic mechanism.

forward and backward catalytic chemical reaction rate constant, respectively.

The equilibrium constant K of eqn (2) can be expressed as:

$$K = \frac{k_f}{k_b} = \frac{[M_O][P_R]}{[M_R][P_O]} = \exp\left[\frac{nF}{RT}(E_p^o - E_M^o)\right] \qquad (3)$$

The mechanism described by eqn (1) and (2) is similar to an electrocatalytic reaction. For redox proteins, the reduced form can be accumulated, so the catalytic reaction may proceed in both directions if the formal potential of the mediator and the protein are close to each other. This also differs from the mediating electrocatalytic mechanism for most redox enzymes, which shows irreversibility due to the specific interaction between the enzymes and the substrates. According to eqn (2) the mediators perform not only the reduction but also the oxidation of proteins (Fig. 5). The only difference, however, are the values of the forward and backward catalytic rate constants. The relationship between k_f and k_b is expressed by eqn (3).

Theory

If M_O and M_R do only dissolve to a negligible extent, the sum of Γ_{M_O} and Γ_{M_R} is constant at the electrode surface

$$\Gamma_{M_O} + \Gamma_{M_R} = \Gamma^* \tag{4}$$

Suppose the reaction takes place so rapidly that the Nernst equation holds,

$$E = E_M^o + \frac{RT}{nF} \ln \frac{\Gamma_{M_O}}{\Gamma_{M_R}} \tag{5}$$

where

$$\frac{\Gamma_{M_O}}{\Gamma_{M_R}} = \exp\left[\frac{nF}{RT}(E - E_M^o)\right] = p \tag{6}$$

By eqns (4) and (6), we obtain

$$\Gamma_{M_R} = \Gamma^*/(1 + p) \tag{7}$$

$$\Gamma_{M_O} = p\Gamma^*/(1 + p) \tag{8}$$

The rate of the catalytic reaction is

$$-\frac{d\Gamma_{M_R}}{dt} = k_f \Gamma_{M_R}(C_{P_O})_s - k_b \Gamma_{M_O}(C_{P_R})_s \tag{9}$$

Numerical results

Figure 6 shows the influence of the formal potentials of mediators on the cyclic voltabsorptometric curves. Under the condition that $E_M^o < E_P^o$, the equilibrium is the constant $K > 1$, and the mediator catalyzes the reduction reaction of the proteins. But the catalytic oxidation can also proceed with a relatively slow rate. With decreasing difference between E_M^o and E_P^o, K appraoches 1 and k_b nearly equals k_f. So the rate of the reoxidation reaction increases. When $E_M^o = E_P^o$, the rate of the oxidation reaction is as fast as that of the catalytic reduction reaction. It can be inferred that if $E_M^o > E_P^o$, then $K < 1$ and $k_f < k_b$, and the absorbance of the reduction process will change more slowly than that of the oxidation process. The loop formed by the oxidation and reduction semi-curves results from the diffusion effect in the thin layer cell. Even if both k_f and k_b are so large that the kinetic factors are negligible, the reduction and oxidation semi-curves can not completely

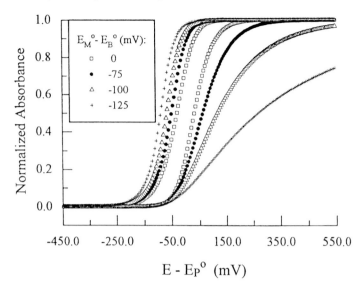

Figure 6. Influence of formal potential of the mediatiors on the cyclic voltabsorptometric curves. $k_f\Gamma^* = 1 \cdot 10^{-4}$ cm/s, $\Delta t = 0.05$ s, V/A = 0.01 cm, $D = 1 \cdot 10^{-6}$ cm^2/s, $D_M = 0.49$, sweep rate 1 mV/s.

overlap. In Figure 2(A) and (B), the formal potential of methylene blue (− 0.19 V, vs Ag/AgCl) is more negative than those of both Hb and Mb (− 0.16 V and − 0.15 V vs Ag/AgCl, respectively). Therefore, the catalytic reduction proceeds more rapidly than the catalytic oxidation. But in Figure 2(C), the formal potential of janus green is ca. + 0.15 V (vs ag/AgCl), more positive than that of cytochrome c (+ 0.064 V, vs Ag/AgCl) and the result is just on the contrary.

When the difference between E_M^o and E_P^o is quite large, only unidirectional catalytic reaction proceeds. If $E_M^o \ll E_P^o$, protein molecules can quickly be reduced when the potential is shifted to negative directions, but the reoxidation process is very slow. If a second mediator with $E_{M2}^o > E_P^o$, which can catalyze the oxidation reaction of proteins, is adsorbed on the electrode surface to construct a catalytic bi-mediator layer, both reduction and oxidation reaction of proteins proceed rapidly. Figure 7 shows cyclic voltabsorptometric curves under mono- and bi-mediator modifications. The mediator with $E_M^o < E_P^o$ facilitates reduction and the mediator with $E_M^o > E_P^o$ facilitates oxidation of proteins. This is in agreement with the experimental results shown in Figure 3. The formal potential of BPR is ca + 0.24 V (vs Ag/AgCl), much more positive than that of hemoglobin, and it catalyzes not the reduction but the oxidation of Hb. When the BPR modified electrode is immersed into the solution mainly containing Hb(III), no reaction can be detected. BCB (− 0.21 V vs Ag/AgCl) catalyzes both reduction and oxidation of Hb, but the reduction reaction is faster than the

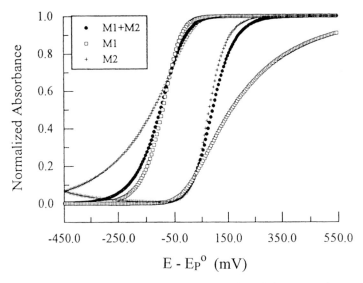

Figure 7. Influence of bi-mediator modification on the cyclic voltabsorptometric curves. $E^{o}_{M1}-E^{o}_{p} = 50$ mV, $k_{f1}\Gamma_{1}^{*} = 5 \cdot 10^{-6}$ cm/s, $E^{o}_{M2} - E^{o}_{p} = -50$ mV, $k_{f2}\Gamma_{2}^{*} = 1 \cdot 10^{-6}$ cm/s, $\Delta t = 0.05$ s, $V/A = 0.01$ cm, $D = 1 \cdot 10^{-6}$ cm^2/s, $D_{M} = 0.49$, sweep rate 1 mV/s.

oxidation. When BPR/BCB bi-mediator modification is used, the oxidation of Hb is accelerated.

Direct electrochemistry of cytochrome C

Cytochrome c is a small electron transfer protein with a molecular weight of ca 12,000 D. The realization of direct electron transfer of cytochrom c is a great breakthrough in direct electrochemistry of redox proteins. Until now numerous investigations have been done on this protein. However, which structure of modifiers can act as an efficient promoter and how this protein conveys electrons with the modified electrode is still not clear.

Promoters with a single functional group

After studying over 50 kinds of small organic molecule modifiers, it was concluded that efficient promoters should possess "X~Y" bifunctional group structure (Armstrong et al., 1988; Allen et al., 1984), where group X anchored the promoter to the electrode surface and group Y interacted with lysine residues of cytochrome c to form a suitable orientation of protein molecules for electron transfer.

However, it was found that molecules with a single functional group, such as carbazole, pyridine and thiophene (Qu et al., 1994; Zhou et al.,

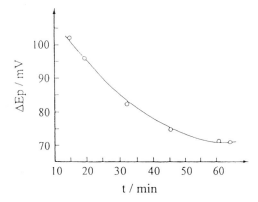

Figure 8. Dependence of the peak separation ΔEp in the cyclic voltammograms of cyto-chrome c on the dipping times used in the prepartion of thiophene-modified gold electrodes.

1992 and 1993) could also accelerate the electrochemical reaction of cyto-chrome c. Long adsorption times (1 h for thiophene and 12 h for pyridine and carbazole) were necessary for preparing modified electrodes. The thio-phene-modified electrode itself gives no redox peaks. When cytochrome c is present in the solution, a pair of well-defined redox peaks is obtained at + 0.01 V (vs SCE). Figure 8 shows the relationship between peak to peak potential separation, ΔEp, and the adsorption time. With the adsorption time increasing, ΔEp appreciably decreases. Long adsorption time is not only necessary for promoters with weak adsorption ability to reach stable adsorption saturation, but also possibly necessary for adsorbed promoters to change their orientation, which in this case is favorable for the elec-tron transfer of cytochrome c. Therefore, the reaction mechanism of cyto-chrome c on molecules with single functional group modifed gold electrode will undoubtedly help to understand the relationship among the structures, the adsorption abilities and the reorientations of modifiers on the electrode surface, as well as the electron transfer of cytochrome c on modified elec-trodes.

Adsorption mechanism on 4-pyridyl derivative CMEs

Allen et al. considered that the binding or adsorption of cytochrome c with 4,4'-bipyridine (PyPy) on the modified gold electrode is rapid and reversible because it cannot be detected by the film transfer method (Albery et al., 1981; Eddowes and Hill, 1982b. However, Hinnen and Niki detected the irreversibly adsorbed cytochrome c on the 4,4'-dithiodipyridine (PySSPy) modified gold electrode surface using infrared and UV-vis electroreflec-tance spectroscopic techniques (Niwa et al., 198; Hinnen and Niki, 1989; Sagara et al., 1990, 1991), although the small current of the adsorbed cyto-

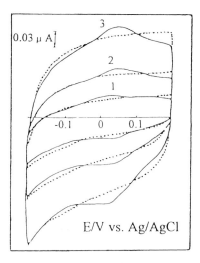

Figure 9. Cyclic voltammograms of cytochrome c adsorbed on gold electrode under various sweep rates. 4 mM phosphate buffer soluton (pH 7.0); sweep rate: 1. 20, 2. 50, 3. 100 mV/s.

chrome c was hidden by the large charging current in voltammetry. Three types of adsorption models were proposed for the irreversible adsorption of cytochrome c on the 4-pyridyl dervative modified gold electrode (Sagara et al., 1990). The authors found that cytochrome c was bound to PySSPy and trans-1,2-bis(4-pyridyl)ethylene (PyC = CPy), but no binding to 4-mercaptopyridine (PySH) or PyPy was detected.

It was found that the difference between these two viewpoints results from the effect of supporting electrolyte on the adsorption of cytochrome c (Xie and Dong, 1994; Dong et al., 1995).

Figure 9 shows the cyclic voltammetric curves of adsorbed cytochrome c, prepared by the film transfer method, on PySSPy modified gold electrode in 4.0 mM phosphate buffer solution (Dong et al., 1995). The redox peaks become more apparent with increasing sweep rate, and the peak to peak potential separation is just 30 mV under the sweep rate of 100 mV/s, possessing adsorption characteristics. This proves that cytochrome c irreversibly adsorbs on the modified gold electrode at lower electrolyte concentration and is in accordance with the conclusion of Hinnen and Niki (1989).

Figure 10 shows the influence of the concentration of electrolyte on the redox peaks of adsorbed cytochrome c on PySSPy modified gold electrode (Dong et al., 1995). With increasing concentration or electrolyte, peak currents of adsorbed proteins decrease gradually and eventually vanish. In this case adsorption of cytochrome c on modified gold electrode may be considered reversible, which is in accordance with the results of Allen et al.

Also, we found that it is the cations rather than the anions of electrolyte that affect the adsorption behavior of cytochrome on PySSPy modified gold electrode (Dong et al., 1995). Cations at high concentration efficient-

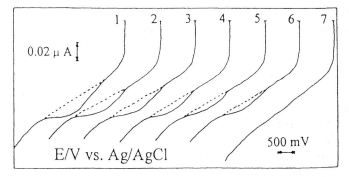

Figure 10. Current-potential curves of cytochrome c adsorbed on PySSPy modified gold electrodes after immersed into various concentrations of supporting electrolytes. KNO_3(mM): 1.0, 2.5, 3.10, 4.10, 5.30, 6.50, 7.100; sweep rate: 100 mV/s; potential range: $+0.18 \sim -0.14$ V (vs Ag/AgCl)

ly compete for the electron-rich nitrogen atoms of the promoter pyridine with cytochrome c, and this may result in the desorption of cytochrome c; thus the adsorption behavior becomes reversible.

Therefore, adsorption of cytochrome c on 4-pyridyl derivative modified gold electrode fits the irreversible adsorption model of Hinnen and Niki (1989) at lower electrolyte concentration and changes to the reversible adsorption model of Allen et al. (1984) when the electrolyte concentration increases.

Heme proteins bound to cryo-hydrogel membranes

Many redox proteins are bound to biological lipid membranes *in vivo.* Therefore, the electrochemical behavior of the immobilized proteins is a good model for showing biological functions of redox proteins. On the other hand, if we want to construct usable molecular electronic devices, we have to utilize the electrode reaction of immobile phase. To achieve efficient and stable electron transfer between redox protein immobilization by a proper membrane at the electrode is a key step.

Here we report a kind of cryo-hydrogel membrane (CHM), which is formed by polyvinyl alcohol and soluble cellulose under $-4\,°C$ for 24 hours. This CHM has been proved to be a good carrier to construct enzyme electrodes (Dong and Guo, 1994, 1995). In the CHM the direct electron transfer of cytochrome c (Chen et al., 1995b), myoglobin (Niu et al., 1995a, b), and even cytochrome c oxidase (Chen et al., under revision), which is the terminal protein of the electron transfer chain of mitochondrial respiratory process, was realized.

It has been reported that in this CHM there are three different states of water, i.e. "non-freezing" water, "bound" water and "free" water (Dong and Guo, 1994; Feng, 1989). "Non-freezing" water molecules interact strongly with hydrogel polymer and do not freeze even at temperatures below $-40\,°C$. "Bound" water molecules weakly interact with the cryo-hydrogel polymer, and reorient their molecules nearby the hydrophilic groups of CHM. So the formed ice crystals are not regular hexagonal piles, which leads to the decrease of the melting point of the ice. "Free" water molecules are in a state like water molecules in aqueous solution. Therefore, this high water-containing structure of CHM forms a favorable hydrophilic microenvironment to maintain the biological activity of biomolecules.

Apart from this, the CHM possesses a polyhydroxyl structure. The hydroxyl groups and carboxyl groups on the backbone of CHM can easily form hydrogen bonds with the amino acid residues of the proteins. This elastic interaction tends to stabilize the biological activity of proteins due to the hydroxyl groups holding or substituting for the "bound" water which is essential for the retention of the tertiary structures of proteins and subsequently the activities of proteins (Gibson and Wordward, 1992).

During the gradual freezing formation process of CHM, the active conformation of the proteins in aqueus solution is maintained in the membrane, and eventually the proteins are "locked" in the three-dimensional interpenetrating network of CHM with their bioactivity. The interaction between proteins and CHM also eliminates the significant interference from the denatured proteins. So, well-defined redox waves have been obtained on this protein membrane electrode using the commercially available protein sample without purification.

Myoglobin

When myoglobin is entrapped in the CHM, it gives a very stable and well-defined redox wave (Fig. 11). No significant change in the voltammogram is observed during continuous measurement. The formal redox potential is ca -0.158 V (vs Ag/AgCl), which is in good agreement with previous reports (Heineman et al., 1979; Taniguchi et al., 1992). This result indicates that the voltammetric response is due to the redox reaction of native myoglobin. The peak to peak separation is 93 mV at 5 mV/s.

The linear relationship between the cathodic peak current and the square root of the sweep rate indicates a semi-infinite diffusion process. The CHM swells as immersed into buffer solution due to the expansion of the polymer chain by absorbing water into its skeleton. The electron transfer process of the proteins in CHM probably proceeds by the movement of free protein molecules or the exchange of electrons between the neighbouring proteins in CHM or both of them, like in other polymer modified electrodes. The observed diffusion coefficient of protein molecules in CHM is much smaller than that in aqueous solution.

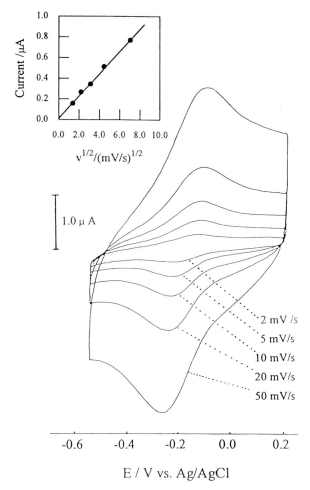

Figure 11. Cyclic voltammograms of myoglobin in a phosphate buffer (pH 6.5) solution under a N$_2$ atmosphere at different sweep rates at 25 °C · 2.0 mM myoglobin at CHM/GC electrode.

Figure 12 shows a pH-controlled conformational equilibrium of myo-globin, which is reflected by variation of its redox potential at the CHM/GC electrode within a pH range of 0~13.8:

$$\text{Acid}_{pH>5.4}\text{Natural}_{>010.4}\text{Basic I}_{>13.2}\text{Basic II}$$
$$\text{State} \underset{<1.2}{\rightleftarrows} \text{State} \underset{<6.8}{\rightleftarrows} \text{State} \underset{<11.3}{\rightleftarrows} \text{State}$$

$$I_1\text{ State} \qquad I_2\text{ State} \qquad I_3\text{ State}$$

States I$_1$, I$_2$ and I$_3$ represent the intermediate states of myoglobin during its conformational conversion from acid state to basic state. When myoglobin

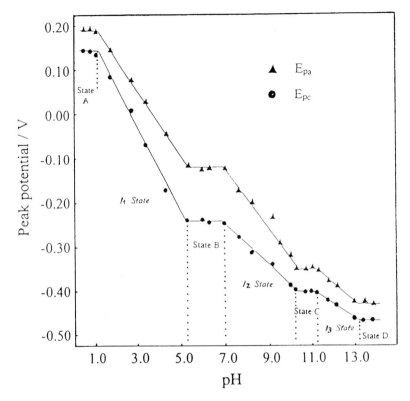

Figure 12. Effect of pH of the buffer solution of the anodic peak potential (E_{pa}) and cathodic peak potential (E_{pc}) of myoglobin at CHM/GC electrode.

is in the three intermediate states I_1, I_2 and I_3, Ep ~ pH curves give slopes of about 60 mV/pH, indicating that one H^+ ion is involved in each conformational conversion of myoglobin. It should be noted that all the mentioned conformation conversions of myoglobin are reversible. They are the co-effects of the bindings of myoglobin with CHM, myoglobin with the electrode, CHM with the electrode, etc.

Cytochrome c

Cytochrome c immobilized by CHM on a graphite electrode also gives a pair of well-defined redox peaks (Fig. 13) at + 70 mV (vs Ag/AgCl), which is close to the result of + 62 mV measured by spectrosocopic titration (Heineman et al., 1975). This also shows that cytochrome c immobilized by CHM maintains its natural conformation. The linear relationship between the peak current and the square root of the sweep rate denotes that it is a semi-infinite diffusion process.

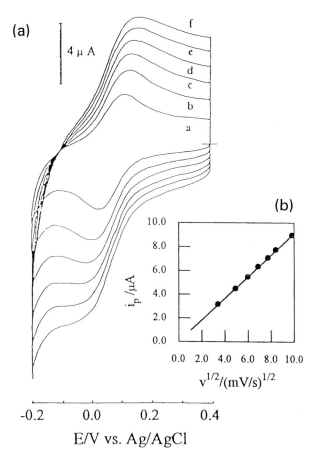

Figure 13. Cyclic voltammograms of cytochrome c entraped in the thin membrane modified graphite electrode. pH 6.3 (a) Changing with sweep rate v: a = 10, b = 20, c = 30, d = 40, e = 50, f = 60 mV/s. (b) Relationship between peak current i_p and square root of sweep rate $v^{1/2}$.

Before the hydrogel is frozen to form the CHM, cytochrome c molecules can move freely in the hydrogel. If there is adequate time for cytochrome c molecules to move and denature on the electrode surface before they are frozen and immobilized by the CHM, (e. g. by using much thicker hydrogel which takes longer to freeze) the adsorption peaks are observed. Under these particular conditions the redox peaks of both the native and the adsorbed cytochrome c can be obtained in CHM.

We also found that in CHM some cytochrome c molecules are in a "free" state, not bound to the backbone of CHM. These molecules can be easily moved out from CHM by rinsing the modified electrodes with water. This results in a slight decrease of the first pair of redox peaks. But the peak current eventually reaches a constant value, and this proves that some molecules are immobilized within the CHM.

Acknowledgements
We gratefully appreciate the support of the National Natural Science Foundation of China.

References

Albery, M.J., Eddowes, M.J., Hill, H.A.O. and Hillman, A.R. (1981) Mechanisms of the reduction and oxidation reaction of cytochrome c at a modified gold electrode. *J. Am. Chem. Soc.* 103:3904–3910.

Allen, P.M., Hill, H.A.O. and Walton, N.J. (1984) Surface modifiers for the promotion of direct electrochemistry of cytochrome c. *J. Eletroanal. Chem.* 178:69–86.

Armstrong, F.A., Hill, H.A.O. and Walton, N.J. (1988) Direct electrochemistry of redox proteins. *Acc. Chem. Res.* 21:407–413.

Bard, A.J. and Faulkner, L.R. (1980) *Electrochemical methods: fundamentals and applications.* John Wiley & Sons, New York, pp 406–413.

Bartlett, P.W. and Farington, J. (1989) The electrochemistry of cytochrome c at a conducting polymer electrode. *J. Electroanal. Chem.* 261:471–475.

Betso, S.R., Klapper, M.H. and Anderson, L.B. (1972) Electrochemical studies of heme proteins: coulometric, polarographic, and combined spectroelectrochemical methods for reduction of the hemeprosthetic group in cytochrome c. *J. Am. Chem. Soc.* 94: 8197–8204.

Caselli, M., Monica, M.D. and Portacci, M. (1991) Electrochemical and spectroscopic study of cytochrome c immobilized in polypyrrole films. *J. Electroanal. Chem.* 319: 361–364.

Chen, T., Dong, S. and Xie, Y. (1995a) Electrochemical reaction of redox proteins speeded up by the dye modified electrodes. *Electrochemistry (Chin.)* 1:125–135.

Chen, T., Guo, Y. and Dong, S. (1995b) Voltammetry of cytochrome c entrapped in hydrogel membrane on graphite electrode. *Bioelectrochem. Bioenerg.* 37:125–130.

Chen, T., Guo, Y. and Dong, S. (1995) Direct electron transfer between immobilized cytochrome c oxidase by cryo-hydrogel and a graphite electrode. subitted to *J. Electroanal. Chem.*

Dong, S. (1991) Advanced electroanalytical chemistry in China (2): chemically modified electrodes, spectroelectrochemistry, bioelectrochemistry and microelectrodes. *Denki Kagaku* 59:664–672.

Dong, S. and Chu, Q. (1992) Study on electrode process of myoglobin at a polymerized azure A film electrode. *Acta Chim. Sin.* 50:589–593.

Dong, S. and Chu, Q. (1993a) Study of the electrode process of hemoglobin at a polymerized azure A film electrode. *Electroanalysis* 5:135–140.

Dong, S. and Chu, Q. (1993b) Study on electrode proces of myoglobin at a polymerized toluidine blue film electrode. *Chin. J. Chem.* 11:12–20.

Dong, S. and Guo, Y. (1994) Organic phase enzyme electrode operated in water-free solvents. *Anal. Chem.* 66:3895–3899.

Dong, S. and Guo, Y. (1995) An organic phase enzyme electrode based on an apparent direct electron transfer between a grahite electrode and immobilized horseardish peroxidase. *J. Chem. Soc. Chem. Commun.* 483–484.

Dong, S. and Song, S. (1991) Direct electron transfer reaction of hemeproteins promoted by methylene blue modified electrode. *Acta Chim. Sin.* 49:493–497.

Dong, S. and Zhu, Y. (1989) Study on the electrode-process of brilliant cresyl blue by optically transparent thin-layer spectroelectrochemistry. *J. Electroanal. Chem.* 263:79–86.

Dong, S. and Zhu, Y. (1990) The reversible electron transfer reaction of myoglobin at brilliant cresyl blue modified platinum gauze electrode. *Acta Chim. Sin.* 48:566–570.

Dong, S., Zhu, Y. and Song, S. (1989) Electrode processes of hemoglobin at a platinum electrode covered by brilliant cresyl blue. *Bioelectrochem. Bioenerg.* 21:233–243.

Dong, S., Niu, J. and Cotton, T.M. (1995) Ultraviolet/visible spectroelectrochemistry of redox proteins. *In*: K. Sauer (ed.): *Biochemical spectroscopy*, Vol. 246 of *Methods in Enzymology.* Academic Press, Inc., Orlando, pp 701–732.

Dong, S., Che, G. and Xie, Y. (1995) Chemically Modified Electrodes. (Chin.) Science Press, Beijing, pp 456–457.

Eddowes, M.J. and Hill, H.A.O. (1977) Novel method for the investigation of the electrochemistry of metalloproteins: cytochrome c. *J. Chem. Soc. Chem. Common.* 771–772.

Eddowes, M.J. and Hill, H.A.O. (1982a) Factors influencing the electron-transfer rates of redox proteins. *Faraday Discuss. Chem. Soc.* 74:331–341.

Eddowes, M.J. and Hill, H.A.O. (1982b) Binding as a prerequisite for rapid electron transfer reactions of metalloproteins. *Adv. Chem. Ser.* 201:173–197.

Feng, Y.D. (1989) Study on state of water in cryo-hydrogel. *J. Sichuan Univ. Natural Science Edition* 26:470–473.

Gibson, T.D. and Wordward, J.R. (1992) Protein stabilization in biosensor systems. *In:* P.G. Edelman and J. Wang (eds.): *Biosensors and chemical senors.* American Chemical Sciety, Washington, DC, pp 40–55.

Gorton, L., Csoregi, E., Dominguez, E., Emneus, J., Jonsson-Pettersson, G., Marko-Varga, G. and Persson, B. (1991) Selective detection in flow analysis based on the combination of immobilized enzymes and chemically modified electrodes. *Anal. Chim. Acta* 250:203–248.

Hahn, C.E.W., Hill, H.A.O., Ritchie, D. and Sear, J.W. (1990) The electrochemistry of proteins entrapped in Nafion. *J. Chem. Soc. Chem. Common.* 125–126.

Heinemann, W.R., Norris, B.J. and Goelz, J.F. (1975) Measurement of Enzyme E^{O} values by optically transparent thin layer electrochemical cells. *Anal. Chem.* 47:79–84.

Heinemann, W.R., Meckstroth, M.L., Norris, B.J. and Su, C.-H. (1979) Optically transparent thin layer electrode techniques for the study of biological redox systems. *Bioelectrochem. Bioenerg.* 6:577–585.

Heinemann, W.R., Hawkridge, F.M. and Blount, H.N. (1984) Spectroelectrochemistry at optically transparent electrodes. II. Electrodes under thin-layer and semi-infinite diffusion conditions and indirect coulometric titrations. *In:* A.J. Bard (ed): *Electroanalytical Chemistry.* Vol. 13, Marcel Dekker, New York, pp 1–113.

Hill, H.A.O. (1987) Bioelectrochemistry. *Pure Appl. Chem.* 59:743–748.

Hinnen, C. and Niki, K. (1989) Roles of 4,4′-bipyridyl and bis(4-pyridyl) disulphide in the redox reaction of cytochrome c adsorbed on a gold electrode. Spectroelectrochemical studies. *J. Electroanal. Chem.* 264:157–165.

Laviron, E. (1979) General expression of the linear potential sweep voltammogram in the case of diffusionless electrochemical systems. *J. Electroanal. Chem.* 101:19–28.

Niu, J., Guo, Y. and Dong, S. (1995a) A cryo-hydrogel immobilized protein electrode for direct electron transfer of myoglobin. *Chin. Chem. Lett.* 6:421–424.

Niu, J., Guo, Y. and Dong, S. (1995b) The direct electrochemistry of cryo-hydrogel immobilized myoglobin at a glassy carbon electrode. *J. Electroanal. Chem.* 399:41–46.

Niwa, K., Furukawa, M. and Niki, K. (1988) IR reflectance studies of electron transfer promoters for cytochrome c on a gold electrode. *J. Electroanal. Chem.* 245:275–285.

Ohtani, M. and Ikeda, O. (1993) Direct electrochemistry of horse heart cytochrome c at the γ-alumina-coated electrode. *J. Electroanal. Chem.* 354:311–317.

Oliver, B.N., Egekeze, J.O. and Murray, R.W. (1988) "Solid-state" voltammetry of a protein in a polymer solvent. *J. Am. Chem. Soc.* 110:2321–2322.

Qu, X, Lu, T., Dong, S. Zhou, C. and Cotton, T.M. (1994) Electrochemical reaction of cytochrome c at gold electrodes modified with thiophene containing one functional group. *Bioelectrochem. Bioenerg.* 34:153–156.

Sagara, T., Niwa, K., Sone, A., Hinnen, C. and Niki, K. (1990) Redox reaction mechanism of cytochrome c at modified gold electrodes. *Langmuir* 6:254–262.

Sagara, T., Murakami, H., Igarashi, S., Sato, H. and Niki, K. (1991) Spectroelectrochemical study of the redox reaction mechanism of cytochrome c at a gold electrode in a neutral solution in the presence of 4,4′-bipyridyl as a surface modifier. *Langmuir* 7:3190–3196.

Salamon, Z. and Tollin, G. (1991) Interfacial electrochemistry of cytochrome c at a lipd bilayer modified electrode: effect of incorporation of negative charges into the bilayer on cyclic voltammetric parameters. *Bioelectrochem. Bioenerg.* 26:321–334.

Song, S., and Dong, S. (1988) Spectroelectrochemistry of the quasi-reversible reduction and oxidation of hemoglobin at a methylene blue absorbed modified electrode. *Bioelectroanal. Bioenerg.* 19:337–346.

Song, S., Zhang, W. and Dong, S. (1990) Electrocatalysis of polypyrrole-methylene blue film modified electrode for direct redox reaction of cytochrome c. *Chin. Sci. Bull.* 35:1961–1964.

Taniguchi, I., Toyosawa, K., Yamaguchi, H. and Yasukouchi, K. (1982) Voltammetric response of horse heart cytochrome c at a gold electrode in the presence of sulfur bridged bipyridines. *J. Electroanal. Chem.* 140:187–193.

Taniguchi, I., Watanabe, K. Tominaga, M. and Hawkridge, F.M. (1992) Direct electron transfer of horse heart myoglobin at an indium oxide electrode. *J. Electroanal. Chem.* 333:331–338.

Veyama, S., Wada, O., Miyamoto, M., Kawakubo, H., Inatomi, K. and Isoda, S. (1993) Vectorial electron transfer in flavolipid/cytochrome c heterolayer. *J. Electroanal. Chem.* 347: 443–449.

Xie, Y. and Dong, S. (1994) Binding of cytochrome c with 4-pyridyl derivatives modified on the gold electrode. *Electroanalysis* 6:567–570.

Zak, J., Porter, M.D. and Kuwana, T. (1983) Thin-layer electrochemical cell for long optical path length observation of solution species. *Anal. Chem.* 55:2219–2222.

Zhan, R., Song, S., Liu, Y. and Dong, S. (1990) Mechanisms of methylene blue electrode processes studied by *in situ* electron paramagnetic resonance and ultraviolet-visible spectroelectrochemistry. *J. Chem. Soc. Faraday Trans.* 86:3125–3127.

Zhang, W., Song, S. and Dong, S. (1990) Heterogeneous electron transfer of cytochrome c facilitated by polypyrrole and methylene blue polypyrrole film modified electrodes. *J. Inorg. Chem.* 40:189–195.

Zhou, C., Qu, X., Lu, T., Dong, S. and Cotton, T.M. (1992) Promotion effect of pyridine for the direct electrochemistry of cytochrome c at gold electrodes. *Chin. Chem. Lett.* 3:133–134.

Zhou, C., Qu, X., Lu, T., and Dong, S. (1993) Direct electrochemistry of cytochrome c at the carbazole modified gold electrode. *Chin. J. Appl. Chem.* 10(4):82–83.

Zhu, Y. and Dong, S. (1990a) Rapid redox reaction of hemoglobin at methylene green modified platinum electrode. *Electrochim. Acta* 35:1139–1143.

Zhu, Y. and Dong, S. (1990b) Rapid oxidation and reduction of hemoglobin at a bifunctional dye of janus green modified electrode. *Bioelectrochem. Bioenerg.* 24:23–31.

Zhu, Y. and Dong, S. (1991) Rapid electrochemical oxidation of hemoglobin at a dye modified electrode. *Bioelectrochem. Bioenerg.* 26:351–357.

Zhu, Y. and Dong, S. (1992) Effect of the modified layer structure of a bifunctional organic compound modified electrode of the electrode reaction of cytochrome c. *Chin. J. Catalysis* 13:209–215.

Frontiers in Biosensorics I
Fundamental Aspects
ed. by F. W. Scheller, F. Schubert and J. Fedrowitz
© 1997 Birkhäuser Verlag Basel/Switzerland

Electron transfer via redox hydrogels between electrodes and enzymes

I. Katakis[1] and A. Heller[2]

[1]Department of Chemical Engineering, Universitat Rovira i Virgili, E-43006 Tarragona, Spain;
[2]Department of Chemical Engineering, The University of Texas at Austin, Austin, TX, 78712, USA

Summary. The molecular "wiring" of redox enzymes provides the basis for amperometric enzyme and affinity sensors. The key features of the useful redox polymers are hydrophilicity and flexibility.

In search of a transduction scheme

The idea of using the selectivity imparted by nature on biological molecules for chemical analysis of complex mixtures has been around for most of this century. An optically monitored reaction of glucose oxidase was one of the first analytical applications of this concept (Bentley, 1963). In the last 30 years the idea has been revived after the realisation by Updike and Hicks (1967) of an enzyme electrode. During this period, numerous configurations have appeared using various biological molecules for analytical purposes, integrating the sensing systems on electrode surfaces (see for example Silverman and Brake, 1970; Davies and Mosbach, 1974; Janata, 1975; Kulys et al., 1980; Cass et al., 1984; Belli and Rechnitz, 1986; Bartlett and Whitaker, 1987; Degani and Heller, 1989; Bartlett et al., 1991; Kuwabata et al., 1995 for representative original works and comprehensive reviews). Although the electrochemical systems present practical advantages (mainly simplicity of operation, and low cost of production and detection devices), recently sensors based on optical and/or gravimetric detection have appeared (Cush et al., 1993; Ramsden, 1993; Ngeh-Ngwainbi et al., 1986; Suleiman and Guilbault, 1994) as the optical detection systems become less expensive and more user friendly while sustaining their low detection levels. A common motif in all such reseach is that of transduction: the biological recognition reaction or interaction is translated into an electro/chemical or optical signal that can be measured. Although there is no such thing as the "perfect transducer" the concept of electrical "wiring" of biomolecules with redox polymers has gained acceptance as a technique to effect an efficient transduction, and as a direct and simple system for the study of biorecognition reactions. This concept is depicted in Figure 1 where it is seen that its application to a biosensor consists of three parts: The biorecognition chemistry (biomolecule), the electron abstraction and propagation

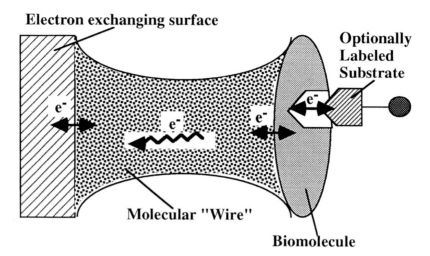

Figure 1. Schematic of the "wiring" transduction of biomolecular recognition reactions.

chemistry (redox polymer) and the retaining surface (although we will generally refer to electrodes and electrochemical detection, the said components can also be immobilized on an optical fiber or other measuring surface). The advantage that this general configuration offered to biosensor research, when it was first introduced (Degani and Heller, 1989), was that it provided a compact, manufacturable system with no leachable components. It was found in the following years, that a number of factors-properties of the redox macromolecules contributed to the efficiency of the process of "wiring" and that these could be not only investigated and quantified, but also rationally manipulated. With such knowledge, a number of designs targeted for various applications became possible. Here we provide an update on the level of our understanding of the "wired" enzyme system and on advances in the way of specific applications.

What happens with electrons received

Before considering the properties of the "wires" that render them efficient transducers we discuss the fate of the passing electrons. This step can be the limiting factor of the response of biosensors and of their performance characteristics. First, a look is due at the nature of the molecular "wires" (Heller 1990, 1992). Of the various redox macromolecules that can be utilized as transducers, the most successful ones to date are complexes of osmium, depicted in Figure 2. In systems with films thicker than a few monolayers, an electron "diffusion" process is necessary for electrical communication between biological molecule and the electron collecting or providing sur-

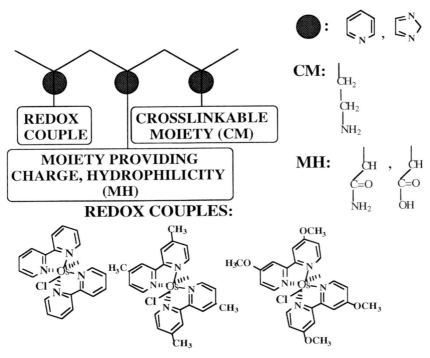

Figure 2. General representation of molecular "wires" with the polymer backbones, side chains, and redox couplex used to date.

face. This process, governed by a phenomenological electron diffusion coefficient described by Dahms (1968) and Ruff and Friedrich (1971) D_e, can be described as:

$$R_{j_{ox}} + R_{j+1_{red}} \xrightarrow{D_e} R_{j_{red}} + R_{j+1_{ox}}$$

where R_j and R_{j+1} are the electron-exchanging relays of the redox polymer.

The answer to the question of the factors influencing the propagation of electrons through this monolithic film can be complex and the results of groups studying the electron diffusion in redox films need to be discussed. Fritsch-Faules and Faulkner (1989, 1992) and Blauch and Savéant (1992, 1993) pointed out that there are three processes leading to electron diffusion in redox polymers: Percolative electron hopping, long-range chain motion, and short-range chain motion. The process can be limited by the diffusion of counterions, required for maintenance of electroneutrality in the redox film. In the redox hydrogels composed of the family of macromolecules similar to those shown in Figure 2, the electron diffusion process involves electron-transferring collisions between chain segments carrying redox centers, i.e. physical motion of the segments. A series of studies (Aoki and

Table 1. Apparent electron diffusion coefficients (D_e) [$cm^2 s^{-1}$] for various molecular "wires" at pH 7.0

POs	: m = 5,	n = 0,	o = 1
POs2	: m = 0.8,	n = 0,	o = 1.2
POs3	: m = 2,	n = 0,	o = 1
POsMe1	: m = 4,	n = 1,	o = 1, R = CH_3
POsMe3	: m = 2,	n = 3,	o = 1, R = CH_3
POsAm1	: m = 3.4,	n = 0.6	o = 1, R = $C_2H_4NH_2$
POsAm9	: m = 4.9,	n = 9,	o = 1, R = $C_2H_4NH_2$

Polymer	D_e	Reference
POs	2.3×10^{-9}	Aoki et al. (1995)
POs2	9.6×10^{-9}	Katakis (1994)
POs3	7.9×10^{-9}	Aoki et al. (1995)
POsMe1	9.2×10^{-9}	Aoki et al. (1995)
POsMe3	3.8×10^{-8}	Aoki et al. (1995)
POsAm1	3.7×10^{-9}	Ye (1994)
POsAm9	1.2×10^{-8}	Ye (1994)

Electron diffusion coefficients determined with interdigitated electrodes in films crosslinked with 6 wt% PEG.

Heller, 1993; Ye 1994; Katakis, 1994; Aoki et al., 1995) aimed at defining the limiting steps and to control the rate of electron diffusion through structural modifications of the redox polymers and the monolithic film indicated that the control of electron diffusion is dictated by the mobility of the polymer chains and thus probably by segmental motion within the crosslinked redox polymer networks. The apparent diffusion coefficients of some members of the family of redox polymers are listed in Table 1. Hydration of the network, i.e., its swelling, increases with the density of the positive charge on the chains, and results in higher electron diffusion coefficients. Charge and swelling increase when the density of Os-redox centers is higher or when more of the nitrogens are quaternized. However, at high osmium loadings the density of crosslinkable functions is reduced (as more available crosslinking sites are complexed with osmium) and the films are not well retained. This fact has prompted consideration of polymeric backbones such as those of poly(vinylimidazole) (Ohara et al., 1993a; Rajagopalan et al., 1994; Ohara et al., 1994) with a high density of crosslinkable nitrogens and also lower redox potentials. Amperometric enzyme electrodes based on such polymers show high current densities reaching about 1 mA cm^{-2} for glucose sensors made with glucose oxidase. The current densities reached, though high, can be limited by electron diffusion, by electron transport between the "wire" and the enzyme, or by enzyme turnover.

At higher charge density and swelling not only the electron diffusion coefficient, but also the permeability of the water soluble substrates are increased. The water-swollen materials used as redox wires are superior to the rigid ones like the poly(vinyl ferrocenes) that swell much less and were first tried by Cenas et al. (1983). Recent efforts in "wiring" redox enzymes by other groups as well, focus on this fact (Calvo et al., 1994; Hale et al., 1991; Willner et al., 1994; Chen et al., 1993).

Although high electron diffusion coefficients have been achieved through hydration and segmental motion of the polymer chains, as can be observed from Table 1, the electron diffusion coefficients are still two orders of magnitude lower than the diffusion coefficient of a freely diffusing mediator. It can be said that in "wired" systems one achieves a reagentless non-leaching system at the price of a two orders of magnitude lower electron diffusion. The question is if the limit of improvement of the rate of electron diffusion has been reached. We believe that this is not so, and that an improvemet by at least one order of magnitude is possible.

Redox enzymes in redox hydrogels

The latest crystallographic information on glucose oxidase (Hecht et al., 1993a, b), an enzyme that has been widely used in the study of biosensors, shows that it is a compact structure with very little access of the solvent to the FAD cofactor, unless through a narrow channel of 7Å in diameter. It is difficult to imagine that the osmium couple which, associated with the polymeric backbone should have a hydrodynamic dimension of at least 12 Å, is capable of "penetrating" in this access channel to effect electron transfer. However, all experimental evidence shows that the minimization of the distance for electron tunnelling effects a formidable increase in the rate of electron transfer. According to recent results by Wuttke et al. (1992) and Casimiro et al. (1993), the effective electron tunnelling distance can be reduced not only through physical approach of the redox partners, but also through contacts at the surface of redox proteins to the endings of conduction pathways through their bulk. Although some redox enzymes will not have such conducting pathways when their function is to guard electrons and not to transfer them indiscriminately, contact between the osmium complexes and near-surface sites of proteins might explain why the "wired" enzymes are electroreduced or oxidized. Yet, the explanation might lie in a closer contact between the redox partners, effected by the penetration of the redox polymer through the oligosaccharide layer of enzymes like glucose oxidase.

The investigation of what makes the communication between the enzymes and the redox polymers more efficient, has led us to try to quantify and manipulate the efficiency of adduct formation between the two redox molecules (Katakis, 1994; Katakis et al., 1994). The results prove that the stronger the bonding in the adduct, the higher the rate of electron transfer from the enzymatic cofactors to the "wiring" redox couple (Table 2). Although the strength of bonding in an adduct does not result exclusively of simple electrostatic interaction, opposite charges on the enzyme and redox polymer usually result in efficient electron transfer. This points to another

Table 2. Relation between electron transfer rates from oxidases to molecular "wires" and the strength of bonding in their adducts

ENZYME	pI	Polycation Equiv. Charge = + 18.95		Zwitterion Equiv. charge = – 9.85	
		Rel. electron transfer rate	Rel. adduct bonding	Rel. electron transfer rate	Rel. adduct bonding
Glucose Oxidase	3.8	1	1	0	< 0.14
Glucose Oxidase	5.0	0.09	0.07	0.02	0.06
Cholesterol Oxidase	6.8	0.0005	0.05	0.009	> 0.06
Cholesterol Oxidase	4.8	0.01	0.07	0.007	< 0.03
Lactate Oxidase	3.6	0.56	0.79	0.006	< 0.09

Equivalent charge of "wire" is given in moles per unit molecular weight in the reduced form. The pI 5.0 glucose oxidase variant was produced by modification with polyamines of the oligosaccharide "coat" of the enzyme. The two cholesterol oxidases were from different microbial sources. The adduct bonding was determined by isoelectric focusing, and the electron transfer rate by the maximum current obtained for thin films electrodes. Both are normalised per unit of enzymatic activity in the adduct (moles of electrons produced per unit time) and the maximum value was taken as unity.

reason of why the polycationic molecular "wires" depicted in Figure 2 have been successful in transferring electrons from redox enzymes that are in their majority poyanions at pH 7 (a pH where most of the evaluation and application of biosensors are done).

Although the adduct formation can slow the leaching of an enzyme from the redox hydrogel, it is impossible to produce robust and stable long-term enzyme electrodes, if the film is not crosslinked. We have studied a number of crosslinkers to effect such crosslinking. The choice of poly(ethylene glycol) diglycidyl ether (PEG) by Gregg and Heller (1991 a, b) resulted from its stability in aqueous solutions, its reactivity with crosslinkable primary amines and its flexibility and hydrophilicity. Polyfunctional aziridines (PAZ) were used as crosslinkers in the case of enzymes having amines that play an important functional role in the catalytic process and need to be left unreacted. Through use of these water-soluble crosslinkers hydrogels that were "open" and flexible were made. Added functions on the redox polymers such as primary amines are necessary for crosslinking and also increase the charge (and thus strength of bonding of adducts) and the flexibility-hydrophilicity of the polymers. The interesting fact is that even without further derivatization of the electrode surface to increase the attachment of the film the glucose electrodes were robust and stable for a week in continuous operation at $37\,^{\circ}C$, and H_2O_2 sensing electrodes made with a thermostable peroxidase were operated for prolonged periods at $45\,^{\circ}C$ and for hours at $65\,^{\circ}C$ (Vreeke et al., 1995 a).

Redox-enzyme based biosensors without leachable components

The "wired" enzyme electrodes provide for the first time a system truly reagentless having no leachable components. This property is a strict requirement for in vivo operation of biosensors and is more than desirable for operation in flow systems.

A disadvantage of the early "wired" enzyme electrodes was that the sensing potential was on the other of 400 mV vs SCE (Gregg and Heller, 1990; Psihko et al., 1990; Katakis and Heller, 1992). It is not trivial to change the potential of the osmium couples attached to the polymeric backbone because of the necessary engagement of a coordination site of the polymer and the pair-type occupation of the rest by bidentate ligands. However, first by the use of poly(vinyl imidazole) by Ohara et al. (1993 a) and next with the modification of the ligands with electron-donating groups, such as methyl and methoxy (Ohara et al., 1994; Taylor et al., 1995), it was possible to lower the redox potential of the "wire" to – 50 mV vs SCE and the sensing potential to + 50 mV in glucose and lactate sensors made with the respective oxidases. At more negative potentials catalytic oxygen reduction is observed. At this sensing potential only ascorbate among all of the commonly present non-specifically oxidizable compounds might be electrooxidized and interference-free glucose and lactate electrodes can be built.

In reality the electrooxidation of interferants is a complex function of the relative rates of the kinetic and diffusion processes of substrate and interferants in the films. It has been shown for example (Ohara et al., 1994), that essentially interference-free sensors can be made even when operated at relatively high potentials, when the right balance between the kinetic and diffusional processes is stricken. A practical solution of the specificity problem would exist if the rate of electron transfer between the enzyme and the "wire" were higher than that between the oxidized relays and interfering

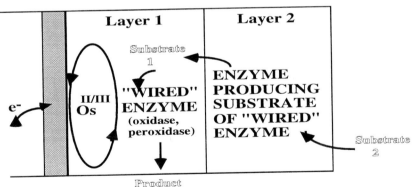

ANODES : Any Oxidase (one-enzyme electrode)
          Diaphorase / Any NAD(P)-dependent Dehydrogenase
          Sarcosine Oxidase / Creatinase
          Cholesterol Oxidase / Cholesterol Esterase
          Choline Oxidase / Choline Esterase
CATHODES: Peroxidase / Any Oxidase

Figure 3. Possible mono-enzyme and bi-enzyme biosensor configurations based on molecular "wires".

Table 3. Enzyme electrodes based on "wired" enzymes

SUBSTRATE	ENZYME 1	ENZYME 2	j_{max}	X-LINKER	$t_{1/2}$ (h)
Glucose	Glucose oxidase	–	1000	PEG	118
Lactate	Lactate oxidase	–	600	PEG	10
L-α Glycerophosphate	L-α glycerophosphate oxidase	–	700	PEG	18
Glutamate	Glutamate oxidase	–	12	PEG	7
Sarcosine	Sarcosine oxidase	–	110	PAZ	3
Theophylline	Theophylline oxidase	–	2	PAZ	6
Cholesterol	Cholesterol oxidase	–	1	PEG	2
D-alanine	D-amino acid oxidase	–	7	PAZ	2
Hydrogen Peroxide	Peroxidase	–	1000	PEG, Ox	> 200
Creatine	Sarcosine oxidase	Creatinase	30	PAZ	7
D-tyrosine	Peroxidase	D-amino acid oxidase	20	PEG, Ox	10
Ethanol	Peroxidase	Alcohol oxidase	17	PEG, Ox	3
Choline	Peroxidase	Choline oxidase	90	PEG, Ox	10

Enzyme 1 is the "wired" enzyme. J_{max} is the highest current density in $\mu A\ cm^{-2}$ observed at saturating substrate concentrations with the best molecular "wire". Except in the cases of lactate, glycerophosphate, glutamate, D-amino acid oxidase, and peroxidase, these values were not optimized. The crosslinkers are PEG: poly(ethylene glycol) diglycidyl ether, PAZ: polyfunctional aziridine, Ox: through oxidation of oligosaccharides on the surface of the enzyme. The $t_{1/2}$ are half-life times of the electrodes in continuous operation under saturating substrate concentrations, usually under vigorous stirring conditions (rotating electrodes at 1000 rpm) at 37 °C, without any stabilizing overlayers. This characteristic was not optimized except for glucose oxidase and peroxidase.

substances. Our kinetic analysis suggests that in a number of oxidase sensors the desired rate is however two or more orders of magnitude lower for the "wires" than for the fastest freely diffusing mediators. Better interference rejection could be based on "wires" that accept electrons from e.g. reduced glucose oxidase at higher rates. These are likely to be redox polymers that are even more flexible and hydrophilic.

Nevertheless, even with the present art, a number of mono enzyme and bienzyme electrodes (Katakis and Heller, 1992; Wang and Heller, 1993; Ohara et al., 1993b; Vreeke and Heller, 1994; Katakis et al., 1996) depicted in Figure 3 and listed in Table 3 with some of their characteristics, were built.

An important consequence of this type of rational design of electrodes is that their performance properties can be tailored. This means that the dynamic range of the electrodes can be changed, their efficiency in competing with oxygen for enzyme reoxidation (in the case of oxidases) can be improved, and their stability extended by changing the rate-limiting steps or their relative significance for response. For example, as seen in Figure 4 the apparent K_m of a sarcosine or a glucose sensor can be increased almost fourfold by increasing the concentration of "wire" in the film; the dynamic range or stability of a glucose sensor can be extended by an order of magnitude through usage of a suitable diffusion limiting membranes (Ye et al., 1994); and, as seen in Figure 5, the oxygen competition in a glycerophosphate electrode can be essentially eliminated by design. This flexibility in the "wired" electrodes provided the attaining of specific performance characteristics for particular applications.

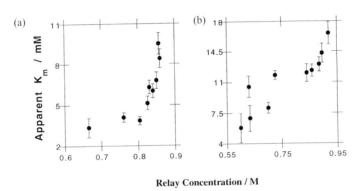

Figure 4. Effect of wire-bound relay concentration in the film on the apparent K_m of (a) sarcosine oxidase and (b) glucose oxidase electrodes. The relay concentration was calculated using the dry film thickness and the electrochemically measured quantity of redox couples. Error bars represent standard errors of the slopes of Eadie-Hofstee plots used for apparent K_m calculation.

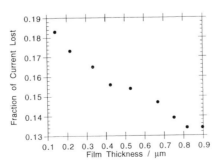

Figure 5. Fraction of current lost as function of film thickness for a L-α glycerophosphate oxidase electrode when switching the solution saturation from Ar to O_2. The film thickness is the measured by profilometry dry film thickness.

Affinity sensors

Because enzymes, that are used as labels of bioaffinity reagents, such as peroxidases are readily "wired" by the redox hydrogels various bioaffinity sensors can be built. With these, the occurrence of a bioaffinity reaction is observed as a change in the electrical current. The current flows upon electrooxidation or electroreduction of a substrate by the "wired" enzyme that labels the affinity reagent. For example, the avidin-biotin reaction is sensed by incorporating in the redox hydrogel avidin and using biotinylated horseradish peroxidase as the avidin's complement. When the biotynilated peroxidase binds the avidin, the enzyme is "wired" and a current flows, i.e. H_2O_2 is electroreduced to water. Both dissolved avidin and dissolved biotin competitively inhibit the binding of the biotinylated peroxidase and prevent the flow of current as demonstrated by Vreeke et al. (1995b).

The same principle was applied in the sensing of oligonucleotides. One oligonucleotide was terminally labeled with the peroxidase while the complimentary one was bound to the redox hydrogel through its opposite end. Hybridization resulted in the flow of current (de Lumley-Woodyear et al., 1996).

Subcutaneously implanted glucose sensors

The subcutaneously implantable glucose sensors were designed for comfort, one-point in vivo calibration and extended life (Csöregi et al., 1994, 1995). The sensors are flexible and plastically deformable 250-μm diameter gold microwires, insulated with a 20-μm thick polyimide. These wires are cut and electrochemically etched to form a recess, shielded by the polyimide. The gold wire tip at the bottom of the recess is coated with the "wired" glucose oxidase sensing layer, a glucose-transport limiting polymer film and a biocompatible polymer film (Quinn et al., 1995). The second layer defines the dynamic range (1 mM–30 mM) and improves the stability, and the outer layer prevents fouling, thrombogenesis, and immune reaction. The total amount of material in the three layers is 2.2 μg. The sensors were designed and operated in rats, with external Ag/AgCl electrodes on the skin, for 1 week.

Conclusions

It has been shown that the concept of electrochemical wiring through redox polymers results in versatile biosensors, enzyme electrodes, and affinity sensors being two examples of these. Emphasis was placed on the design of the molecular wires forming redox hydrogels where flexibility, hydrophilicity, and fast redox properties are of essence. The achievement of implantable glucose electrodes based on this concept is the direct consequence of the versatility of the electron-conducting hydrogels.

Acknowledgements
A.H. thanks the National Science Foundation, the Office of Naval Research, The Welch Foundation and the National Institutes of Health for financial support. I.K. gratefully acknowledges a startup grant of the University Rovira i Virgili and the autonomous government (Generalitat) of Catalonia (grant #94 255A2) and a grant from the Environment ministry (Departament de Medi Ambient) of the same (contract #2292). Dr. Ling Ye read and improved the manuscript.

References

Aoki, A. and Heller, A. (1993) Electron Diffusion Coefficients in Hydrogels Formed of Cross-Linked Redox Polymers. *J. Phys. Chem.* 97:11014–11019.

Aoki, A., Rajagopalan, R. and Heller, A. (1995) Effect of Quaternization on Electron Diffusion Coefficients for Redox Hydrogels Based on Poly(4-vinylpyridine). *J. Phys. Chem.* 99:5102–5110.

Bartlett, P.N. and Whitaker, R.G. (1987/88) Enzyme Electrodes. *Biosensors* 3:359–379.

Bartlett, P.N., Tebbutt, P. and Whitaker, R.G. (1991) Kinetic Aspects of the Use of Modified Electrodes and Mediators in Bioelectrochemistry. *Prog. Reaction Kin.* 16:55–155.

Blauch, D.N. and Savéant, J.-M. (1992) Dynamics of Electron Hopping in Assemblies of Redox Centers. Percolation and Diffusion. *J. Am. Chem. Soc.* 114:3323–3329.

Blauch, D.N. and Savéant, J.-M. (1993) Effects of Long-Range Electron Transfer on Charge Transport in Static Assemblies of Redox Centers. *J. Phys. Chem.* 97:6444–6450.

Belli, S. and Rechnitz, G. (1986) Prototype Potentiometric Biosensor Using Intact Chemoreceptor Structures. *Anal. Lett.* 19:403–416.

Bentley, R. (1963) Glucose Oxidase. *In*: P.D. Boyer, H. Lardy and K. Myrback (eds.): *The Enzymes* Vol. 7. Academic Press, London, pp 567–586.

Calvo, E.J., Dalilowicz, C. and Diaz, L. (1994) A New Polycationic Hydrogel for Three-Dimensional Enzyme Wired Modified Electrodes. *J. Electroanal. Chem.* 369:279–282.

Cass, A.E.G., Davis, G., Francis, G.D., Hill, H.A.O., Aston, W.J., Higgins, I.J., Plotkin, E.V., Scott, L.D.L. and Turner, A.P.F. (1984) Ferrocene-Mediated Enzyme Electrode for Amperometric Determination of Glucose. *Anal. Chem.* 55:667–671.

Casimiro, D.R. Wong, L.-L., Colón, J.L., Zewert, T.E., Richards, J.H., Chang, I.-J., Winkler, J.R. and Gray, H.B. (1993) Electron Transfer in Ruthenium/Zinc Porphyrin Derivatives of Recombinant Human Myoglobins. Analysis of Tunneling Pathways in Myoglobin and Cytochrome c. *J. Am. Chem. Soc.* 115:1485–1489.

Cenas, N.K., Pocius, A.K. and Kulys, J.J. (1983) Electron Exchange between Flavin and Heme-Containing Enzymes and Electrodes Modified by Redox Polymers. *Bioelectrochem. Bioenerg.* 11:61–67.

Chen, C.J., Liu, C.C. and Savinell, R.F. (1993) Polymeric Redox Mediator Enzyme Electrodes for Anaerobic Glucose Monitoring. *J. Electroanal. Chem.* 348:317–338.

Csöregi, E., Quinn, C., Lindquist, S.-E., Schmidtke, D., Pishko, M., Ye, L., Katakis, I., Hubbel, J. and Heller, A. (1994) Design, Characterization, and One-Point *in Vivo* Calibration of a Subcutaneously Implanted Glucose Electrode. *Anal. Chem.* 66:3131–3138.

Csöregi, E., Schmidtke, D.W. and Heller, A. (1995) Design and Optimization of a Selective Subcutaneously Implantable Glucose Electrode Based on "Wired" Glucose Oxidase. *Anal. Chem.* 67:1240–1244.

Cush, R., Cronin, J.M., Stewart, W.J., Maule, C.H., Molloy, J. and Goddard, N.J. (1993) The Resonant Mirror: A Novel Optical Biosensor for Direct Sensing of Biomolecular Interactions Part I: Principle of Operation and Associated Instrumentation. *Biosens. Bioelectr.* 8:347–352.

Dahms, H. (1968) Electronic Conduction in Aqueous Solution. *J. Phys. Chem.* 72:362–364.

Davies, P. and Mosbach, K. (1974) The Application of Immobilized NAD in an Enzyme Electrode and in Model Enzyme Reactor. *Biochim. Biophys. Acta* 370:329–338.

de Lumley-Woodyear, T., Campbell, C.N. and Heller, A. (1995). Direct Enzyme-Amplified Electrical Recognition of a 30-Base Model Oligonucleotide. *J. Am. Chem. Soc.* 118: 5504–5505.

Degani, Y. and Heller, A. (1989) Electrical Communication between Redox Centers of Glucose Oxidase and Electrodes via Electrostatically and Covalently Bound Redox Polymers. *J. Am. Chem. Soc.* 111: 2357–2358.

Fritsch-Faules, I. and Faulkner, L.R. (1989) A Microscopic Model for Diffusion of Electrons by Successive Hopping among Redox Centers in Networks. *J. Electroanal. Chem.* 263: 237–248.

Fritsch-Faules, I. and Faulkner, L.R. (1992) Use of Microelectrode Arrays to Determine Concentration Profiles of Redox Centers in Polymer Films. *Anal. Chem.* 64: 1118–1126.

Gregg, B.A. and Heller, A. (1990) Cross-Linked Redox Gels Containing Glucose Oxidase for Amperometric Biosensor Applications. *Anal. Chem.* 62: 258–263.

Gregg, B.A. and Heller, A. (1991a) Redox Polymer Films Containing Enzymes. 1. A Redox-Conducting Epoxy Cement: Synthesis, Characterization, and Electrocatalytic Oxidation of Hydroquinone. *J. Phys. Chem.* 95: 5970–5975.

Gregg, B.A. and Heller, A. (1991b) Redox Polymer Films Containing Enzymes. 2. Glucose Oxidase Containing Enzyme Electrodes. *J. Phys. Chem.* 95: 5976–5980.

Hale, P.D., Lan, H.L., Boguslawsky, L.I., Karan, H.I., Okamoto, Y. and Skotheim, T.A. (1991) Amperometric Glucose Sensors Based on Ferrocene-Modified Poly(Ethylene Oxide) and Glucose Oxidase. *Anal. Chim. Acta* 251: 121–128.

Hecht, H.J., Schomburg, D., Kalisz, H.M. and Schmid, R.D. (1993a) The 3D Structure of Glucose Oxidase from *Aspergillus niger*. Implications for the Use of GOD as a Biosensor Enzyme. *Biosens. Bioelectr.* 8: 197–203.

Hecht, H.J., Kalisz, H.M., Hendle, J. and Schmid, R.D. (1993b) Crystal Structure of Glucose Oxidase from *Aspergillus niger*. Refined at 2.3 Å Resolution. *J. Mol. Biol.* 229: 153–172.

Heller, A. (1990) Electrical Wiring of Redox Enzymes. *Acc. Chem. Res.* 23: 128–134.

Heller, A. (1992) Electrical Connections of Enzyme Redox Centers to Electrodes. *J. Phys. Chem.* 96: 3579–3587.

Janata, J. (1975) An Immunoelectrode. *J. Am. Chem. Soc.* 97: 2914–2921.

Katakis, I. (1994) *Development and Analysis of Operation of Enzyme Electrodes Based on Electrochemically "Wired" Oxidoreductases*, Ph.D. Thesis. The University of Texas at Austin, pp 30–73.

Katakis, I. and Heller, A. (1992) L-α-Glycerophosphate and L-Lacatate Electrodes Based on the Electrochemical "Wiring" of Oxidases. *Anal. Chem.* 64: 1008–1013.

Katakis, I., Ye, L. and Heller, A. (1994) Electrostatic Control of the Electron Transfer Enabling Binding of Recombinant Glucose Oxidase and Redox Polyelectrolytes. *J. Am. Chem. Soc.* 116: 3617–3618.

Katakis, I., Vreeke, M., Ye, L., Aoki, A. and Heller, A. (1996) Electron Conducting Adducts of Water-Soluble Redox Polyelectrolytes and Enzymes. *In*: Advances in Molecular and Cell Biology. JAI Press, Inc., Greenwich, CT, USA, pp 389–407.

Kulys, J.J., Samalius, A.S. and Svirmickas, G.J.S. (1980) Electron Exchange Between the Enzyme Active Center and Organic Metal. *FEBS Lett.* 114: 7–13.

Kuwabata, S., Okamoto, T., Kajiya, Y. and Yoneyama, H. (1995) Preparation and Amperometric Glucose Sensitivity of Covalently Bound Glucose Oxidase to (2-Aminoethyl)ferrocene on an Au Electrode. *Anal. Chem.* 67: 1684–1690.

Ngeh-Ngwainbi, J., Foley, P.H., Kuan, S.S. and Guilbault, G.G. (1986) Parathion Antibodies on Piezoelectric Crystals. *J. Am. Chem. Soc.* 108: 5444–5460.

Ohara, T.J., Rajagopalan, R. and Heller, A. (1993a) Glucose Electrodes Based on Cross-Linked [Os(bpy)₂Cl]$^{+/2+}$ Complexed Poly(1-vinylimidazole) Films. *Anal. Chem.* 65: 3512–3517.

Ohara, T.J., Vreeke, M.S., Battaglini, F. and Heller, A. (1993b) Bienzyme Sensors Based on "Electrically Wired" Peroxidase. *Electroanalysis* 5: 825–831.

Ohara, T.J., Rajagopalan, R. and Heller, A. (1994) "Wired" Enzyme Electrodes for Amperometric Determination of Glucose or Lactate in the Presence of Interfering Substrates. *Anal. Chem.* 66: 2451–2457.

Pishko, M.V., Katakis, I., Lindquist, S.-E., Ye, L., Gregg, B. and Heller, A. (1990) Direct Electrical Communication between Graphite Electrodes and Surface Adsorbed Glucose Oxidase/Redox Polymer Complexes. *Angew. Chem.* 102: 109–111.

Quinn, C.P., Pathak, C.P., Heller, A. and Hubbell, J.A. (1995) Photo-crosslinked Copolymers of 2-Hydroxyethyl Methacrylate, Poly(ethylene glycol) and Ethylene Dimethacrylate for Improving Biocompatibility of Biosensors. *Biomaterials* 16: 389–396.

Rajagopalan, R., Ohara, T.J. and Heller, A. (1994) Electrical Communication between Glucose Oxidase and Electrodes Based on Poly(vinylimidazole) Complex of Bis(2,2'-bipyridine)-N,N'-dichloroosmium. *In*: A.M. Usmani and N. Akmal (eds.): *Diagnostic Biosensor Polymers*. ACS Symposium Series 556, Washington D.C., pp 307–317.

Ramsden, J.J. (1993) Partial Molar Volume of Solutes in Bilayer Lipid Membranes. *J. Phys. Chem.* 97:4479–4485.

Ruff, I. and Friedrich, V.J. (1971) Tranfer Diffusion. I. Theoretical. *J. Phys. Chem.* 75: 3297–3301.

Silverman, H.P. and Brake, J.M. (1970) Method of determining microbial populations, enzyme activities, and substrate concentrations by electrochemical analysis. *U.S. Patent* 3,506,544.

Suleiman, A.A. and Guilbault, G.G. (1994) Recent Developments in Piezoelectric Immuno-sensors. *Analyst* 119:2279–2282.

Taylor, C., Kenausis, G., Katakis, I. and Heller, A. (1995) "Wiring" of Glucose Oxidase within a Hydrogel made with Poly(vinil imidazole) Complexed with Os(4-4'-dimethoxy 2-2'-bipyri-dine)Cl. *J. Electroanal. Chem.* 396:511–515.

Updike, S.J. and Hicks, G.P. (1967) The Enzyme Electrode. *Nature* 214:986–988.

Vreeke, M.S. and Heller, A. (1994) Hydrogen Peroxide Electrodes Based on Electrical Connec-tion of Redox Centers. *In*: A.M. Usmani and N. Akmal (eds.): *Diagnostic Biosensor Poly-mers*. ACS Symposium Series 556, Washington D.C., pp 180–193.

Vreeke, M.S., Yong, K.T. and Heller, A. (1995a) A Thermostable Hydrogen Peroxide Sensor Based on Wiring of Soybean Peroxidase. *Anal. Chem.* 67:4247–4249.

Vreeke, M.S., Rocca, P. and Heller, A. (1995b) Direct Electrical Detection of Dissolved Biot-nylated Horseradish Peroxidase, Biotin, and Avidin. *Anal. Chem.* 67:303–306.

Wang, D.L. and Heller, A. (1993) Miniaturized Flexible Amperometric Lactate Probe. *Anal. Chem.* 65:1069–1073.

Willner, I., Lapidot, N., Riklin, A., Kasher, R., Zahavy, E. and Katz, E. (1994) Electron-Trans-fer Communication in Glutathione Reductase Assemblies: Electrocatalytic, Photocatalytic, and Catalytic Systems for the Reduction of Oxidized Glutathione. *J. Am. Chem. Soc.* 116:1428–1441.

Wuttke, D.S., Bjerrum, M.J., Winkler, J.R. and Gray, H.B. (1992) Electron-Tunneling Pathways in Cytochrome c. *Science* 256:1007–1015.

Ye, L. (1994) *Characterization and Development of Glucose Electrodes Based on Enzyme-Containing Redox Hydrogels,* Ph.D. Thesis. The University of Texas at Austin, pp 36–56.

Ye, L., Katakis, I., Schummann, W., Schmidt, H.-L., Duine, J.A. and Heller, A. (1994) Enhan-cement of the Stability of Wired Quinoprotein Glucose Dehydrogenase Electrode. *In*: A.M. Usmani and N. Akmal (eds.): *Diagnostic Biosensor Polymers*. ACS Symposium Series 556, Washington D.C., pp 34–40.

Frontiers in Biosensorics I
Fundamental Aspects
ed. by F. W. Scheller, F. Schubert and J. Fedrowitz
© 1997 Birkhäuser Verlag Basel/Switzerland

Direct redox communication between enzymes and electrodes

T. Ikeda

Department of Agricultural Chemistry, Kyoto University, Sakyo-ku, Kyoto 606, Japan

Summary. In this mini-review bioelectrocatalysis based on direct electron transfer of redox enzymes at electrodes is described. Bacterial membrane-bound flavocytochrome enzyme and quinocytochrome enzymes adsorb spontaneously on the surfaces of carbon and metal electrodes. The adsorbed enzymes communicate with the electrodes at the heme c moiety to catalyze the electrolytic oxidation of the substrates. The catalytic mechanism is discussed and an application to a biosensor is demonstrated.

Introduction

Direct electron transfer between cytochrome c and electrodes was first demonstrated at a tin-doped indium oxide electrode by Yeh and Kuwana (1977) and at 4-4′ bipyridyl-modified gold electrodes by Eddows and Hill (1977) on the time scale of cyclic voltammetry. It was found in the same year that cytochrome c_3 from *Desulfovibrio vulgaris* MIYAZAKI produced a reversible voltammetric wave at a mercury electrode as studied in detail by Niki et al. (1979). Since then, a great number of papers have appeared dealing with direct electron transfer of cytochrome c and other redox proteins at bare and promoter-modified electrodes as reviewed by Armstrong et al. (1988) and Armstrong (1992).

From the biosensoric point of view, redox enzymes are more interesting than redox proteins; direct electron transfer between redox enzymes and electrodes allows bioelectrocatalysis of the oxidation or reduction of the substrates, that is, the acceleration of electrochemical reaction of the substrates in the absence of additional redox molecules serving as electron transfer mediators. Tarasevich et al. (1979) reported unmediated bioelectrocatalysis using laccase extracted from fungi. They showed that oxygen reduction at pH 5.0 started from 1.2 V vs NHE at the carbon black electrode on which laccase had been immobilized by irreversible adsorption. They discussed the mechanism of the bioelectrocatalytic reaction; however, the catalytic mechanism is quite involved and is not fully clarified yet. Lee et al. (1984) also observed catalytic current for the reduction of oxygen to water at laccase-coated pyrolytic graphite electrodes. The bioelectrocatalysis behavior was similar to but not exactly the same as that reported by Tarasevich et al. (1979). The potential for the catalytic oxygen reduction was

0.74 V vs NHE at pH 4.7, which contrasted with the potential of 1.27 V reported in the latter paper.

Horseradish peroxidase adsorbed on a graphite electrode catalyzes the reduction of H_2O_2 to water by the mechanism of direct unmediated electron transfer as discussed by Ruzgas et al. (1995). These authors have indicated the importance of the peroxidase-catalyzed reduction of H_2O_2 for detecting H_2O_2 produced in the reactions driven by oxidases when constructing amperometric biosensors. Armstrong and Lannon (1987) have demonstrated that cytochrome c peroxidase also catalyzes the reduction of H_2O_2 by a mechanism of direct electron transfer when aminoglycoside-modified edge-oriented pyrolytic graphite electrodes are used as the working electrodes. Padock and Bowden (1989) have shown that cytochrome c peroxidase adsorbed on an unmodified edge-oriented pyrolytic graphite electrode is also active to produce a catalytic current for the reduction of H_2O_2 via direct electron transfer from the electrode to the adsorbed enzyme. The catalytic function of the adsorbed cytochrome c peroxidase hase been studied in more detail by Scott et al. (1992).

Diaphorase form *Bacillus stearothermophilus* and ferredoxin-NADP$^+$ reductase from spinach leaves catalyze the oxidation of NADH and NADPH, respectively, at carbon and metal electrodes in the absence of external redox compounds (Kobayashi et al., 1992). However, the catalytic currents are rather small. More recently, two papers have appeared reporting direct bioelectrocatalysis of a glucose oxidation by glucose oxidase. Alvarez-Icaza and Schmid (1994) have reported the observation of a current response to glucose at an aminophenylboronic acid-modified electrode on which glucose oxidase was immobilized. Jiang et al. (1995) have prepared an electrode having glucose oxidase immobilized at a self-assembled monolayer by using 3,3′ dithiobis-sulfosuccinimidylpropionate, which is thiol-cleavable and assembles onto a gold surface. They have observed an increase in the oxidation current at the electrode on addition of glucose using cyclic voltammetry. The oxidation current starts to appear at the potential close to –0.6 V vs Ag/AgCl, while the half-wave potential of the wave due to the immobilized enzyme itself is located at about –0.28 V, a potential much more positive than –0.6 V. This is difficult to understand from the thermodynamic and kinetic point of view.

Available data on the electron transfer of redox proteins and enzymes at electrodes indicate that the electron transfer rate depends on the states of the electrode surface and the nature of proteins. There have been a number of attempts to modify the surface state of an electrode appropriately as mentioned above. Kulys (1986) has attempted to use organic salt conducting electrodes to realize bioelectrocatalysis based on the mechanism of electron transfer between the enzyme-active centers and organic salt. One possible mechanism is that electron transfer occurs via mediators formed in the layer near the electrode surface as a result of slight dissolution of the organic salts, but the electron transfer mechanism still remains unclear.

Degani and Heller (1987) changed the nature of glucose oxidase by introducing a sufficient number of electron-relaying centers to the enzyme, whereby the chemically modified enzyme becomes active to catalyze the electrochemical oxidation of glucose by electrical communication between the active center of the enzyme and an ordinary electrode through the relaying centers. An extension of this line of study was made by Riklin et al. (1995). They removed flavine adenine dinucleotide (FAD) cofactors from glucose oxidase, and reconstituted the apoprotein with modified FADs carrying redox-active ferrocene groups. In this way, electrical contact between an electrode and the modified enzyme in solution is enhanced, producing a catalytic wave for the oxidation of the substrate. This is not direct redox communication between electrodes and native enzymes, but it is interesting in view of the electron transfer reactions operating in multi-component enzymes of biological membranes.

Flavocytochrome enzymes and quinocytochrome enzymes are such multi-component enzymes consisting of more than two subunits; one of the subunits contains flavin (FAD) or pyrroloquinoline quinone (PQQ), and the others contain heme redox centers. The FAD or PQQ is the site to accept electrons from the substrates, and the electrons reach the heme redox centers through the enzyme molecules. We expected the heme groups of the multicomponent enzymes to serve as built-in mediators in direct bioelectrocatalysis, and in fact found that gluconate dehydrogenase, a membrane-bound flavocytochrome, adsorbed on carbon and gold electrodes produces a catalytic current for the oxidation of D-gluconte (Ikeda et al., 1988). Guo et al. (1989, 1990) have also demonstrated bioelectrocatalytic oxidations of p-cresol and sulfide, respectively, based on the direct electrochemistry of p-cresolmethylhydroxylase and sulfide: cytochrome c oxidoreductase in solution, both of which are flavocytochromes. Sucheta et al. (1992) have obtained a catalytic oxidation wave for succinate and reduction wave for fumarate at a pyrolytic graphite electrode on which succinate dehydrogenase was immobilized by irreversible adsorption; succinate dehydrogenase is a multicomponent enzyme which consists of two different subunits containing one covalently bound FAD and three Fe-S clusters. In this mini-review, we describe our own works concerning the redox communication between electrodes and the multicomponent enzymes from bacterial membranes.

Materials and methods

Enzymes

D-gluconate dehydrogenase (GADH; EC 1.1.99.3): This enzyme is a bacterial membrane-bound oxidoreductase that catalyzes the oxidation of D-gluconate (GlcA) to 2-keto D-gluconate in vivo at cytoplasmic membrane surfaces to give electrons to ubiquinone in the membrane. The enzyme is composed of three subunits with molar masses of 6600 g, 50,000 g, and 22,000 g, each of which contains covalently bound FAD, heme c and Fe-S clusters, respectively. FAD and heme c are considered to be the sites which react with GlcA and ubiquinone, respectively, allowing unidirectional electron flow through the enzyme from GlcA in solution to ubiquinine in the membrane. GADH was isolated from *Pseudomonas fluoescens* FM-1 by the method of Matsushita et al. (1982) and was obtained in solution (145 µg ml^{-1}) in 0.01 M phosphate buffer (pH 6.0) containing 0.1% Triton X-100, 10 mM sodium gluconate and 5 mM MgCl$_2$. The purity of the GADH preparation was checked by polyacrylamide gel electrophoresis, gel electrophoresis with sodium dodecyl sulfate and visible spectrophotometry; all these methods gave the same results as those reported for GADH by Matsushita et al. (1982). The GADH solution was stored in quantities of 0.2 ml at –30 °C. A batch of the solution was kept in a refrigerator at 5 °C before use and was used within three days, during which time the enzyme activity did not decrease.

Alcohol dehydrogenase (ADH; no EC number): This enzyme is a bacterial membrane-bound oxidoreductase that catalyzes the oxidation of ethanol to acetaldehyde. It consists of three subunits of molar masses of 85,000 g, 49,000 g and 14,400 g; the first subunit contains tightly bound PQQ and heme c, the second two heme c, and the third no redox group. PQQ is the site for reacting with the substrate, and all heme c can accept electrons from the PQQ. ADH was isolated from *Gluconobacter suboxydans* (IFO 12528) by the method of Ameyama et al. (1982), and was obtained in solution (145 µg ml^{-1}) in 0.1 M phosphate buffer (pH 6.0) containing 0.1% Triton X-100. The purity of the ADH preparation was checked in a similar manner; the enzyme was stored at -30 °C.

Fructose dehydrogenase (FDH; EC 1.1.99.11): This enzyme is also a membrane-bound oxidoreductase containing PQQ and heme c as redox active sites, PQQ being the site for reacting with D-fructose to produce 5-keto fructose. FDH preparation from *Gluconobacter* sp., Grade **III**, 29 units mg^{-1}, Lot 85710 was obtained from Toyobo Co. Stock solutions of FDH were prepared by dissolving 3 mg of the FDH preparation in 1 ml of McIlvaine buffer (pH 6.0) containing 0.1% Triton X-100. The FDH concentration of the stock solution was determined spectrophotometrically with reference to the spectrum given by Ameyama et al. (1981). Ubiquinone, which might be contained in the FDH preparation was determined by the method of Redfearn (1967), and was not detected. The stock solution was used within 2 weeks of preparation, during which time the FDH activity remained unchanged when stored at 5 °C.

Electrodes

Carbon paste electrodes (GPEs; geometrical surface area, 0.09 cm^2) were prepared as described by Ikdeda et al. (1985). Disks of pyrolytic graphite

(PGEs; Union carbide Corp.) held on the end of glass tubing with heat-shrinkable Teflon tubing were used as pyrolytic graphite electrodes; the mounted disks were freshly cleaved to expose an area of the basal plane of 0.19 cm^2. Glassy carbon (GCEs; 3.2 mm in diameter) and gold (AuEs; 1.6 mm in diameter) electrodes were obtained from BAS Corp. and platinum electrodes (PtES; 5 mm in diameter) from Toa Electronics Corp. Disks of silver (AgEs; 2 mm in diameter) glued to the ends of glass tubing with epoxy cement were used as silver electrodes. The GCE, AuE, PtE, and AgE were polished to a mirror-like finish with 0.05 mesk alumina, and were rinsed with water in an ultrasonic bath. A gold-plated platinum disk electrode (Au-PtE; 5 mm in diameter) and a platinized platinum disk electrode (Pt-PtE; 5 mm in diameter) were prepared using PtEs according to conventional procedures immediately before their modification with enzymes. A mercury disk electrode (Hg(Au)E; 2 mm in diameter) was pepared by dipping a gold-plated platinum electrode into mercury for a short time.

Preparation of enzyme-coated electrodes

Enzyme-coated electrodes were prepared by dip coating. The electrodes were kept in enzyme solutions for a given time (t_{exp}), usually 30 s, at 5 °C, and then rinsed with a solution of the same composition as the buffer to be used for the electrochemical measurements. Preparation of electrodes for use as a fructose sensor was the following: A 10-μl portion of the FDH stock solution, which contained 2.2×10^{-12} mol FDH, was syringed onto the surface of a carbon paste electrode. The solvent was allowed to evaporate and then the electrode surface was covered with a dialysis membrane. The whole electrode was covered with a nylon net to give it physical strength.

Electrochemical measurements

Electrochemical measurements were made in a buffer solution stirred with a magnetic stirrer unless stated otherwise. A three-electrode system was used, in which an Ag/AgCl/saturated KCl electrode and a platinum coil were used as the reference electrode and the counter electrode, respectively. In this paper, all potentials are referred to the Ag/AgCl reference electrode. Electroreflectance (ER) measurements were using the procedure described performed by Sagara et al. (1991).

Results and discussion

GADH

Adsorption of GADH on electrodes
Figure 1(A) shows cyclic voltammograms obtained with a GADH-coated CPE in an acetate buffer (pH 5.0) at 5 °C. The voltammogram obtained in the absence of GlcA (broken line) is very similar to that obtained with a bare CPE, and no wave attributable to the redox reaction of GADH is observed. However, when the solution contained GlcA, a characteristic voltammetric wave with a hump at about 20 mV appears (full line). No such wave was obtained with a bare CPE in the same solution (Ikeda et al., 1988). The results indicate that the voltammetric wave is due to the GADH-catalyzed electrolytic oxidation of GlcA. Voltammograms of almost the same magnitude were obtained, provided that the measurements were finished within 3 h of the preparation of the GADH-coated electrode, suggesting that the absorbed GADH remained active for at least 3 h when the electrode was stored at 5 °C in a buffer at pH 5.0 (Ikeda et al., 1993 a).

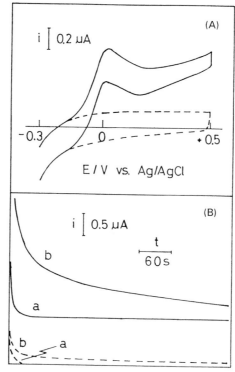

Figure 1. (A) Cyclic voltammograms recorded with a GADH-coated CPE in solution of acetate buffer (pH 5.0) (broken line) and in the solution containing 20 mM GlcA (full line). Scan rate, 5 mV s⁻¹. (B) Chronoamperograms recorded at 0.2 V (curve a) and 0.8 V (curve b) in the buffer solution (broken line) and in the solution containing 20 mM GlcA (full line).

Figure 1 (B) shows chronoamperograms recorded with GADH-coated CPEs at 0.2 V (curve a) and 0.8 V (curve b). A constant current is obtained at 0.2 V when correction is made for the base current. On the other hand, the curent measured at 0.8 V decreases gradually, indicating denaturation or desorption of the adsorbed GADH at the positive potential. Time-dependent decrease in the current magnitude were observed even at 0.2 V when the measurements were carried out at elevated temperatures. The time-dependent current decreases can be analyzed in terms of first-order kinetics as

$$I = I_0 \exp(-k_d t) \tag{1}$$

where I_0 is the current at the time $t = 0$ and k_d is a coefficient expressing the rate of decrease.

The plot of $\ln I$ against t for the data in curve b in Fig. 1(b) gives a line from which k_d is calculated to be $6.3 \times 10^{-4} \mathrm{s}^{-1}$. Values of k_d at 0.1 V were measured at different pH values and temperatues. The pH dependence was similar to the one of the activity of the enzyme in solution, and the Arrhenius activation energy was in the order of magnitude usually obtained for the denaturation of proteins. The results suggest that the time-dependent decrease in the current magnitude is primarily due to the denaturation of the GADH-adsorbed on the carbon paste electrode, although desorption of GADH from the electrode might also contribute to the decrease. The denaturation at 5 °C was slow at pH values from 5.0 and 6.0, as long as the applied potential was no more than 0.5 V. The catalytic current measured under these conditions remained unchanged for at least the first 40 min. The same was true for currents measured with other GADH-coated electrodes.

Cyclic voltammograms of GADH-coated electrodes
GADH-coated electrodes prepared with GCE, PGE, AuE, AgE, Au-PtE, Pt-PtE and Hg(Au)E produced anodic currents for the enzyme-catalyzed oxidation of GlcA. Figure 2 shows cyclic voltammograms recorded with (A) a GADH-coated PG (basal plane) electrode, (B) a GADH-coated AuE, (C) a GADH-coated Au-PtE and (D) a GADH-coated AgE in buffer (pH 5.0) containing 20 mM GlcA at 5 °C. The positive and negative scan voltammograms on each cyclic voltammogram are superimposable when corrected for the base current obtained in the absence of GlcA. The corrected voltammograms have characteristic shapes similar to that obtained with the GADH-coated CPE. Carbon electrodes often produce surface redox waves, owing to their quinone-like group, which may mediate electron transfer between the electrode and the redox site of the immobilized enzyme. We have observed such surface waves at about 0.2 V with a PGE (Fig. 2(A)). However, it is unlikely that the redox group producing the surface wave is a mediator, since the catalytic current begins to appear

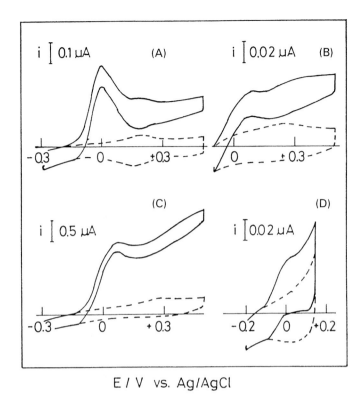

E / V vs. Ag/AgCl

Figure 2. Cyclic voltammograms recorded with GADH-coated electrodes prepared with (A) PG (basal plane) E, (B) AuE, (C) AuPtE, and (D) AgE in the buffer (pH 5.0) solution (broken line) and in the solution containing 20 mM GlcA (full line). Scan rate, 5 mVs⁻¹.

at a much more negative potential than the peak potential of the wave (Fig. 2(A)) and since very similar current-potential curves are obtained with GADH-coated metal electrodes for the oxidation of GlcA (Figs. 2 (B) – (D)). The results strongly suggest that the catalytic current originated from direct electron transfer between these electrodes and the GADH adsorbed on them.

GADH has three kinds of redox active sites: FAD, heme c, and Fe-S clusters. That the redox potential E_0 of heme c (– 97 mV Matsushita et al., 1982) is close to the half-wave potentials (– 40 to –60 mV) of the voltammograms in Figures 1 (A) and 2 suggests that the redox site for reaction with the electrode is heme c. This means that intramolecular electron transfer occurs from FAD, the site reacting with the substrate, to the heme c during the electrocatalytic reaction. The rate of the intramolecular electron transfer may depend on the electrode potential, which may account for the unusual shape of the voltammograms (Figs. 1 (A) and 2). However, another possibility that Fe-S clusters also take part in the electrode reaction with a different potential dependence cannot be ruled out.

Bioelectrocatalytic behavior of GADH-coated electrodes

The current measured at a fixed potential attained a steady state only seconds after the addition of GlcA to the solution. The steady state current I measured at 0.1 V increases to approach saturation as the concentration of c_{GADH} increased. The results obtained with a GADH-coated GCE are shown in Figure 3(A). The saturation tendency is typical of those seen in enzyme kinetics and the plot of c_{GADH}/I against, c_{GADH}, (Hanes-Woolf plot in enzyme kinetics) is a straight line as shown in Figure 3(B). Accordingly, the steady state current can be expressed by

$$I = I_{max} c_{GADH}/[K_m + c_{GADH}] \tag{2}$$

with

$$I_{max} = n F A k_{cat}[GADH]_{ad} \tag{3}$$

where n, F and A are the number of electrode, the Faraday constant and the surface area of the electrode, respectively, and k_{cat}, and K_m are the catalytic constant and Michaelis constant for the reaction of adsorbed GADH. $[GADH]_{ad}$ is the surface concentration of the adsorbed GADH. The I_{max}/A and K_m values of the GADH-coated GCE were calculated from the slope and intercept of the plot in Figure 3(B). The I_{max}/A and K_m values of the

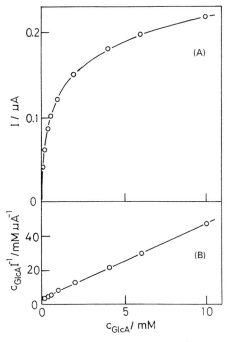

Figure 3. (A) Dependence of the steady state current I on the concentration of GlcA (c_{GlcA}) measued at 0.1 V with a GADH-coated GCE and (B) the plot of c_{GlcA}/I against c_{GlcA}.

Table 1. I_{max}/A and K_m values of the enzyme-electrochemical reaction at GADH-coated electrodes

Electrode	$I_{max}/A^a/\mu A\ cm^{-2}$	K_m^a/mM	n^b
CPE	1.8	0.88	1
GCE	1.9 ± 0.4	0.92 ± 0.07	19
PGE[c]	1.3 ± 0.4	0.98 ± 0.01	4
AuE	4.4 ± 0.5	0.75 ± 0.02	2
Au-PtE	8.6 ± 0.9	0.87 ± 0.04	6
AgE	1.6 ± 0.3	0.83 ± 0.01	2
PtE	0.02	1.1	1
Pt-PtE	16.1	1.1	1
Hg(Au)E	0.27	1.0	1

[a] Values are at 0.1 V vs. Ag/AgCl, except those at 0.5 V for PtE and Pt-PtE. [b] Number of runs.
[c] Basal plane.

other GADH-coated electrodes were calculated in a similar manner. The results are summarized in Table 1. The K_m values obtained with these electrodes are similar to each other and to the Michaelis constant of 0.8–2.3 mM reported for the GADH reaction in solution by Matsushita et al. (1982), which indicates that the affinity of the enzyme to the substrate is affected slightly by the adsorption and not by the electrode material onto which GADH is adsorbed. The I_{max}/A values differ depending on the electrode material; the GADH-coated AuE, Au-PtE and Pt-PtE give large I_{max}/A values. On the other hand, the GADH-coated PtE gives a very small I_{max}/A values. The reason for the difference in the I_{max}/A values is not clear, thought the amount of adsorbed GADH may depend on the electrode materials. It is interesting that the GADH-coated Hg(Au)E gives a small but noticeable I_{max}/A value. It has been considered that proteins are easily unfolded on a mercury electrode by adsorption. Since GADH is a membrane-bound protein, it may be relatively stable in the adsorbed state on mercury having hydrophobic nature.

ADH

ADH is an another example of multicomponent enzyme that is capable of direct electron transfer at an electrode to catalyze the electrolytic oxidation of the substrate. The catalytic mechanism of this enzyme has been studied in more detail than that of GADH (Ikeda et al., 1993 b).

Cyclic voltammograms of ADH-coated electrodes

Figure 4 shows cyclic voltammograms obtained with an ADH-coated Au-PtE in (A) a phosphate buffer (pH 6.0) and (B) the buffer containing 10 mM ethanol at 5 °C. A clear anodic current appears when the solution contains ethanol, though waves attributable to the surface redox reaction of the adsorbed ADH are not observed. The voltammogram is independent of the potential scan rate from 2 to 50 mV s^{-1} when correction is made for the

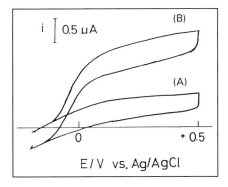

Figure 4. Cyclic voltammograms recorded with an Au-PtE in (A) a phosphate buffer solution (pH 6.0) and (B) the solution containing 10 mM ethanol. Scan rate 5 mVs⁻¹.

base current, and is also independent of the stirring rate of the solution at and above 200 rev min⁻¹. The results indicate that the voltammetric wave is due to the catalytic reaction of ADH coated on the electrode and that the concentration polarization of ethanol in the solution during the electrolysis is negligibly small. The ADH coated on the electrode remain active and is stable for the first 40 min as long as the positive potential scan dose does not exceed + 0.6 V, which is enough to allow a quantitative study of the electrocatalysis. ADH-coated electrodes prepared at different t_{exp} values between 30 s and 10 min produce voltammetric waves of almost the same magnitude, indicating that 30 s is enough for the maximum amount of ADH to be adsorbed. When McIlvaine buffer is used, the same voltammogram as that in phosphate buffer is obtained.

Figure 5 shows the pH dependence of the voltammograms of ADH-coated Au-PtE at several pH values, where correction is made for the base current. The voltammogram shifts to positive potentials as the solution pH decreases. The wave height is almost the same between pH 4.0 and 6.0 and decreases significantly at pH 3.0. This pH dependence of the wave height is very similar to the pH dependence of the enzyme activity of ADH which has a optimum pH at about 5.5. The Arrhenius activation energy is calculated from the temperature dependence of the wave height at pH 6.0 between 5 and 25 °C to be 10.8 kJ mol⁻¹, which is to the order of magnitude of the activation energy for enzymatic reactions. When a current is measured at a fixed potential, a steady state is attained seconds after the addition of ethanol to the solution. The steady-state current I measured at 0.2 V increases with increasing concentrations of ethanol, c_{EtOH}, to approch saturation as illustrated in Figure 6. All these experimental results confirm that the current is due to the catalytic action of ADH coated on the electrode and that the current magnitude is determined by the rate of the calalytic reaction.

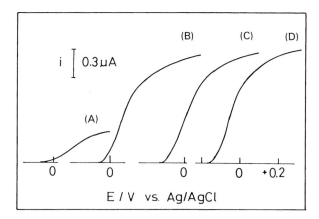

Figure 5. Corrected current-potential curves of ADH-coated Au-PtES at pH (A) 3.0, (B) 4.0, (C) 5.0, and (D) 6.0. The curves were obtained by subtracting the base current from the voltammograms obtained in the presence of 10 mM ethanol.

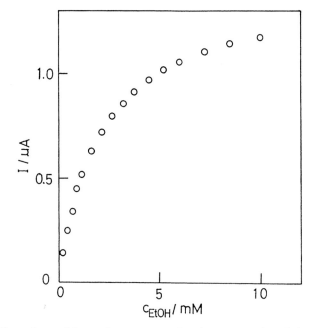

Figure 6. Dependence of the steady state current I on the concentration of ethanol c_{EtOH}. I was measured at 0.2 V with an ADH-coated Au-PtE at pH 6.0

Effects of electrode materials

ADH-coated electrodes prepared with AuE, AgE, GCE, and PGE (basal plane) produce similar voltammetric waves to that obtained with the ADH-coated AuPtE (Fig. 4). Figure 7 shows the currents measured at fixed potentials with (A) an ADH-coated AuE, (B) an ADH-coated AgE, (C) an ADH-coated GCE, and (D) an ADH-coated PGE (basal plane). All the electrodes produce anodic currents and attain steady states seconds after the addition of ethanol to the solution. Interestingly, the steady-state currents are affected differently by the attention of Triton X-100 to the solution; Triton X-100 is a surfactant employed for solubilizing ADH from the bacterial membranes. The steady-state current is greatly decreased with the ADH-coated GCE (Fig. 7(C)) and the ADH-coated PGE (basal plane, Fig. 7(D)), whereas the decrease is small with the ADH-coated AuE (Fig. 7(A)) and there is no decrease with the ADH-coated AgE (Fig. 7(B)). The results show that ADH on the carbon electrodes is easily desorbed from the electrode surface in the presence of Triton X-100, whereas ADH can remain adsorbed on the metal electrodes.

Observation of ER spectrum of heme c of adsorbed ADH

The amount of a protein of the size of ADH adsorbed in a monomolecular layer on an electrode is considered to be in the order of 10^{-12} mol. This

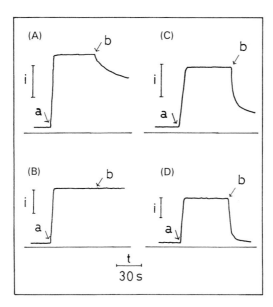

Figure 7. Current response at a fixed potential of ADH-coated electrodes prepared with (A) AuE, (B) AgE, (C) GCE, and (D) PG (basal plane) E in buffer of pH 6.0. Ethanol was added to the solution at the points denoted a, and then Triton X-100 was added at the points denoted b to make the surfactant concentration 0.1%. The potential was fixed at (A, C, and D) 0.2 V and (B) 0.1 V. The concentration of ethanol was (A und B) 1.5 mM and (C and D) 0.5 mM; $i = 0.01$ μA.

amount of an enzyme is not enough to produce a surface redox wave clearly distinguishable from the base current obtained with a bare electrode. ER measurement is a powerful technique which provides information on the redox state of the substance adsorbed in a monomolecular layer on an electrode as demonstrated by Sagara et al. (1991). When the ER spectrum is measured with an ADH-coated AuE at 0.0 V in buffer of pH 6.0, small signals due to heme c appear in the range 390–590 nm. A spectrum of adsorbed heme c is obtained by subtracting from the spectrum at 0.0 V a spectrum at – 0.2 V, at which potential such signals as those at 0.0 V are not observed. The difference spectrum is shown in Figure 8. A negative peak is observed at 406 nm and positive peaks at 427, 520, and 552 nm. These wavelengths are very similar to those demonstrated by Sagara et al. (1991) of the peaks observed with cytochrome c coadsorbed with 4,4′-bipyridyl on a gold electrode; the relative magnitudes of the peaks at 406 and 427 nm are also similar to the ER spectrum of the adsorbed cytochrome c. Heme c of ADH has absorption peaks at 417, 522, and 553 nm in the reduce form and at 409 nm in the oxidized form, which are very similar to those of mammalian cytochrome c: 415, 521, and 550 nm in the reduced form and 409 nm in the oxidized form. Thus, we may conclude that heme c of the ADH coated on an AuE dose exchange electrons with the electrode.

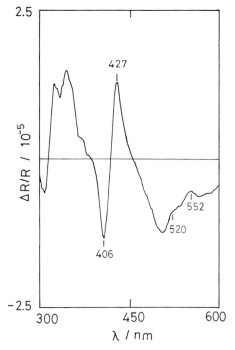

Figure 8. An ER spectrum of heme c of an ADH-coated AuE in a buffer of pH 6.0 with a modulation of 70 mV at 8.0 Hz.

Figure 9 shows ER voltammograms at pH 6.0, both imaginary and real parts, measured at 405 nm. The ER voltammograms produce waves with peak potentials at 2 ± 35 mV in both components with a width at half-height of 176 ± 10 mV, where the measurements are taken using presumed base lines given by the broken lines in Figure 9. Assuming a one-electron process, we can calculate the rate constant k_s of the electrode reaction to be 30 s^{-1} from the ratio of the ER voltammetric wave heights of the imaginary and real parts according to the method by Sagara et al. (1991). When an ER voltammogram is measured at pH 4.0, the waves in the imaginary and real components appear with the peak potentials 39 ± 20 mV and 54 ± 35 mV, respectively. The result that the waves at pH 4.0 appear at more positive potentials than the waves at pH 6.0 is in line with the fact than the cyclic voltammetric wave due to the catalytic oxidation of ethanol shifts to positive potentials as pH decreases.

A mechanism of direct bioelectrocatalysis at an multicomponent enzyme-coated electrode

The above-mentioned experimental results show that GADH and ADH, membrane-bound multicomponent enzymes containing heme c, adsorb spontaneously on electrodes and are active in catalyzing the electrolytic oxidation of the substrates. In the case of an ADH-coated AuE, the site to exchange electrons with the electrode is heme c of the adsorbed enzyme. Heme c is probably the site for exchanging electrons with the electrodes in

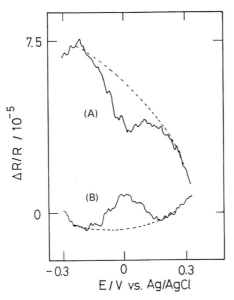

Figure 9. ER voltammogram at 405 nm of an ADH-coated AuE in a buffer of pH 6.0 with a modulation of 70 mV at 8.0 Hz and a sweep rate of 2 mV s^{-1}, (A) real part and (B) imaginary part.

the other ADH-coated electrodes and the GADH-coated electrodes, because the voltammetric waves due to the bioelectrocatalysis start to appear at the potentials close to the redox potentials of heme c of ADH and GADH.

We assume the following reaction scheme for the bioelectrocatalyis reaction:

$$S + [E(ox)]_{ad} \underset{k_{-1}}{\overset{k_1}{\rightleftharpoons}} [S-E]_{ad} \tag{4}$$

$$[S-E]_{ad} \overset{k_2}{\longrightarrow} [E(red)]_{ad} + P \tag{5}$$

$$[E(red)]_{ad} \underset{k_{be}}{\overset{k_{fe}}{\rightleftharpoons}} [E(ox))]_{ad} \tag{6}$$

S and P are the substrate and product respectively; $[E(ox)]_{ad}$, $[E(red)]_{ad}$, and $[S-E]_{ad}$ are the oxidized and reduced forms of the absorbed enzyme, and the enzyme-substrate complex respectively; k_1, k_{-1}, and k_2 are the rate constants of the enzymatic reactions given by eqns. (4) and (5); k_{fe} and k_{be} are the rate constant of the electrode reaction of the adsorbed enzyme given by eqn. (6). It should be noted that GADH and ADH have two kinds of redox active groups each; FAD or PQQ the site for reacting with the substrate and heme c which can accept electrons from the FAD or PQQ. The step of intramolecular electron transfer from the FAD or PQQ to heme c is included in eqn. (4).

The steady-state current for reactions (4)–(6) can be expressed by

$$I = \frac{I_{max}}{(1 + k_{be}/k_{fe}) K_m + (1 + k_2/k_{fe}) (S)} \tag{7}$$

with

$$I_{max} = nFA k_2 [Enz]_{ad} \tag{8}$$

and

$$K_m = (k_{-1} + k_2)/k_1 \tag{9}$$

where (S) and $[Enz]_{ad}$ are the concentrations of the substrate in solution at the surface of the enzyme-coated electrode and the total surface concentration of the adsorbed enzyme, respectively.

At a potential where a limiting current is obtained, eqn. (7) can be simplified to

$$I = I_{max}/[K_m + (S)] \tag{10}$$

This is identical with eqn. (2) and k_{cat} in eqn. (3) corresponds to k_2.

Table 2. K_m and I_{max}/A values of bioelectrocatalytic reaction at ADH-coated electrodes

Electrode	$(I_{max}/A)/$ $\mu A\ cm^{-2a}$	K_m/mm^a	Number of runs
AuE	5.4 (0.7)	1.5 (0.2)	4
Au-PtE	5.3 (0.9)	1.0 (0.3)	2
AgE	1.3 (0.1)	1.4 (0.1)	2
GCE	0.75	1.0	1
PG(basal)E	0.27 (0.09)	1.0	1
PG(polished)E	0.18 (0.07)	1.2 (0.1)	2

Numbers in parentheses are standard deviations.
[a] The data were obtained from the measurements at 0.2 V except for AgE (0.1 V) in buffer of pH 6.0 a 5 °C.

In the following, we describe the analysis of the experimental results obtained with ADH-coated electrodes based on the mechanism assumed above. I_{max}/A and K_m values can be calculated from the data like those in Figure 6 by plotting c_{EtOH}/I against c_{EtOH}. Here, (S) in eqn. (10) can be equated to c_{EtOH}, since the concentration polarization of ethanol is negligibly small under the experimental conditions. The I_{max}/A and K_m values thus obtained are summarized in Table 2. The K_m values obtained with the different ADH-coated electrodes agree well with each other and with the Michaelis constant, 1.6 mM, for the ADH reaction in solution reported by Ameyama and Adachi (1982). Considering that Michaelis constant expresses affinity of an enzyme to be substrate, we may say that the affinity of ADH to ethanol is affected little by the absorption on AuE.

The quantitiy I_{max}/A is a function of both k_2 and $[Enz]_{ad}$ (eqn. (8)). The I_{max}/A values differ depending on the electrode materials. The largest value is obtained with the ADH-coated gold electrodes, and the rather small value with the ADH-coated PGE. Armstrong et al. (1989) have suggested that electron transfer of cytochrome c occurs only at oxygen-functionalized electroactive sites on the surfaces of pyrolytic graphite, and that a basal-plane electrode has a low surface density of electroactive site compared with an edge-plane electrode. We measured the bioelectrocatalytic current using a polished basal-plane electrode that had a high density of the oxygen-functionalized sites. The results are included in Table 1. The I_{max}/A value is even smaller than that measured at a basal-plane electrode, suggesting that ADH, in contrast to cytochrome c, prefers hydrophobic surfaces. This is consistent with the fact that ADH absorbs more strongly on a hydrophobic surface of gold than a graphite electrode and that ADH undergoes little structural change compared with cytochrome c when adsorbed on a bare gold electrode as indicated by Sagara et al. (1991).

The current-potential curves corrected for the base current as illustrated in Figure 5 can be simulated by eqn (7) combined with the Butler-

Volmer formulation for k_{fe} and k_{be}:

$$k_{fe} = k_s \exp[(1-\alpha)\,(nF/RT)\,(E-E_0')] \tag{11}$$

$$k_{be} = k_s \exp[(-\alpha nF/RT)(E-E_0')] \tag{12}$$

where E_0', k_s, and α are the formal potential, the rate constant at this potential, and the transfer coefficient respectively. R and T have their usual meanings. For example, using the experimental values $K_m = 1.5$ mM and (S) = 10 mM at a ADH-coated AuE, and assuming $\alpha = 0.5$, we obtain the best fit with E_0' and k_2/k_s values of -22.4 ± 2.3 mV and 0.68 ± 0.05, respectively. The E_0' value is located in the potential region where ER voltammogram of heme c apears (Fig. 9). This supports the idea that heme c is the site donating electrons to the electrode during the bioelectrocatalytic reaction. It should be noted that ADH has three heme c groups; one is in the subunit containing PQQ and the other two are in another subunits. However, only one broad wave centered at about 0 V is observed on the ER voltammograms (Fig. 9). The result that the width at half-height of the ER voltammogram is somewhat larger than that calculated by simulation with the assumption of a one-electron process might indicate that two or three heme c groups take part in the electrode reaction in the potential region where the ER voltammetric waves appear. The k_2/k_s value of 0.68 gives a k_2 value of 20^{-1} by the use of the k_s value of 30 s^{-1} calculated from the ER voltammograms. This k_2 value is much smaller than the catalytic constant 482 s^{-1} of the native ADH in solution calculated from the specific activity given by Ameyama and Adachi (1982). The amount of adsorbed ADH is calculated from the k_2 value by eqn 8 with the I_{max}/A value of 5.4 μAcm^{-2} to be 1.4×10^{-12} mol cm^{-2}, which seems to be a value acceptable for a monomolecular layer adsorption of the enzyme.

Although details of the electron transfer reaction of heme groups of the adsorbed ADH are not clear, we may postulate an oriented adsorption of ADH with the PQQ moiety facing toward the solution as illustrated in Figure 10. The oriented adsorption is anticipated from the indication that ADH in vivo is partially buried in the cytoplasmic membranes of bacteria with the PQQ moiety exposed to the periplasmic space and with the heme c moiety within the membranes. The PQQ moiety is presumed to be hydrophilic, and is allowed to stay apart from the electrode of hydrophobic nature, while the heme c moiety is allowed to be in close contact with the electrode surface. Thus, electrons can flow from the substrate to the electrode through the enzyme during the bioelectrocatalytic reaction, in which heme c acts as a built-in mediator. The small k_2 value compared with the catalytic constant of ADH suggests a conformational change of ADH by the adsorption. We have expected that chemical modification of the surfaces of gold and mercury electrodes with alkanethiols allows the adsorption of ADH on the modified surfaces in its native state. We have

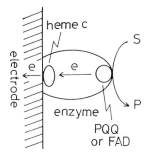

Figure 10. Schematic representation of bioelectrocatalysis at an enzyme-coated electrode with heme c serving as a built-in mediator: S substrate, P product.

made a preliminary experiment with the gold and mercury electrodes modified with n-alkanethiols ($CH_3(CH_2)_nSH$, n = 3, 5, and 11) using ADH from *Acetobacter aceti*, a membrane-bound quinocytochrome c enzyme with properties similar to those of ADH from *Gluconobacter suboxidans* (Yanai et al., 1994). However, the alkanethiol layers had an inhibitory effect rather than a promotion effect on the bioelectrocatalytic reaction.

FDH

FDH is also a quinocytochrome c enzyme which catalyzes the electrolytic oxidation of the substrate based on a mechanism of direct bioelectrocatalysis. In this section, we describe an applied aspect of an FDH-coated electrode as an amperometric fructose sensor (Ikeda et al., (1991). A dialysis membrane covered FDH-coated carbon paste electrode, a film-FDH-CPE, is used in this section throughout, the preparation of the film-FDH-CPE being described in the experimental section.

Effects of applied potential, solution pH, and temperature
The film-FDH-CPE shows an anodic current response to fructose when the electrode immersed in a buffer solution is kept at a fixed potential and when fructose is added to the solution. Figure 11 shows the time dependence of the current response measured at 0.5 V and at pH 4.5. The 90% level of the steady-state current, I, is attained 50 s after the addition of fructose in the solution. This time dependence is longer than those at the GADH- and ADH-coated electrodes mentioned above, indicating that the time dependence is governed by the permeation rate of the substrate in the dialysis membrane. Figure 12 shows the dependence of I on the potential applied to the electrode. I starts to appear at 0 V and increased with increasing positive potential to approach a limiting current.

The magnitude of I is sensitive to the pH of the solution; the maximum current response is obtained at around pH 4.5–5.0, close to the optimum

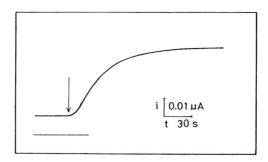

Figure 11. Time dependence of the current response of the film-FDH-CPE. At the point indicated by the arrow, fructose was added to a buffer solution of pH 4.5 to make 0.2 mM in fructose concentration. The horizontal line shows the current zero level.

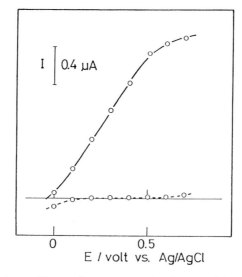

Figure 12. Dependence of the steady state current I on th epotential applied to the film-FDH-CPE. I was measured at 15 °C in pH 4.5 McIlvaine buffer (broken line) and the buffer containing 20 mM fructose (full line).

pH of FDH, pH 4.0. When the film-FDH-CPE devices are used for measurements at pH less than 3 or above 7, the electrodes become inactive owing to the denaturation of the enzyme. Accordingly, control of the solution pH is critical for the film-FDH-CPE to be used as a sensor. The magnitude of I increases with increasing temperature in the range 5 to 25 °C with the temperature coefficient of 2.3% at 25 °C. The magnitude at 25 °C remains unchanged during the course of continuous measurements for 2 h. At above 30 °C, however, I decreases gradually; the higher the temperature, the greater the rate of decrease. In contrast, no decreae in I is observed during more than 24 h of continuous measurements at low temperatures such as 5 °C.

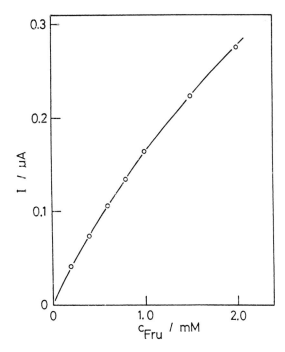

Figure 13. Calibration graph for fructose obtained with the film-FDH-CPE. c_{Fru} the concentration of fructose.

Reproducibility, selectivity, and long-term stability
Measurements have been made in air-saturated buffer solutions of pH 4.5 at 0.5 V and at 25°C. Figure 13 shows the calibration graph for fructose in the range 0.2 to 2.0 mM. The reproducibiltiy of the current responce at 2 mM fructose is 1.9% (n = 10) coefficient of variation. Oxygen has no effect on the current response: deaeration by passing nitrogen into the solution causes no change of the current magnitude. No interference is observed on addition of other sugars (2 mM), such as glucose, galactose, sucrose, lactose, maltose, xylose, and arabinose. When I for 2 mM fructose is measured with the film-FDH-CPE every day, and when the electrode is stored when not in use at 5°C in pH 4.5 McIlvaine buffer containing 0.1% Triton X-100, the magnitude of I decrease gradually to 30% of the original value after 2 weeks. The current, however, remains unchanged during the course of continuous measurements for 2 h even when the measurements are done after 10 days.

Measurements of fructose in fruits
Using the film FDH-CPE we have measured the concentration for fructose in an apple and a lemon. The results are given in Table 3 together with the results obtained by the F-Kit method; F-Kit for fructose measurements was

Table 3. Comparison of the present methods with the F-Kit method

Sample	Present methods		F-kit method (mM)
	Film-FDH-CPE (mM)	Film-ASOD/FDH-CPE (mM)	
Apple	422 ± 4 ($n = 3$)	420 ± 3 ($n = 3$)	420 ($n = 1$)
Lemon	92.7 ± 4.1 ($n = 3$)	60.6 ± 2.5 ($n = 3$)	58.9 ($n = 1$)

n, number of runs.

purchased from Boehringer Mannheim Yamanouchi Co. In the case of an apple, the present method gives a result agreeing with that obtained by the F-Kit method, while it gives a rather high value in the case of a lemon, compared with the value by the latter method. This is because the lemon contains ascorbic acid in a high concentration of 1.7 mM, which is 2.9% of the concentration of fructose. Ascorbic acid is directly oxidizable at the film-FDH-CPE at 0.5 V, and consequently interferes with the measurement of fructose. This inferference can be eliminated by using the film FDH-CPE which contains ascorbate oxidase (ASOD) in the enzyme layer; ASOD catalyzes the oxidation of ascorbic acid by oxygen to produce dehydroascorbic acid, which is electrochemically inactive at 0.5 V. The results obtained with this electrode (Table 3) agree well with those by the F-Kit method.

Conclusion

GADH, a flavocytochrome enzyme, and ADH and FDH, quinocytochrome enzymes, are capable of redox communication between ordinary electrodes and the enzymes adsorbed on them to produce catalytic currents for the oxidation of the substrates. The experiments with ADH at a gold electrode give evidence that heme c is the site to communicate with the electrode. Present results indicate that the use of multicomponent enzymes having different sites for accepting and donating electrons is promising to achieve bioelectrocatalysis based on direct redox communication between an electrode and the enzyme immobilized on it. This type of redox communication will also be possible with an enzyme having only one redox active cofactor, provided that the enzyme donates electrons to the electron acceptor of the enzyme at a site different from the site receiving electrons from the substrate. Ferredoxin-FNR reductase is an example of such an enzymes, and produces a catalytic current, though the magnitude of the current is very small as shown by Kobayashi et al. (1992).

Acknowledgements
This work ws supported in part by Grants-in-Aid from the Ministry of Education, Science and Culture, Japan.

References

Alvarez-Icaza, M. and Schmid, R.D. (1994) Observation of direct electron transfer from the active center of glucose oxidase to a graphite electrode achieved through the use of mild immobilization. *Bioelectrochem. Bioenerg.* 33:191–199.

Ameyama, M. and Adachi, O. (1982) Alcohol dehydrogenase from acetic acid bacteria, membrane-bound. *Methods Enzymol.* 89:450–457.

Ameyama, M., Shinagawa, E., Matsushita, K. and Adachi, O. (1981) D-Fructose dehydrogenase of Gluconobacter industrius: purification, characterization and application of enzymatic microdetermination of D-fructose. *J. Bacteriol* 145:814–823.

Armstrong, F.A. (1992) Dynamic electrochemistry of iron-sulfur proteins. *Adv. in Inorg. Chem.* 38:117–163.

Armstrong, F.A. and Lannon, A.M. (1987) Fast interfacial electron transfer between cytochrome c peroxidase and graphite electrodes promoted by aminoglycosides: Novel electroenzymatic catalysis of H_2O_2 reaction. *J. Am. Chem. Soc.* 109:7211–7212.

Armstrong, F.A., Hill, H.A.O. and Walton, N.J. (1988) Direct electrochemistry of redox proteins. *Acc. Chem. Res.* 21:407–413.

Armstrong, F.A., Bond, A.M., Hill, H.A.O., Psalti, S.M.I. and Zoshi, C.G. (1989) A microscopic model of electron transfer at electroactive sites of molecular dimensions for reduction of cytochrome c at basal- and edge-plane graphite electrodes. *J. Phys. Chem.* 93:6485–6493.

Degani, Y. and Heller, A. (1987) Direct electrochemical communication between chemically modified enzymes and metal electrodes. 1. Electron transfer from glucose oxidase to metal electrodes via electron relays, bound covalently to the enzyme. *J. Phys. Chem.* 91:1285–1289.

Eddows, M.J. and Hill, H.A.O. (1977) Novel method for the investigation of the electrochemistry of metalloproteins: cytochrome c. *J. Chem. Soc. Chem. Commun.* 770–772.

Guo, L.H., Hill, H.A.O., Lawrence, G.A., Sanghera, G.S. and Hopper, D.J. (1989) Direct unmediated electrochemsitry of the enzyme p-cresolmethylhydroxylase. *J. Electroanal. Chem.* 266:379–396.

Guo, L.H., Hill, H.A.O., Hopper, D.J., Lawrence, G.A. and Sanghera, G.S. (1990) Direct voltammetry of the *Chromatium vinosam enzyme*, sulfide, cytochrome c oxidase (flavocytochrome d_{552}). *J. Biol. Chem.* 265:1958–1963.

Ikeda, T., Hamada, H., Miki, K. and Senda, M. (1985) Glucose oxidase-immobilized benzoquinone-carbon paste electrode as a glucose sensor. *Agric. Biol. Chem.* 49:541–543.

Ikeda, T., Fushimi, F., Miki, K. and Senda, M. (1988) Direct bioelectrocatalysis at electrode modified with D-gluconate dehydrogense. *Agric. Biol. Chem.* 52:2655–2658.

Ikeda, T., Matsushita, F. and Senda, M. (1991) Amperometric fructose sensor based on direct bioelectrocatalysis. *Biosens. Bioelectron.* 6:299–304.

Ikeda, T., Miyaoka, S. and Miki, K. (1993 a) Enzyme-catalyzed electrochemical oxidation of D-Gluconate gluconate at electrodes coated with D-gluconate dehydrogenase, a membrane-bound flavohemoprotein. *J. Electroanal. Chem.* 352:267–278.

Ikeda, T., Kobayashi, D., Matsushita, F., Sagara, T. and Niki, K. (1993 b) Bioelectrocatalysis at electrodes coated with alcohol dehydrogenase, a quinohemoprotein with heme c serving as a built-in mediator. *J. Electroanal. Chem.* 361:221–228.

Jiang, L., McNeil, C.J. and Cooper, J.M. (1995) Direct electron transfer reaction of glucose oxidase immobilized at a self-assembled monolayer. *J. Chem. Soc. Chem. Commun.* 1293–1295.

Kobayashi, D., Ozawa, S., Mihara, T. and Ikeda, T. (1992) Flavoenzyme-catalyzed electrochemical oxidation of NADH and NADPH in the absence of external mediators. *Denki Kagaku* 60:1056–1062.

Kulys. J.J. (1986) Enzyme electrodes based on organic metals. *Biosensors* 2:3–13.

Lee, Chi-Woo, Gray, H.B., Anson, F.C. and Malmstron, Bo G. (1984) Catalysis of the reduction of dioxygen at graphite electrodes coated with fungal laccase A. *J. Electroanal. Chem.* 172:289–300.

Matsushita, K., Shinagawa, E. and Ameyama, M. (1979) D-gluconate dehydrogenase from bacteria. 2-Keto-D-gluconte-yielding, membrane bound. *Methods Enzymol.* 89:187–193.

Niki, K., Yagi, I., Inokuchi, H. and Kimura, K. (1979) Electrochemical behavior of cytochrome c_3 of desulfovibrio vulgaris, strain Miyazaki, on the mercury electrode. *J. Am. Chem. Soc.* 101:3335–3340.

Paddock, R.M. and Bowden, E.F. (1989) Electrocatalytic reduction of hydrogen peroxide via direct electron transfer from pyrolytic graphite electrodes to irreversibly adsorbed cytochrome c peroxidase. *J. Electroanal. Chem.* 260:487–494.

Redfearn, E.R. (1967) Isolation and determination of ubiquinone. *Methods. Enzymol.* 10: 381–384.

Riklin, A., Katz, E., Willner, I., Stockers, A. and Buckmann, A.F. (1995) Improving enzyme-electrode contacts by redox modification of cofactors. *Nature* 376:672–675.

Ruzgas, T., Gorton, L., Emneus, J. and Marko-Verga, G. (1995) Kinetic models of horseradish peroxidase action on a graphite electrode. *J. Electroanal. Chem.* 391:41–49.

Sagara, T., Murakami, H., Igarashi, S., Sato, H. and Niki, K. (1991) Spectroelectrochemical study of the redox reaction mechanism of cytochrome c at a gold electrode in a neutral solution in the presence of 4,4'-bipyridyl as a surface modifier. *Langmuir* 7:3190–3196.

Scott, D.L., Paddock, R.M. and Bowden, E.F. (1992) The electrocatalytic enzyme function of adsorbed cytochrome c peroxidase on pyrolytic graphite. *J. Electroanal. Chem.* 341: 307–321.

Sucheta, A., Ackrell, B.A.C., Cochran, B. and Armstrong, F.A. (1992) Diode-like behavior of a mitochondrial electron-transfer enzyme. *Nature* 356:361–362.

Tarasevich, M.R., Yaropolov, A.I., Bogdanovskaya, V.A. and Varfolomeev, S.D. (1979) Electrocatalysis of a cathodic oxygen reduction by laccase. *Bioelectrochem. Bioenerg.* 6:393–403.

Yanai, H., Miki, K., Ikeda, T. and Matsushita, K. (1994) Bioelectrocatalysis by alcohol dehydrogenase form *Acetobacter aceti* adsorbed on bare and chemically modified electrodes. *Denki Kagaku* 62:1247–1248.

Yeh, P. and Kuwana, T. (1977) Reversible electrode reaction of cytochrome c. *Chemistry Letters* 1145–1148.

Frontiers in Biosensorics I
Fundamental Aspects
ed. by F. W. Scheller, F. Schubert and J. Fedrowitz
© 1997 Birkhäuser Verlag Basel/Switzerland

Microbiosensors using electrodes made in Si-technology

R. Hintsche, M. Paeschke, A. Uhlig and R. Seitz

Fraunhofer Institut für Siliziumtechnologie, D-25524 Itzehoe, Germany

Summary. The combination of electrochemical transducers made in silion technology with chemical and biochemical components has been used to manufacture miniaturized sensor structures. Three different types of sensors have been developed and optimized for practical use: (i) an ion-selective sensor, (ii) a glucose enzyme sensor, (iii) a redox-amplifying sensor for immunosensing. The immunodetection based on the redox recycling of mediator molecules is shown for low and high molecular weight analytes. The sensors have been integrated with miniaturized fluidic components and combined with sensor-related electronics and a common microcontroller.

Introduction

There are several arguments for using silicon technology to manufacture transducers for biosensors. Thin film technology enables the arrangement of multiple transducer structures in micrometer and submicrometer dimensions. On the other hand, the combination with bulk micromachining is useful to miniaturize well-known transducer principles for novel applications.

For sensors used in medical applications, it is important to build in small-sized sensors as near as possible to the site of signal origin. This is directly related to the demand of low voltages and low power consumption. These features are also required for portable, analytical devices as well as in industrial process control and field measurements in the environmental. Many efforts have been made and several approaches have been described to integrate silicon-made electrochemical transducers with biocomponents such as enzymes or immunoproteins. In several articles the present state and relevant references have been summarized (Göpel 1994; Scheller et al. 1992; Hintsche et al. 1994). Field effect transistors (Izquierdo and de Castro, 1994), and thin film electrodes are mostly used as electrochemical transducers. The thin film metal electrodes allow the arrangement of very complex constructions. Thus the integration of structures in μm and nm shapes open the way to new sensor principles, such as the electrochemical method of redox recycling (Bard et al. 1986), which has been adapted to silicon technology using thin film noble metal electrodes of close proximity (Niwa et al. 1990; Hintsche et al. 1994).

Another advantage of silicon technology is the possibility to realize a large number of equal of different electrodes or multiple arrays with high precision at the same chip. These electrode arrays were used for multicomponent detection and signal enhancement. The electrodes are manufactured in a batch process at a low price and show a high variability of the shape by simple changing of only one lithographic mask, by which a high volume production can be achieved.

With bulk micromachining, e.g. wet etching holes, cavities or lattices can be realized (Sibbald and Shaw, 1987; Knoll et al. 1994; Uhlig et al. 1995). The planar structures are useful for covering, whereas the three-dimensional structures can be covered with functional organic and biochemical layers, where adhesive, polymeric and enzymatic layers have been combined (Hintsche et al. 1989; Urban et al. 1992; Koudelka et al. 1993; Uhlig et al. 1995).

The practical application of these electrodes in liquid handling setups can be realized by inserting small chips into the analyte stream or by stacking the planary transducers in flow-through devices. Due to the aqueous nature of the samples, the chemical and electrical resistance in specification of the bulk and insulating material and for the metals used is quite different compared to common IC-technology. The moisture must be excluded from the non-active sensor areas by covering the chips with special insulating and protecting layers. Furthermore, the electrical connection and mechanical integration of the sensor element in a fluidic system containing pumps, valves and sample preparation units has to be managed. Some approaches have been shown to develop a fully integrated microanalytical system (v.d. Berg and Bergfeld, 1995).

Here, we describe the construction, manufacturing and application of a potentiometric potassium sensor and an amperometric glucose sensor and their application in catheters. Also the use of highly integrated arrays of interdigitated electrodes as transducers in amplifying immunosensors is shown. Modular coupling of an immunoreactor and the detection electrodes offers a high flexibility for the detection of different analytes by changing only the respective immunocomponents. The immunoassays are produced on commercial polymeric beads by standard procedures.

Technology

The transducers for the different types of electrochemical sensors can be manufactured with one technological basis: the standard silicon technology (Benecke, 1990; Wise and Najati, 1991). Thus different technologies are provided, such as evaporation, sputtering of insulation layers or chemical functionalized membranes, photolithography or anistropic and isotropic etching techniques. These technological tools can be variably combined, resulting in planar or three-dimensional sensors or in flow-through elements, as shown in Figure 1.

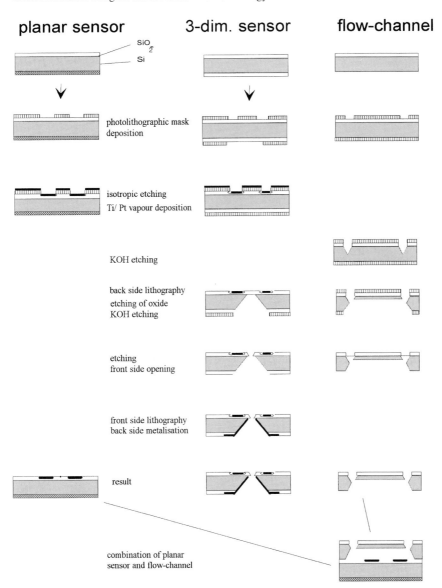

Figure 1. Scheme of processing different sensor devices.

Planar transducers require only few technological steps (Fig. 1, left column). After thermal oxidation of the silicon wafer, a photo resist is structured by UV-light through a mask. By removing those parts of the resin, which were not exposed to the light, the structure of the resin is realized on the wafer. The standard photolithography is limited to dimensions of 1–20 μm, whereas UV-photolithography with photostepper is used for structural resolution down to 0.6 µm. The fabrication of electrode arrangements with smaller dimensions requires the more expensive electrode-beam technique (Griffith et al., 1994, Reimer et al., 1995; Paeschke et al., 1995a). In a next step a thin metal film is deposited on the whole wafer. Typical metals are platinum, gold, silver or iridium. The metals are structured by dissolution of the resin. Lift-off technology is used to planarize the transducer surface.

For three-dimensional sensors (Fig. 1, central column), the metal deposition is combined with bulk-etching steps by both sides wafer processing (Knoll et al. 1994; Uhlig et al. 1995). Planar metal electrodes are realized on both sides of the wafer with lift-off technology. After back-side lithography a cavity is formed by etching the silicon wafer in KOH-solution. This cavity is then opened to the front side through reactive ion etching (RIE) structuring of the oxide and time controlled etching the silicon in tetramethylammonium hydroxide (TMAH) solution. After metallization of the cavity side walls a three-dimensional sensor is manufactured, where chemical sensitive membranes can be immobilized.

In the right column of Figure 1, the principal fabrication steps of a flow-through element are shown as an example for flow devices (Fiehn et al., 1995; Paeschke et al., 1995b), enzyme reactors (Murakami et al., 1993) or separation columns (Manz et al., 1995). Here, three-dimensional structuring and both sides wafer processing is used without a metallization step. After oxidizing and photolithographical patterning of a resin first openings are etched in the front side of the wafer. A rectangular channel is structured at the back side. The channel is open to the front side. The complete structure can be covered with a chemically inert isolation or immobilization layer.

The combination of a planar sensor with a flow-through element leads to a part of a mimiaturized total analytical system, which is called µTAS (Haemmerli et al., 1992; Verpoorte et al., 1994). The flow-through element is used to ensure a defined contact between sensor and sample solution. Inlet and outlet of the sample solutions are realized through the openings, whereas the solution passes the active sensor part, thus flushing the channel. The flow channel can be pasted or anodically bonded on the transducer surface (Shoji and Esashi, 1995).

Modular portable analytical systems

Electrochemical biosensors are used in portable analytical systems, e.g. from the companies i-STAT, USA, Medisense, USA and EKF, Germany (see Pfeiffer, this volume). These applications demonstrate the benefit of direct signal recognition without analyte treatment. Connected with standardized digital electronics, small portable analytical systems have been realized.

We have developed a modular system for the use of potentiometric, amperometric and multichannel voltammetric thin film sensors and sensor arrays. The system is based on common transducer technology and common microcontroller (µc) with data acquisition, display and key control. The sensor responses are processed by sensor related electronics. In Figure 2 the electrochemical principles: potentiometric (a), amperometric with two electrode configuration (b), amperometric with potentiostat (c) and biopotentiostat (d), are shown.

The potentiometric circuit amplifies only the concentration-dependent potential and is connected to the analogue to digital converter (ADC) for the data acquisition. (a) The corresponding single channel measurement is applied for the amperometric sensor with fixed reference. (b) This tpye is employed for the amperometric enzyme sensors (glucose, lactate), where hydrogen peroxide can be measured at 600 mV vs Ag/AgCl. The resulting current is converted to a potential via current follower for the further data processing. For the voltammetric sensors, a potentiostat with feedback and potential control is used. The working potential is adjustable and is generated by a µc-controlled 10-bit digital to analogue converter (DAC). Step potentials with different scan rates are software controlled. Bipotentiostats are used for interdigitated array electrode (IDA) based immunosensors. Redox recyling requires simultaneously on oxidation and a reduction potential. These potentials are generated also via two DAC channels. The measuring currents, converted to potentials, are averaged in the controller to one enhanced signal.

ACD and DAC are used as system interface. The digital hardware, consisting of power supply, display, clock, device and flow system driver, can be used for all sensor types. Additionally, software tools, such as data acquisition, digital filter or calibration routines are handled by one computer configuration.

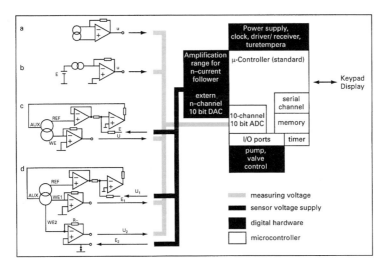

Figure 2. Scheme of the modular sensor control unit.

Potentiometric potassium chips in catheter application

In critical care the blood potassium concentration is an important parameter in characterizing the patient's condition (Oesch et al., 1986). *In vivo* monitoring of K^+ in whole blood opens new dimensions in diagnostic (Knoll et al., 1994; Lindner et al., 1993; Reinhoudt et al., 1994). For this purpose, a new potassium sensitive microsensor with an integrated micro reference electrode was developed using silicon technology (Hintsche et al., 1995a; Uhlig et al., 1995a). Furthermore, the transducer can be combined with the three different membrane types commonly used for ion selective electrodes. Thus, the whole spectrum of ions measurable with the ion selective technique can be detected. For the purpose of miniaturization, the ion selective electrode as well as the reference electrode are integrated on one 1×5 mm silicon chip (Fig. 3). Potassium detection is carried out by potentiometric measurement of the potential difference between a reference electrode and a potassium selective electrode. Potassium selectivity is realized using a neutral carrier (valinomycin or BME-44), which is fixed in a polymeric structure (PVC-COOH, silicon and polyurethane). The ion selective membrane is cast into a cavity. The surface of the Si-cavity is oxidized and then covered completely with a thin metal layer. For better mechanical stability of the membrane and to improve the solid state contact, the opening to the front side is structured as a lattice. After casting the membrane in the cavity, the back side of the chip is totally covered with an electrically insulating silicon adhesive (Dow Corning).

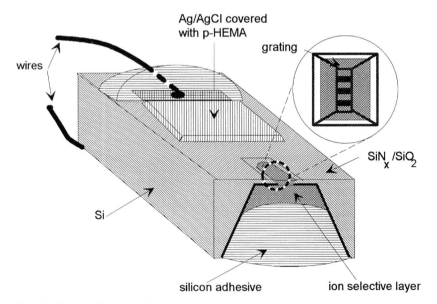

Figure 3. Scheme of the potassium selective electrode.

   The reference electrode is located on the front side of the chip. To stabilize the reference signal against chloride concentration fluctuations of the sample, the Ag/AgCl-composite-reference element is covered with a photolithographically structured hydrogel layer of poly-hydroxyethyl-methacrylate (p-HEMA). This layer serves as a diffusion barrier as well as a chloride buffer. The overcome problems resulting from a long-time contact of the sensor with protein-containing sample solution, the sensor is brought in contact with the sample only for short-time intervals. Immediately after the stabilization of the measuring signal, the chip is flushed with a Ringer solution of constant potassium concentration.

   The sensors were characterized with respect to slope, drift, long-term stability and measuring performance in biological fluids. The slope of the ISE was tested in a concentration range of 10^{-4} to 10^{-1} in decadic concentration steps. The slopes (n = 5 for each membrane type) DE/DpM were 50 ± 4 [mV/pK$^+$] for the PCV-COOH membrane, 56 ± 3 [mV/pK$^+$] for the silicon membrane and 56 ± 3 [mV/pK$^+$] for the polyurethane membrane. Thus, the sensors show near-Nernstian behavior (theoretical slope = 59 mV/pK$^+$). After 5 hours of continuous electrolyte contact, a drift of less than 0.2 mV/h can be achieved for all types of sensors tested. Long-term stability of the sensors was tested by continuously changing the concentration of K$^+$ between two concentrations, c_1 and c_2. For consecutive concentration changes over 100 h, the calculated E = E(c2) – E(c1) – values vary by only 2 %. Figure 2 demonstrates a whole blood measurement with the sensor under flow through conditions. The sensor is flushed with Ringer-solution containing 1×10^{-3} mol/LK$^+$ and was brought alternately in contact with the heparinized whole blood sample. Thus the *in vivo* measurement is simulated. As shown in this figure, the signals are stable and reproducible. The same performance of the sensor has been found using human serum or urine.

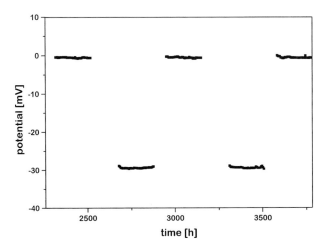

Figure 4. Measurement of potassium in whole blood (detail).

Amperometric enzyme sensor

For the on site analysis chemical parameters, especially in critical care, a miniaturized chip sensor system has been developed and implemented in a common catheter system (Gumprecht, 1991; Hintsche, 1994). Based on the principle of amperometric thin-film noble metal sensors made in silicon technology, enzyme chips for the determination of lactate and glucose have been constructed (Koudelka et al., 1989; Murakami et al., 1986; Hintsche et al., 1991; Roe, 1992; Urban et al., 1992; Koudelka et al., 1993).

The chip size (Fig. 5) is 1×5 mm and fits into the inner diameter of a three-lumen catheter. The platinum working electrode and the Ag/AgCl-reference electrode are placed on one chip side. Active electrodes areas and the connecting pads are defined by structuring an isolating silicon nitride layer. The enzymes, glucose oxidase and lactate oxidase respectively, are entraped into polymeric films, which are directly attached on the chip surface and optimized by chemical surface modification (Hintsche et al., 1989). A special packaging procedure has been developed for direct bonding of an isolated copper wire to the contact pads.

The chip and the long wire can be directly inserted into a lumen of a catheter (20 to 50 cm length). By means of a segmented analyte management, the exposure time of the sensors to blood is reduced to about 2 min/h, assuming that every 15 minutes an analysis has to be carried out. In Figure 6 the measuring setup for the catheter with built-in chemical sensors is shown. The sensor is conditioned during the standby time and is calibrated immediately before each measurement as described for the potassium sensor. The analysis and the signal treatment are controlled by the microcontroller. A smart interface module is located at the end of the catheter and acts as an pre-amplifier for the signal which is transmitted for the computer system.

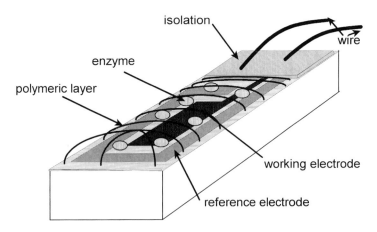

Figure 5. Scheme of the planar glucose sensor.

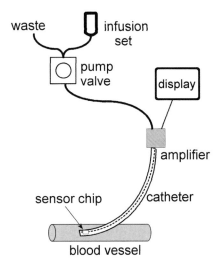

Figure 6. Scheme of the measuring setup.

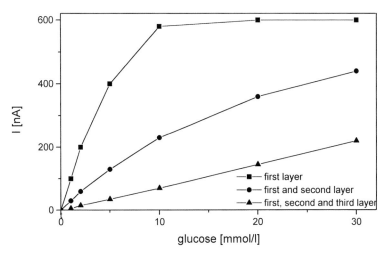

Figure 7. Adjustment of linearity of the glucose response by varying the number of layers on the sensor.

Polyurethane is used for entrapping the enzymes. The films are made by dispensing enzyme solution inaqueous polyurethane dispersions and subsequent covering of the chips. The adjustment of the sensitivity and the diffusion of the analyte molecules into the enzyme-containing membrane were established by repeated covering of thin polyurethane films. After the formation of the first polyurethane/enzyme film by evaporating water and forming a gel-like structure, the second film of the same material was added. In Figure 7 the stepwise change of the calibration graph due to three layers is shown.

In order to realize a well-defined diffusion barrier, we use partial cross-linking by poly-isocyanate as described earlier (Hintsche et al., 1991). This surface cross-linking is a very effective way to adjust sensor properties to the conditions of the analytical problem. In the case of whole blood analysis we have observed that a conditioning and calibration phase of 10 min within two measurements is sufficient to remove all clotted materials from the sensor surface. The aspiration of blood into the chip region of the catheter for about 30 s in combination with the conditioning period allows the use of this metabolite-measuring catheter sensor for more than 90 h.

Redox recycling based immunosensors

In conventional immunoassay techniques the presence of the analyte (haptens or macromolecules) is detected by optical or radioactive readout systems (Langone and Vernazis, 1983). Enzymatic labels are used in ELISA systems, where the enzymatic activity is determined mainly by optically (e.g. colorimetric, fluorimetric, turbudimetric) detectable products (Daniel et al., 1987). On the way to a direct electrochemical readout, electrochemical transducer principles have been tested (Hua et al., 1988; Spener et al., this volume). By the application of suitable enzyme substrates, electrochemically redox active compounds are created after turnover with the respective enzymes. (Niwa et al., 1993; Hintsche et al., 1995b).

Employing the interdigitated electrode array (IDA) structured by silicon technology, electrochemically active mediator substances can be detected in a range which is sensitive enough for the requirements of medical and environmental analysis. The potential of high structural resolution of planar electrode allows novel measurement methods. For the amperometric detection of either enzymatically formed substances or redox-labeled analytes (Niwa et al., 1990; 1993; Wollenberger et al., 1994; Paeschke et al., 1995b).

This measurement is based on the redox recylization between the adjacent microband electrodes of an interdigitated electrode array, with each electrode within the diffusion layer of the other. The process occurs when both the oxidation and the reduction potential of the reversible redox species are applied to pairs of interdigitated electrodes. The detection limit of redox mediators could be lowered to the range of 10^{-9} to 10^{-10} mol/l (Wollenberger et al., 1995; Paeschke et al., 1995c; Niwa et al., 1995). We have measured relevant mediators with these detection limits as shown in Table 1. For this purpose, IDA-electrodes with an electrode width of 1.5 μm and an interelectrode spacing of 0.8 m were used.

Enhancement of the sensitivity and reliability can be obtained by averaging the simultaneous measurements at more than one electrode pair arranged to an array (Paeschke et al., 1996). Here signals of the different electrode pairs are derived simultaneously and averaged by means of a computer.

Table 1. List of reversible mediators studied with redox recycling

Mediator	Ea, mV	Ec, mV	Detection limit, nmol/l	Amplifi-cation factor	Collection efficiency
p-AP	350	−150	5−10	12	0.853
Fe(CN)$_6$[4]	350	−120	30	8	0.98
o-HQ/o-BQ	600	−200	5−10	9	0.923
Ferrocene dicarbocylic acid[1]	600	0	50	8	0.93
Ferrocene lysine[1]	600	0	5−10	13	0.913
Gentisyl aldehyd	650	−50	60	4	0.83
OsbpypyNH[4]	550	−50	500		0.933
PEG-ferrocene[2]	400	100		7	0.94/0.833

1: Au IDA; 2: PEG MW:10/20 T, Pt; 3: flow, 4: Osmiumbipyridyl-NH by I. Katakis, Univ. Tarragona, Spain; HQ: hydroquinone, BQ: benzoquinone.

This increases the signal/noise ratio by the square root of the number of used electrode channels. The redox recycling and the signal averaging together form an amplification cascade which can be further enlarged by the use of enzymatic reactions within the assay. In this case enzyme labels liberate – corresponding to their turnover number – thousands of electro-active mediator molecules per molecule of analyte to be detected.

The integration of the enzymatic amplification mechanism of ELISA systems and interdigitated electrodes within sub-μm silikon structures enables the construction of cheap and portable micro-ELISA's offering some advantages compared to common optical ELISA systems for analytes within the medical and environmental area.

We propose a displacement assay as shown in principle in Figure 8. Covalently with antigen covered polymeric beads can be used to bind anti-body-enzyme conjugates. If the solution to be analyzed contains anti-anti-gen antibodies, a corresponding amount of antibody-enzyme conjugates will be displaced in an immunoreactor. After adding an electrode inactive enzyme substrate electrochemical active redox mediators are liberated in the electrode cell by the enzymes bound to be antibodies. The enzymati-cally liberated electrode-active molecules are detected by the redox recycling as described above. The current respone at the IDA reflects the amount of analyte. The amplification cascade caused by redox recycling, multichannel detection and enzyme catalysis enables high sensitivity.

In order to increase the sensitivity we optimized the enzymatic liberation of p-aminophenol from p-aminophenyl-hexopyranosides and -phosphate. For use in enzyme-analyte conjugates we tested the hydrolytic enzymes a-galactosidase, b-galactosidase, b-glucosidase, a-mannosidase and alka-line and acidic phosphatase with respect to their K_m-value, turnover and detection limit using our readout principle. It was evaluated that the low molecular weight enzymes guarantee highest displacement efficiencies

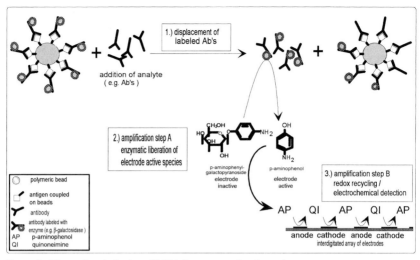

Figure 8. Electrochemical micro-ELISA for macromolecular antigens.

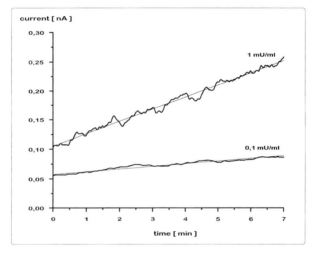

Figure 9. Detection limit of enzyme-antibody conjugate. label: β-galactosidase; antibody: < ciliary neurotrophic factor >; substrate: pAP-β-D-galactopyranoside, 1 mM; phosphate buffer pH 6.7, 0.1 M NaCl; detection: IDA, – 50/+ 300 mV; redox recycling.

and, thereby, the highest sensitivities. Concerning the liberation of the redox mediator p-aminophenol (pAP) we found that it is favorably carried out under slightly acidic conditions which pursue optimal operation of the sub-μm IDA.

We studied the commercially available ciliary neurotrophic factor optical assay (CNTF; Boehringer Mannheim, Germany) consisting of an anti-CNTF-b-galactosidase conjugate, anti-CNTF-antibody and recombinant CNTF. Because of the identical label enzyme, we adapted this ELISA for our electrochemical readout system. Compared to the original optical ELISA, the incubation time for the redox mediator liberation in our approach can be lowerd by a factor of about 5. With respect to the label enzymes it is obvious that 2 decades less volume activity and 4 decades less absolute activity (2 µU anti-CNTF-β-galactosidase/test) are sufficient to detect pAP in the lower nanomolar range with IDA's (Fig. 9) compared to the optical detection in the CNTF-assay.

By combining immunoreactor chambers of about 5 µl volume and IDA flow-through cells of about 2 µl it is possible to electrically read out commercial ELISA formats if appropriate hydrolytic enzymes are used. The corresponding decrease of reagent volumes in the micro-ELISA with a volume of about 5 µl results in great savings of bioreagents and a decrease in the relative detection limit. The modular coupling of the incubation chamber and the detection electrodes offer high flexibility.

For the determination of small molecular weight compounds such as pesticides and other organic pollutants in the environment, a variation of the displacement assay is used. In this case the assay works with direct mediator-labeled antigens as shown in Figure 10. In a first step anti-hapten antibodies covalently bound to small polymeric beads and saturated with redox mediator-labeled haptens have to be prepared. After addition of hapten-containing solution a corresponding amount of labeled haptens will be competitively displaced. By means of redox recycling using IDA's these labeled hapten conjugates displaced into solution can be detected in the IDA-electrode cell.

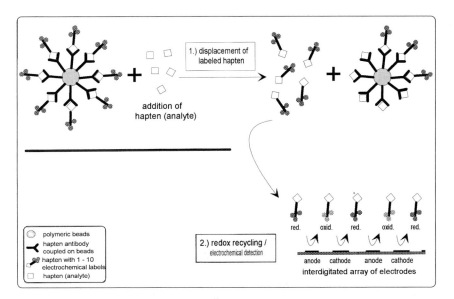

Figure 10. Electrochemical micro-ELISA for haptens.

In order to construct an immunoassay for di- and tri-nitrophenol derivatives we studied their redox labeling and detection limit within the electrochemical displacement assay. A promising candidate for redox labeling is
the ferrocenelysine, which has been synthesized at the University of Ham-
burg, Dept. Biochemistry. For the detection of ferrocene-lysine the anode,
which serves as generator electrode, is potentiostated to 600 mV and the
cathode (collector) to 0 mV. The additon of 10 µmol/l ferrocene-lysine
results in an anodic current of 53 nA within 1 s in flow. A response in the
same extent is indicated simultaneously at the cathode. The collection
efficiency was 91% over the concentration range studies. A concentration
dependence for ferrocene-lysine illustrates the increase of sensitivity
(Fig. 11). The steady state response is enhanced by a factor of 10–15 by the
electrochemical recycling and summing up the anodic and cathodic
currents the amplification is about 25. Thus, the detection limit could be
reduced to 10 nmol/l with reasonable reproducibility. Further improvement
of the sensitivity was obtained by using a multilabeling procedure, where
the antigen is labeled with a small peptide chain carrying two ferrocene-
lysine residues. For this purpose another redox label carrying two m-dihy-
droxy-phenolic residues instead of ferrocene has also been studied. This
multilabeling of haptens as tri-nitrophenol enhances the amplification
cascade of redox recycling and multichannel signal averaging as mentioned
above further.

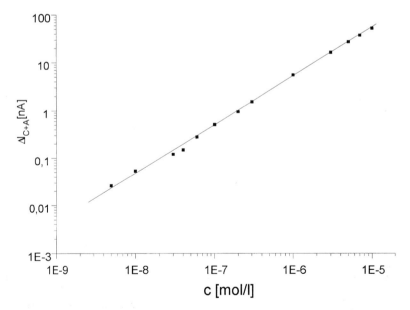

Figure 11. Dose-dependent response to ferrocene-lysine. Sum of generator and collector current
(amplified with redox recycling) of an IDA. Generator: 600 mV, collector: 0 mV, gap between
microband: 1.1 µm.

Preliminary studies of a complete assay with tri-nitrophenol antibodies bund to sepharose beads and complexed with their bis-ferrocenyl labeled antigen showed the usefulness of the proposed electrochemical readout immunoassay.

References

Bard, A.J., Crayston, J.A., Kittlesen, G.P., Shea, T.V. and Wrigton, M.S. (1986) Digital simulation of the measured electrochemical response of reversible redox couples at microelectrode arrays: Consequences arising from closely spaced ultramicroelectrodes. *Anal. Chem.* 58:2321–2331.

Benecke, W. (1990) Silicon micromachining for microsensors and microactuators. *Microelectron. Eng.* 11:73–82.

Daniel, W.C. (1987) General principle of Immunoassay. *In*: W.C. Daniel and M.T. Perlstein (eds): *Immunoassay – a practical guide.* Academic Press Inc., London, pp 1.

Fiehn, H., Howitz, S. and Pham, M.T. (1995) Components and technology for a fluidic-ISFET-microsystem. *In*: A.v.d. Berg and P. Bergfeld (eds): *Micro Total Analysis Systems.* Dordrecht: Kluver Academic Publ., The Netherlands, pp 289–294.

Göpel, W. (1994) *Biosensors & Bioelectronics:* 9/10:601–760.

Griffith, A. and Cooper, J. (1994) Investigation of biological electron transport at gold nanostructures. *Proc. Third World Congress on Biosensors*, No. 4.48 New Orleans.

Gumprecht, W., Schelter, W., Montag, B., Bos, J.H.A., Eijkang, E.P. and Lachmann, B. (1991) Monitoring of blood pO_2 with a thin amperometric sensor. *Technical Digest Transducer'* 91:85–87.

Haemmerli, S., Manz, A. and Widmer, H.M. (1992) Use of sensors in chemical analysis – concept of total analysis system. *VDI-Ber.* 939 (Sensoren):83–86.

Hintsche, R., Neumann, G., Dransfeld, I., Kampfrath, G., Hoffmann, B. and Scheller, F. (1989) Polyurethane enzyme membranes for chip biosensors. *Anal. Letters* 22:2175–2190.

Hintsche, R., Möller, B., Dransfeld, I., Wollenberger, U. and Scheller, F. (1991) Chip biosensors on thin-film metal electrodes. *Sensor. Actuator. B* 4:287–291.

Hintsche, R., Paeschke, M., Wollenberger, U., Schnakenberg, U., Wagner, B. and Lisec, T. (1994b) Micro Electrode arrays and application to biosensing devices. *Biosensors & Bioelectronics* 9:697–705.

Hintsche, R., Kruse, C., Uhlig, A., Paeschke, M., Lisec, T., Schnakenberg, U. and Wagner, B. (1995a) Chemical microsensor systems for medical applications in catheters. *Sensor. Actuator. B* 27:471–473.

Hintsche, R., Paeschke, M., Seitz, R., Wollenberger, U., Bredehorst, R. and Vogel, C.W. (1995b) Immunoanalysis using nm-thinfilm electrodes and redox labels in microsystems. *In: Microfabrication Technology for Research and Diagnostics.* San Francisco; Cambridge Healthtec Institute, MA, USA.

Hua, T.T., Tang, H.T., Lunte, C.F., Halsall, H.B. and Heineman, W.R. (1988) p-Aminophenyl phosphate: an improved substrate for electrochemical enzyme immunoassay. *Anal. Chim. Acta* 214:187–195.

Izquierdo, A. and de Castro, M. (1994) Ion-selective field-effect transistors and ion-selective electrodes as sensors in dynamic systems. *Electroanalysis* 7:505–512.

Knoll, M., Cammann, K., Dumschat, C., Eshold, J. and Sundermeier, C. (1994) Micromachined ion-selective electrodes with polymer matrix membranes. *Sensor. Actuator. B* 21:71–76.

Koudelka, M., Gernet, S. and de Rooij, N.F. (1989) Planar amperometric enzyme based glucose microelectrode. *Sensor. Actuator. B* 18:157–165.

Koudelka-Hep, M., Strike, D.J. and de Rooij, N.F. (1993) Miniature electrochemical glucose biosensors. *Anal. Chim. Acta* 281:461–466.

Langone, J.J. and Vunakis, H. (1983) *Immunochemical techniques, Part A.* Methods in Enzymology, vol. 92. New York: Academic Press.

Lindner, E., Cosofret, V., Ufer, S., Buck, R., Kusy, R., Ash, B. and Nagle, T. (1993) Flexible (Kapton-based) microsensor arrays of high stability for cardiovascular applications. *J. Chem. Farad. Trans.* 89/2:361–376.

Manz, A., Verpoorte, E.M.J., Raymond, D.E., Effenhauser, C.S., Burggraf, N. and Widmer, H.M. (1995) µ-TAS: miniaturized total chemical analysis systems. *In:* A. v.d. Berg and P. Bergfeld (eds): *Micro Total Analysis Systems.* Dordrecht: Kluver Academic. Publ., The Netherlands, pp 5–23.

Murakami, T., Nakamoto, S., Kumura, Jlk Kuriyama, T. and Karube, I. (1986) A microplanar amperometric glucose sensor using an ISFET as a reference electrode. *Anal. Lett* 22: 1973–1986.

Murakami, Y., Takeuchi, T., Yokohama, K., Tamiya, E., Karube, I. and Suda, M. (1993) Integration of enzyme-immobilized column with electrochemical flow cell using micromachining techniques for a glucose detection system. *Anal. Chem.* 65:2731–2735.

Niwa, O., Morita, M. and Tabei, H. (1990) Electrochemical behaviour of reversible redox species at interdigitated array electrodes with different geometries: Consideration of redox recycling and collection efficiency. *Anal. Chem.* 62:447–452.

Niwa, O., Xu, Y., Halsall, H.B. and Heinemann, W.R. (1993) Small-volume voltametric detection of 4-aminophenol with interdigitated array electrodes and its application to electrochemical enzyme immunoassay. *Anal. Chem.* 65:1559–1563.

Niwa, O., Tabei, H., Solomon, B.P., Xie, F. and Kissinger, P.T. (1995) Improved detection limit for catecholamines using liquid chromatography- electrochemistry with a carbon interdigitated array microelectrode. *J. Chromatogr. B. Biomed. Appl.* (1): 21–28.

Oesch, U., Amman, D. and Simon, W. (1986) Ion-selective membrane electrodes for clinical use. *Clin. Chem.* 32:1448–1459.

Paeschke, M., Wollenberger, U., Köhler, C., Lisec, T., Schnakenberg, U. and Hintsche, R. (1995a) Properties of interdigital electrode arrays with different geometries. *Anal. Chim. Acta* 305:126–131.

Paeschke, M., Wollenberger, U., Uhlig, A., Schnakenberg, U., Wagner, B. and Hintsche, R. (1995b) A stacked multichannel amperometric detection system. *In*: A. v.d. Berg and P. Bergfeld (eds): *Micro Total Analysis Systems.* Dordrecht: Kluver Academic. Publ., The Netherlands, pp 249–254.

Paeschke, M., Wollenberger, U., Lisec, T., Schnakenberg, U. and Hintsche, R. (1995c) Highly sensitive electrochemical microsensors using submicrometer electrode arrays. *Sensor. Actuator. B* 27:394–397.

Paeschke, M., Dietrich, F., Uhlig, A. and Hintsche, R. (1996) Voltammetric multichannel measurements using silicon-fabricated microelectrode arrays. *Electroanalysis*: 7/1; in press.

Reimer, K., Koehler, C., Lisec, T., Schnakenberg, U., Fuhr, G., Hintsche, R. and Wagner, B. (1995) Fabrication of electrode arrays in the quarter micron regime for biotechnological applications. *Sensor. Actuator. A* (1–203):66–70.

Reinhoudt, D.N., Engbersen, J.F.J., Brzozka, Z., van den Vlekkert, H.H., Honig, G.W.N., Holterman, H.A.J. and Verkerk, U.H. (1994) Development of durable K$^+$-selective chemically modified field effect transistors with functionalised polysiloxane membranes. *Anal. Chem.* 66:3618–3623.

Roe, J.N. (1992) Biosensor development. *Pharmaceut. Res.* vol. 9, no. 7:835–844.

Scheller, F.W., Heyn, S.P., Wollenberger, U., Pfeiffer, D., Makower, A., Paeschke, M., Neumann, B. and Riedel, K. (1992) Electrochemische Biosensoren – Grundlagen, Anwendungen und Perspektive. *Dechema Monographies* 126:201–218.

Shoji, S. and Esashi, M. (1995) Bonding and assembling methods for realizing a µ-TAS. *In*: A. v.d. Berg and P. Bergfeld (eds): *Micro Total Analysis Systems.* Dordrecht: Kluver Academic. Publ., The Netherlands, pp 165–180.

Sinclair, Y., Hong, J. and Lawrence, K.C.L. (1988) Miniature liquid junction reference electrode with micromashined silicon cavity. *Sensor. Actuator.* 15:337–345.

Uhlig, A., Schnakenberg, U., Lindner, E., Dietrich, F. and Hintsche, R. (1995) Catheter system for potassium measurements in medical application. *Technical Digest Transducer '95, Eurosensors IX*: 469–472.

Uhlig, A., Schnakenberg, U., Lindner, E., Dietrich, F. and Hintsche, R. (1995) Catheter system for potassium measurements in medical application. *Sensor. Actuator. B24–25*, 899–903.

Urban, G., Jobst, G., Keplinger, F., Ascherau, A., Jachimowitz, A. and Kohl, F. (1992) Miniaturized biosensor for integration on flexible polymer carriers. *Proc. Biosensors '92*. Elsevier, Oxford, pp 467–471.

v.d. Berg, A. and Bergfeld, P. (1995) *Micro Total Analysis Systems.* Dordrecht: Kluver Academic. Publ., The Netherlands.

Verpoorte, E.M.J., Schoot, B.H.V., Jeanneret, S., Manz, A. and Rooij, N.F.d. (1994) Silicon-based chemical microsensors and microsystems. *ACS Symp. Ser.* 561 (INTERFACIAL DESIGN A): 244–254.

Wise, K.D. and Najafi, K. (1991) Microfabrication techniques for integrated sensors and micro-systems. *Sience* 254: 1335–1342.

Wollenberger, U., Paeschke, M. and Hintsche, R. (1994) Interdigitated array microelectrodes for the determination of enzyme activities. *Analyst* 119: 1245–1249.

Wollenberger, U., Hintsche, R. and Scheller, F. (1995) Biosensors for analytical microsystems. Microsystem Technology 1: 275–283

Subject Index

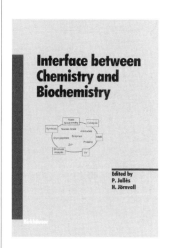

Interface between Chemistry and Biochemistry

Edited by
P. Jollès, *Univ. René Descartes, Paris, France*
H. Jörnvall, *Karolinska Institute, Stockholm, Sweden*

1995. 312 pages. Hardcover • ISBN 3-7643-5081-4
(EXS 73)

The increasing importance of the interface between chemistry and biology is probably the largest change in chemistry in the past 15 years. More and more organic chemists are working on problems dealing with biology. Once considered to be at the very outside edge of either field, interfacial research is poised to move into the mainstream of both disciplines. This merging of two types of approach has resulted in a vigorous research discipline with unprecedented potential to address important biological and chemical problems. A series of examples are developed in this book.

Some analytical aspects are discussed first as the fundamental concepts are not only chemical, but chemistry has provided biochemistry with powerful tools of analysis.

Physico-chemical aspects are devoted to spectrometric studies of nucleic acids as well as lipids, lipases and membrane proteins (receptors). Three chapters are included in the section dealing with enzymes. The part devoted to metalloproteins is mainly directed toward zinc metallochemistry and NMR structural work on zinc proteins.

Chemists have been able to bring to biology their characteristic approach of synthesizing new molecules; three chapters are devoted to peptides, sugar compounds and biocatalysts. Two chapters discuss new active compounds (antibacterial peptides, catalytic antibodies), which are the result of collaboration between chemists and biochemists.

From the Contents:

Chemistry at interfaces and in transport • Enzyme function in organic solvents • Chemistry and biochemistry • Analysis of proteins and nucleic acids • UV and nucleic acids • Synthesis of active compounds • Metalloproteins

Birkhäuser Verlag • Basel • Boston • Berlin

Lysozymes: Model Enzymes in Biochemistry and Biology

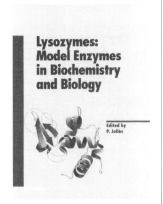

Edited by
P. Jollès
Muséum National d'Historie Naturelle, Paris, France

1996. 464 pages. Hardcover • ISBN 3-7643-5121-7
(EXS 75)

More than seventy years after Flemingís discovery of lysozyme, this enzyme continues to play a crucial role as a model enzyme in protein chemistry, in enzymology, in crystallography, in molecular biology and genetics, in immunology and also in evolutionary biology. The classical representative of this widespread enzyme family is the hen egg-white lysozyme. Chicken (c)-type lysozymes have also been characterized in many other animals including mammals, reptiles and invertebrates. Besides this c-type lysozyme, other distinct types, differing on the basis of structural, catalytic and immunological critera, have been described as well, these in birds, phages, bacteria, fungi, invertebrates and plants. The specificity, however, of all these enzymes is the same: they cleave β (-glycosidic bond between the C-1 of N-acetylmuramic acid and the C-4 of N-acetylglucosamine of the bacterial peptidoglycan.

In this volume special emphasis is placed on results obtained during the last ten years. Lysozymes are by no means merely defence or, in certain cases, digestion enzymes. In fact, peptidoglycan fragments released by the lytic action of this enzyme family can trigger the synthesis of immunostimulating or antibacterial substances, and a host of other unexpected, biological reactions may be provoked by lysozymes as well. As Fleming prophesied: "We shall hear more about lysozyme".

Birkhäuser Verlag • Basel • Boston • Berlin

Bioelectrochemistry: Principles and Practice

Experimental Techniques in Bioelectrochemistry

Bioenergetics

Edited by **P. Gräber,** *University of Freiburg, Germany /* **G. Milazzo†,** *formerly Istituto Superiore di Sanità, Rome, Italy*

Bioelectrochemistry:
Principles and Practice: Vol. 4
1996. Approx. 500 pages. Hardcover. ISBN 3-7643-5295-7

Bioelectrochemistry: General Introduction

Bioelectrochemistry of Cells and Tissues

Edited by **S.R. Caplan, I.R. Miller,** *The Weizmann Institute of Science, Rehovot, Israel /* **G. Milazzo†,** *formerly Istituto Superiore di Sanità, Rome, Italy*

Bioelectrochemistry:
Principles and Practice: Vol. 1
1995. 384 pages. Hardcover
ISBN 3-7643-2687-5

This first volume discusses nonequilibrium thermodynamics and kinetics, particularly enzyme catalysis, for processes and systems in the steady state. Methods of mathematical modeling by means of network simulations are also treated, since they serve to assess the transient behavior of a system on its way to a steady state. Water as a ubiquitous constituent plays an essential role in bioelectrochemical systems, hence its structure is carefully evaluated, both in the pure state and in the ionic hydration shell. Similarly, the interface between water and a membranous or biocolloidal phase is of major importance. The phenomena occurring at such interfaces, including diffuse double layers, as well as binding and adsorption of solutes, are extensively examined.

Edited by **D. Walz,** *University of Basel, Switzerland /* **H. Berg,** *Institute of Molecular Technology, Jena, Germany /* **G. Milazzo†,** *formerly Istituto Superiore di Sanità, Rome, Italy*

Bioelectrochemistry:
Principles and Practice: Vol. 2
1995. 328 pages. Hardcover
ISBN 3-7643-5085-7

The role of electric and magnetic fields in biological systems forms the focus of this second volume in the *Bioelectrochemistry* series. The most prominent use of electric fields is found in some fish. These species generate fields of different strengths and patterns serving either as weapons, or for the purpose of location and communication. Electrical phenomena involved in signal transduction are discussed by means of two examples, namely excitation-contraction coupling in muscles and light transduction in photoreceptors. Also examined is the role of electrical potential differences in energy metabolism and its control.

Edited by **V. Brabec,** *Academy of Sciences of the Czech Republic, Brno, Czech Republic /* **D. Walz,** *University of Basel, Switzerland /* **G. Milazzo†,** *formerly Istituto Superiore di Sanità, Rome, Italy*

Bioelectrochemistry:
Principles and Practice: Vol. 3
1996. 576 pages. Hardcover
ISBN 3-7643-5084-9

The measurements of electrochemical impedance, voltammetric (polarographic) analysis, and spectroelectrochemistry represent a basis for analysing molecules of biological significance in bulk solution and at interfaces. These principles are reviewed in the first four chapters of this volume. The following three chapters demonstrate how these principles are utilized in voltammetric and interfacial analysis of biomacromolecules such as nucleic acids, proteins, polysaccharides and viruses in vitro, in the development of biosensors with electrochemical transducers, and in in vivo voltammetry. The final two chapters are devoted to the principles of electrophoresis used for separation analysis of biomolecules and to the theoretical principles and practical description of the patch-clamp technique.

Approximately 0.05% of all the sunlight reaching the surface of the earth is used by photosynthetic organisms to synthesize organic compounds. All other organisms use these compounds as energy sources for their metabolism. This volume describes the energetics, kinetics and molecular mechanisms of these processes. The volume begins by considering the thermodynamics of open systems, and the global aspects of the biological processes involved. Subsequent chapters focus on the differences between scalar and vectorial chemical reactions in the cell, and the coupling between the different reactions. Energy transduction in bacteria, chloroplasts, and mitochondria as well as the different pathways evolved in order to utilize energy from various external sources are described.

Birkhäuser Verlag • Basel • Boston • Berlin